T0291810

CAMBRIDGE LIBRARY COLLECTION

Books of enduring scholarly value

Physical Sciences

From ancient times, humans have tried to understand the workings of the world around them. The roots of modern physical science go back to the very earliest mechanical devices such as levers and rollers, the mixing of paints and dyes, and the importance of the heavenly bodies in early religious observance and navigation. The physical sciences as we know them today began to emerge as independent academic subjects during the early modern period, in the work of Newton and other 'natural philosophers', and numerous sub-disciplines developed during the centuries that followed. This part of the Cambridge Library Collection is devoted to landmark publications in this area which will be of interest to historians of science concerned with individual scientists, particular discoveries, and advances in scientific method, or with the establishment and development of scientific institutions around the world.

Contributions to Molecular Physics in the Domain of Radiant Heat

Professor of natural philosophy for the Royal Institution between 1853 and 1887, the physicist John Tyndall (1820–93) passionately sought to share scientific understanding with the Victorian public. Reissued here is the collected research he contributed to the *Philosophical Transactions of the Royal Society* and other journals. Published in 1872, it complements Tyndall's *Heat Considered as a Mode of Motion* (1863), which is also reissued in this series. Here each memoir is preceded by a short summary, explaining what he discovered and his reasons for embarking on the investigations in question. Accompanying the detailed descriptions of experimental methods are illustrations of the scientific apparatus used. Tyndall also shows how his work built upon previous research, acknowledging the insights of distinguished scientists such as William Herschel and Macedonio Melloni. In particular, he discusses at length his academic debates with Heinrich Gustav Magnus.

Cambridge University Press has long been a pioneer in the reissuing of out-of-print titles from its own backlist, producing digital reprints of books that are still sought after by scholars and students but could not be reprinted economically using traditional technology. The Cambridge Library Collection extends this activity to a wider range of books which are still of importance to researchers and professionals, either for the source material they contain, or as landmarks in the history of their academic discipline.

Drawing from the world-renowned collections in the Cambridge University Library and other partner libraries, and guided by the advice of experts in each subject area, Cambridge University Press is using state-of-the-art scanning machines in its own Printing House to capture the content of each book selected for inclusion. The files are processed to give a consistently clear, crisp image, and the books finished to the high quality standard for which the Press is recognised around the world. The latest print-on-demand technology ensures that the books will remain available indefinitely, and that orders for single or multiple copies can quickly be supplied.

The Cambridge Library Collection brings back to life books of enduring scholarly value (including out-of-copyright works originally issued by other publishers) across a wide range of disciplines in the humanities and social sciences and in science and technology.

Contributions to Molecular Physics in the Domain of Radiant Heat

A Series of Memoirs Published in the 'Philosophical Transactions' and 'Philosophical Magazine', with Additions

John Tyndall

Cambridge University Press

CAMBRIDGE
UNIVERSITY PRESS

University Printing House, Cambridge, CB2 8BS, United Kingdom

Published in the United States of America by Cambridge University Press, New York

Cambridge University Press is part of the University of Cambridge.
It furthers the University's mission by disseminating knowledge in the pursuit of education, learning and research at the highest international levels of excellence.

www.cambridge.org
Information on this title: www.cambridge.org/9781108067904

This edition first published 1872
This digitally printed version 2014

ISBN 978-1-108-06790-4 Paperback

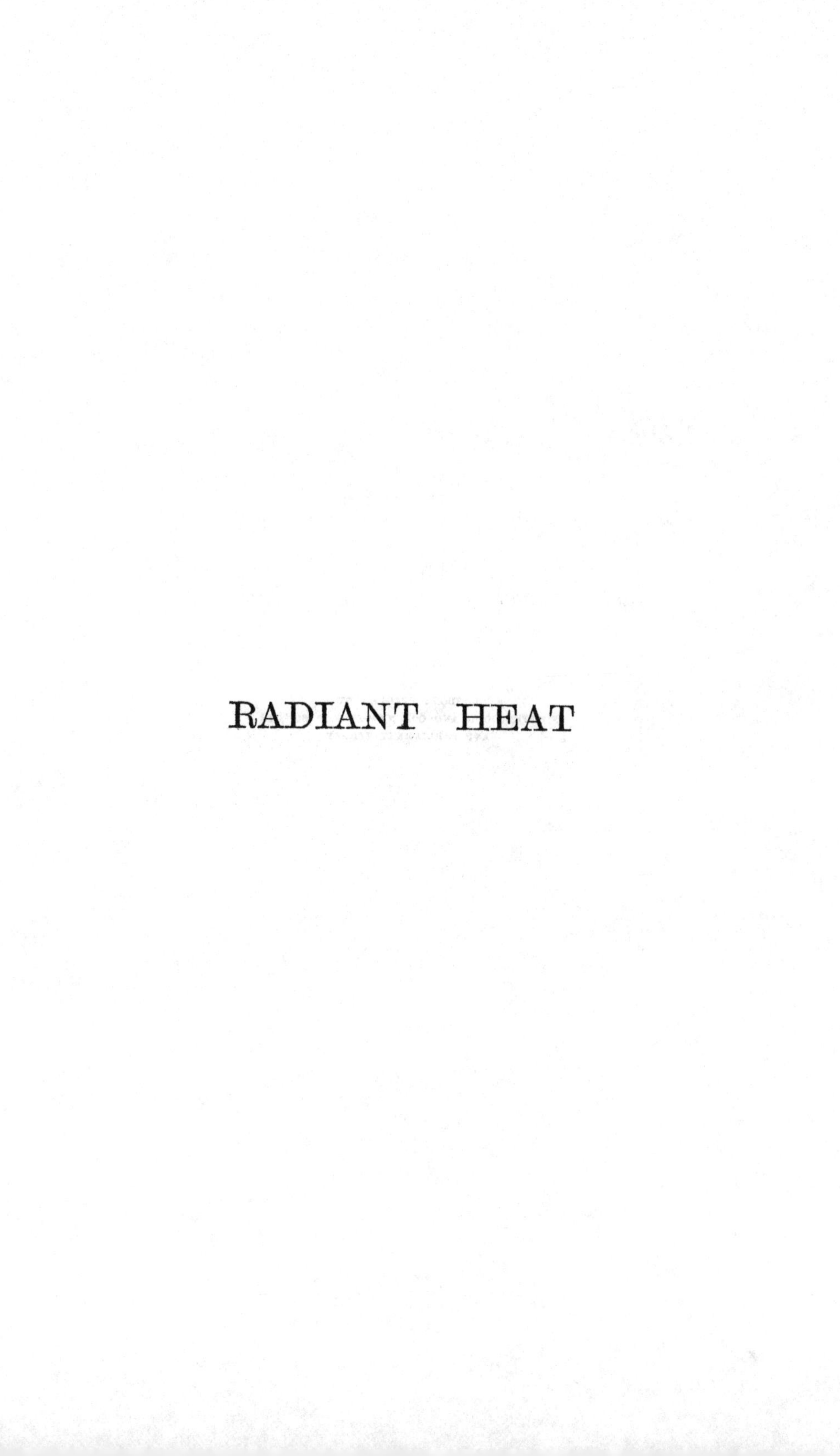

RADIANT HEAT

LONDON: PRINTED BY
SPOTTISWOODE AND CO., NEW-STREET SQUARE
AND PARLIAMENT STREET

FRONTISPIECE

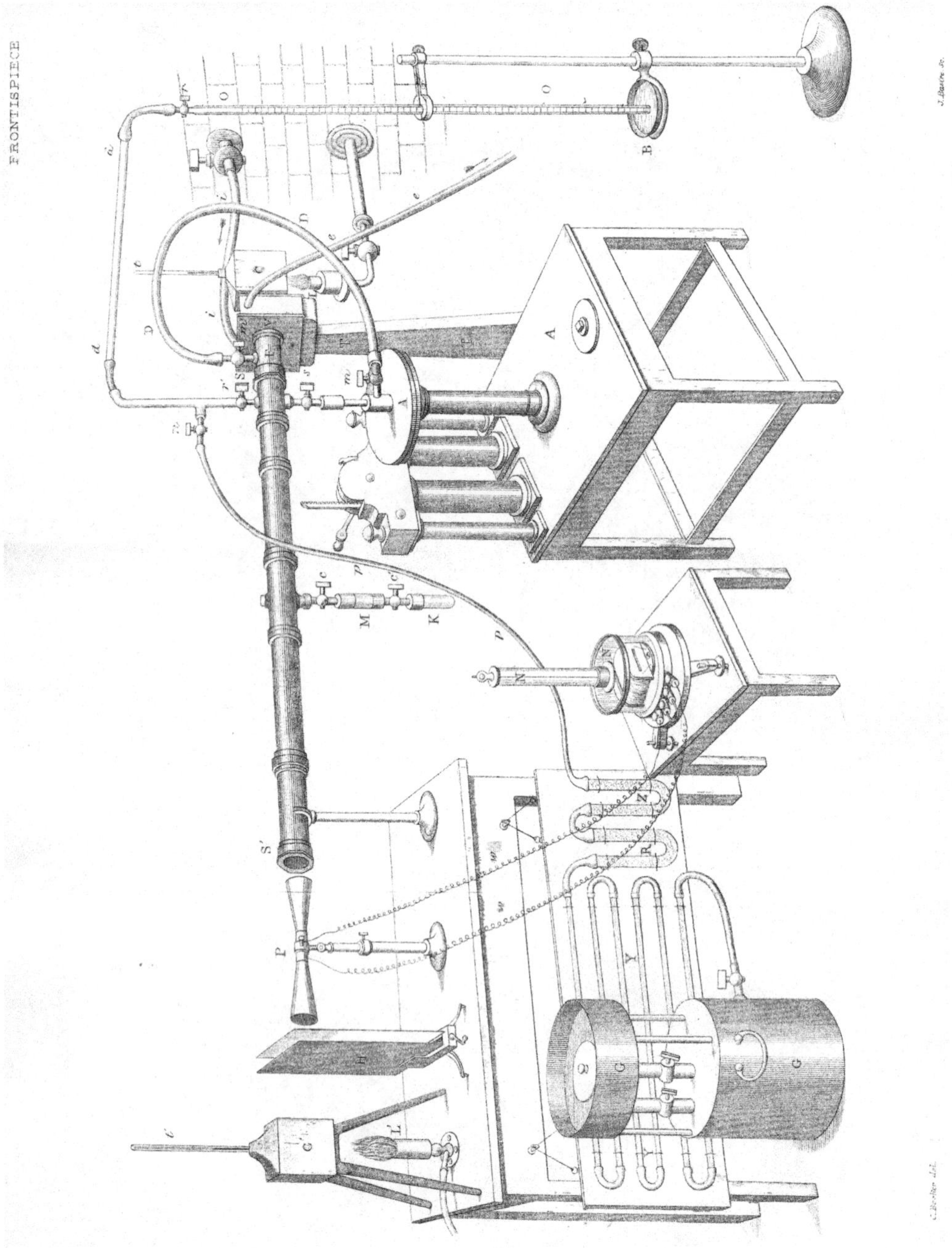

The material originally positioned here is too large for reproduction in this reissue. A PDF can be downloaded from the web address given on page iv of this book, by clicking on 'Resources Available'.

CONTRIBUTIONS

TO

MOLECULAR PHYSICS

IN THE DOMAIN OF

RADIANT HEAT.

A SERIES OF MEMOIRS PUBLISHED IN THE 'PHILOSOPHICAL TRANSACTIONS' AND 'PHILOSOPHICAL MAGAZINE,' WITH ADDITIONS.

BY

JOHN TYNDALL, LL.D. F.R.S.

PROFESSOR OF NATURAL PHILOSOPHY IN THE ROYAL INSTITUTION.

LONDON:
LONGMANS, GREEN, AND CO.
1872.

TO

HENRY BENCE JONES, M.D. D.C.L. F.R.S.

HON. SEC. R. I.

If unswerving devotion to the Royal Institution, firstly, and above all, as a school of original enquiry, and secondly as an organ for the diffusion of scientific knowledge, merit the grateful recognition of its Members and its Professors, then justice ought to require no stimulus from friendship, in associating these Researches with your name.

They were one and all conducted on the spot whence, during sixty years, issued in unbroken succession the labours of Young, Davy, and Faraday. Would that they were more worthy of their immortal antecedents!

JOHN TYNDALL.

Royal Institution:
May 1872.

Erratum

Preface, line 8, *for* Magno-crystallic *read* Magne-crystallic

PREFACE.

In the Preface to the Third Edition of my work on Heat, written in January 1868, the hope was expressed that before the end of that year the original Memoirs which I had contributed to the 'Philosophical Transactions,' and other journals, during the previous eighteen years, would be presented to the scientific public. Hitherto this hope has been only partially fulfilled by the publication of the researches on Diamagnetism and Magno-crystallic Action.

The present volume contains the Memoirs on Radiant Heat, considered as an explorer of Molecular Condition. I have read them over carefully, and have tried to augment their clearness without altering their substance.

In front of each memoir is placed an analysis of its contents, from which the reader can at once learn the nature of the inquiry. I have also added here and there some necessary historic data.

The points of difference between the late Professor Magnus and myself regarding the action of air and that of aqueous vapour on radiant heat are placed in their proper sequence and relation. At the end of the series of Memoirs the discussion is resumed, and brought, I trust, to a fair conclusion.

I ought to inform the reader who desires but a partial or general acquaintance with these researches, that summaries of most of them have been already published in the various editions of my work on Heat.

Finally, I would offer my best thanks to the Council of the Royal Society for the ready courtesy with which they granted me the use of the Plates employed to illustrate these Memoirs in the 'Philosophical Transactions.'

JOHN TYNDALL.

ROYAL INSTITUTION:
May 1872.

CONTENTS.

LIST OF PLATES.

I.

ON THE ABSORPTION AND RADIATION OF HEAT BY GASES AND VAPOURS, AND ON THE PHYSICAL CONNEXION OF RADIATION, ABSORPTION, AND CONDUCTION.

ANALYSIS OF MEMOIR I.

THE researches embodied in the following memoir were begun in the early part of 1859; and the first notice of them is published in the 'Proceedings of the Royal Society' for the 26th of May of that year.

They arose in part from the desire to do for the gaseous form of matter what had been previously so well done by Melloni for the liquid and solid states of aggregation. They were also stimulated by the persuasion that not only the physical but the chemical, in other words the *molecular*, condition of bodies probably played a part previously unrecognised in the phenomena of radiation and absorption.

At the time here referred to, the belief was general that as regards its relation to radiant heat the gaseous form of matter was inaccessible to experiment. Of the published attempts in this direction, two only are known to me, the one by Melloni and the other by Franz. Both are referred to in the 'Introduction' to this memoir, and are there stated, I believe correctly, to have left this field of inquiry 'perfectly unbroken ground.'

The memoir naturally begins with a description of the instruments employed; the difficulty of obtaining a galvanometer-coil free from magnetic action is dwelt upon; and a simple method is proposed of testing the galvanometric purity of copper wire. It is shown that by an experiment of a moment's duration we can satisfy ourselves, before the coil is constructed, whether the wire is fit for galvanometric purposes or not.

Following the methods of observation introduced by Melloni, experiments on air and other gases were executed and recorded. They proved conclusively that to grapple with these gases far more powerful sources of heat than any previously employed would be necessary. Hence arose the problem how to employ such sources, and at the same time to maintain the needle of the galvanometer in a condition as sensitive as if no heat at all were falling upon the thermo-electric pile.

The problem is thus solved:—Two sources of heat are permitted to radiate against the two opposite faces of the pile. However powerful these opposing radiations may be, if they be only equal, they neutralise each other, and permit the needle of the galvanometer to point tranquilly to zero. Between one of these sources and the pile a wide tube is introduced, which can be exhausted or filled at pleasure with any gas. Supposing the opposing forces to be equal when the tube is empty, the introduction of any gas, capable of quenching even an infinitesimal fraction of the heat, would give the opposite source the mastery; a deflection of the galvanometer would follow, and from the magnitude of the deflection the amount of heat quenched by the gas could be immediately deduced.

The final form of the apparatus thus sketched in outline was determined by a long course of tentative experiments. The possibilities of error were numerous, both in the arrangement of the apparatus and in the selection and purification of the substances to be examined. Experiments are recorded which show the infinitesimal action of the elementary gases and the perfectly enormous action of some of the compound gases upon radiant heat. To render these contrasting results secure; to guard against instrumental defects which might readily substitute a delusive for a real action; and to avoid impurities which, though infinitesimal when measured chemically, were found competent in the case of the feebler gases to entirely vitiate the results, some thousands of experiments were executed.

As regards instrumental arrangements, for example, it was proved that to avoid errors arising from convection, and from dynamic heating and chilling, it was absolutely necessary that the gases examined should not come into contact either with the radiating source or with the face of the thermo-electric pile. It was necessary that the heat lost by the one and received by the other should be purely *radiant heat*—a result which could never be calculated upon if contact were permitted with either the pile or the source.

One of the earliest trials with the new apparatus, in which electrolytic oxygen was employed, proved the modicum of ozone which accompanied the oxygen to exercise a far more potent action upon radiant heat than the oxygen itself. In § 19 Memoir II. this result is further developed, the constitution of ozone now generally agreed upon being deduced from these developments.

While the action of oxygen, hydrogen, nitrogen, and air, even at the full pressure of the atmosphere and subjected to the most powerful tests, proved barely measurable, amounting certainly to not more than a fraction of a unit per cent. of the incident heat; experiments on olefiant gas are recorded in which the quantities employed varied from $\frac{1}{10000}$th of an atmosphere to a whole atmosphere, and yielded throughout a perfectly measurable action. The absorption by this gas, under the pressure of an atmosphere, amounted to fully 81 per cent. of the incident radiation.

Experiments on sulphuric-ether vapour are also recorded in which the pressures employed varied from $\frac{1}{500000}$th of an atmosphere to the maximum pressure of the vapour, the quantity of vapour corresponding to the smallest of these pressures being proved capable of producing a measurable effect. Comparing equal pressures up to 5 inches of mercury, sulphuric-ether vapour was found to be more than twice as potent as olefiant gas.

The action of the following vapours at various degrees of density upon radiant heat is afterwards recorded:—Bisulphide of carbon, amylene, iodide of ethyl, iodide of methyl, iodide of amyl, chloride of amyl, benzol, methylic alcohol, formic ether, propionic ether, chloroform, and alcohol.

A curious explosive effect occurring in the cylinders of the air-pump when air is mixed with attenuated bisulphide-of-carbon vapour is referred to and explained.

Chlorine gas is next operated on, and its action in altering the reflective power of the interior surface of the experimental tube is demonstrated. This leads to a searching inquiry whether any of the effects observed with the vapours above mentioned could be due to a diminution of interior reflexion. Two parallel series of experiments executed with two different experimental tubes, the one polished, and the other blackened within, are recorded. In nine cases out of thirteen the order of absorption in both tubes was the same; in fact, the absorptions in the blackened tube multiplied by a certain coefficient were sensibly equal to those in the polished tube. In the case of the four remaining vapours slight deviations from the order of absorption were observed. These it was deemed unnecessary to follow up, *so conclusive was the evidence that the observed effects were really cases of absorption, and by no means to be referred to any alteration of the reflecting surface whether by chemical action or by condensation.*

The actions of other permanent gases than those already referred to are then recorded.

Experiments are next described which illustrate the action of the aqueous vapour of our atmosphere on radiant heat; and considerations follow regarding the influence of an atmosphere like ours upon the temperature of a planet. In former speculations upon this subject the density and height of the atmosphere were dwelt upon by distinguished writers; but it is here pointed out that a comparatively slight change in the variable constituents of our atmosphere, by permitting free access of solar heat to the earth, and checking the outflow of terrestrial heat towards space, would produce changes of climate as great as those which the discoveries of geology reveal.

Thus far our attention has been restricted to the *absorption* of radiant heat by gaseous matter, not only the general fact of absorption, but vast differences of absorptive power being established experimentally. We now come to a series of reciprocal experiments on the *radiation* of heat by gases, which demonstrate not only the general fact of radiation, but that the order of radiation is precisely the same as the order of absorption. As regards both radiation and absorption the elementary gases in the experiments here recorded, stand lowest; olefiant gas highest; and between these extremes stand the other compound gases without any shifting of position.

It is further shown that a film of gas, coating a polished metallic surface, may, both as regards radiation and absorption, be made to do the duty of a coat of varnish, or of lampblack, in increasing the emissive and absorptive power of the surface.

Our knowledge of this subject prior to the foregoing experiments is thus briefly summed up by Melloni. 'On ne connaît encore aucun fait qui démontre directement le pouvoir émissif des fluides élastiques purs et transparents.'—*Annales de Chimie et de Physique*, vol. xxii. p. 494.

Directly bearing upon this portion of the subject is an observation which, for a time, constituted one of the numerous perplexities besetting this inquiry in its earlier stages. A residue of vapour being in the experimental tube, air is permitted to enter; a prompt deflection follows, indicating an increase instead of a diminution of the transmission. The needle then returns, and finally, takes up a position indicating a slightly higher absorption than when the vapour-residue alone was present. On pumping out, the needle at first moves promptly, indicating a diminution instead of an increase of the transmission. After a momentary impulse in this direction, the needle returns to zero. The observation on analysis turned out to be an exceedingly interesting case of gaseous radiation and absorption. When the air entered, the vapour was dynamically heated; it discharged its heat against the pile, and thus apparently augmented the transmission. When the tube was exhausted the vapour was chilled, and the radiation into it from the adjacent face of the pile produced for a moment the deflection due to absorption. In subsequent memoirs, under the name of Dynamic Radiation and Absorption, this subject is fully developed.

In the last section of the memoir an attempt is made to establish a physical connexion between radiation, absorption, and conduction. One of the speculative notions in this section subsequent experience has caused me to modify. Radiation and Absorption are here regarded as the acts of the molecule as a whole, whereas I now hold them to be mainly the work of the constituent atoms of the molecule. Experimental reasons for this change of conception will be given subsequently. The memoir winds up with some supplementary remarks on the thermo-electric pile and galvanometer, intended chiefly for the use of the younger student.

I.

ON THE ABSORPTION AND RADIATION OF HEAT BY GASES AND VAPOURS, AND ON THE PHYSICAL CONNEXION OF RADIATION, ABSORPTION, AND CONDUCTION.

The Bakerian Lecture delivered before the Royal Society, February 7, 1861.*

INTRODUCTION.

THE RESEARCHES on Glaciers which I have had the honour of submitting from time to time to the Royal Society directed my attention in a special manner to the observations and speculations of De Saussure, Fourier, Pouillet, and Hopkins, on the transmission of solar and terrestrial heat through the earth's atmosphere, and gave practical effect to a desire long previously entertained to make the mutual action of radiant heat and gases and vapours of all kinds the subject of experimental inquiry.

Our acquaintance with this department of Physics is exceedingly limited. So far as my knowledge extends, the literature of the subject may be stated in a few words.

From experiments with his admirable thermo-electric apparatus, Melloni inferred that for a distance of 18 or 20 feet the absorption of radiant heat by atmospheric air is perfectly insensible.†

* Received January 10. *Philosophical Transactions* for 1861; *Philosophical Magazine*, vol. xxii. p. 169.

† *La Thermochrose*, p. 136.

With a delicate apparatus of the same kind, Dr. Franz, of Berlin, found that the air contained in a tube 3 feet long absorbed 3·54 per cent. of the heat sent through it from an Argand lamp; that is to say, calling the number of rays which passed through the exhausted tube 100, the number which passed when the tube was filled with air was only 96·46.*

In the sequel it will be shown that the result obtained by Dr. Franz was due to an inadvertence in his mode of observation. These are the only experiments of this nature with which I am acquainted, and they leave the field of inquiry now before us perfectly unbroken ground.†

§ 1.

The Galvanometer and its Defects.—Magnetic Analysis of its Wire.

At an early stage of the investigation I experienced the need of a first-class galvanometer. My instrument was constructed by that excellent workman Sauerwald, of Berlin. The needles are suspended independently of the shade, which is constructed so as to enclose the smallest possible amount of air, the disturbance of aërial currents being thereby practically avoided. The plane glass plate, which forms the cover of the instrument, is close to the needle; so that the position of the latter can be read off with ease and accuracy either by the naked eye or by a magnifying lens.

The wire of the coil belonging to this instrument was drawn from copper obtained from a galvano-plastic manufactory in the Prussian capital; but it was not free from magnetic action.

In consequence of this, when the needles were as perfectly astatic as I could make them they deviated as much as 30° right and left of the neutral line. To neutralise this deflection, a minute magnetic 'compensator' was made use of, by which the needle was gently drawn to zero in opposition to the magnetism of the coil.

But the instrument suffered much in point of delicacy from this arrangement, and accurate quantitative determinations

* Pogg. *Ann.* vol. xciv. p. 342.

† No doubt many experimenters had attempted to establish the action of air upon radiant heat; otherwise the conviction could not have become universal that no such action was discoverable.

with it were unattainable. I therefore sought to replace the Berlin coil by a less magnetic one. Mr. Becker first supplied me with a coil which reduced the lateral deflection from 30° to 3°.

But even this small residue was a source of great annoyance, and for a time I almost despaired of obtaining pure copper wire. I knew that Professor Magnus had succeeded in obtaining it for his galvanometer, but the labour of doing so was immense. He first fused, and had drawn into wire, copper obtained from a galvano-plastic manufactory, but found, after the completion of his coil, its magnetic condition intolerable. 'I have therefore,' he says, 'specially purified my copper in the following manner: A solution of sulphate of copper was saturated with ammonia, until the precipitated oxide was again dissolved. The precipitated oxide of iron was removed by filtration, and, as copper is not easily precipitated electrolytically from an ammoniacal solution, the fluid was evaporated to dryness, and all the ammonia thus drawn off. The sulphate of copper thus purified was dissolved in water and precipitated by the Voltaic current. As it was found impossible to separate the copper in an adherent mass, it was necessary to fuse it. Unhappily a very brittle metal is thus obtained, which cannot be drawn into wires. The metal had to be fused eight times in succession before it was rendered fit for this purpose. This process,' continues Magnus, 'of purifying copper is very troublesome and very costly. Without doubt it would be possible to obtain silver quite as free from magnetism as this wire, and by an easy calculation it might be proved that it would cost considerably less than an equal weight of copper prepared in the foregoing way.' *

* Pogg. *Ann.* vol. lxxxiii. p. 489.

Melloni gives the following account of the formidable nuisance of a magnetic coil: 'Les systèmes astatiques très-sensibles appliqués comme nous venons de l'indiquer aux hélices d'un fil ordinaire de cuivre ou d'argent, présentent presque toujours le fait curieux de ne pouvoir s'arrêter au zéro du cadran; c'est-à-dire que, généralement, les systèmes astatiques doués d'une grande sensibilité ne peuvent s'arrêter dans le plan vertical qui divise l'hélice en deux portions égales, parallèlement à la direction des spires. Lorsqu'on cherche à les amener dans ce plan, en tournant doucement l'hélice vers leur position d'équilibre, on les voit s'écarter aussitôt, à droite ou à gauche, et, après quelques oscillations, se fixer stablement dans une position d'équilibre plus ou moins éloignée du zéro. On mesure aisément cet arc de déviation moyennant un cercle gradué que l'on fixe à la partie supérieure de l'hélice après y avoir pratiqué une ouver-

While pondering over the means of avoiding so formidable a task, the thought occurred to me that a magnet furnished an immediate and perfect test as to the quality of the wire. Pure copper is *diamagnetic*; hence its repulsion or attraction by the magnet would at once declare its fitness or unfitness for the purpose in view.

Naked fragments of the wire furnished by M. Sauerwald were strongly attracted by the magnet. The wire furnished by Mr. Becker, when covered with its green silk, was also attracted, though in a much feebler degree.

I then removed the silk covering from the latter and tested the naked wire. *It was repelled.* The whole annoyance in its case was thus fastened on the green silk; some iron compound had been used in the dyeing of it, and to this the deviation of the needle from zero was manifestly due.

I had the green coating removed and the wire overspun with silk, clean hands being used in the process. A perfect galvanometer is the result. The needle, when released from the action of the current, returns accurately to zero, and is perfectly free from all magnetic action on the part of the coil. In fact, while we have been devising agate plates and other elaborate methods to get rid of the impurities of our galvanometer coils, the means of doing so by magnetic analysis are at hand. Diamagnetic copper wires are readily found. Out of eleven specimens, four of which were furnished by Mr. Becker, and seven taken at random from our laboratory, nine were found diamagnetic and only two paramagnetic.

Perhaps the only defect of those noble instruments with which

ture longitudinale dans le sens du zéro et de la division des spires. La déviation est égale des deux côtés; elle peut aller jusqu'à 10 ou 12 degrés et même davantage, si, en opérant sur un système astatique d'une grande perfection, *on donne une certaine largeur à la fente qui sert à introduire dans l'hélice l'aiguille inférieure du système.* Le phénomène dérive donc du partage du fil en deux masses égales, qui ont chacune un centre d'attraction, vers lequel tendent les pôles des aiguilles aimantées. Ainsi le cuivre, dont ces fils sont ordinairement composés, tout en n'étant pas un métal magnétique par lui-même, opère sur les aiguilles aimantées comme s'il contenait des parcelles de fer. C'est, en effet, le cas du cuivre de commerce; et l'on en devine facilement le motif, lorsqu'on réfléchit à l'imperfection des procédés de raffinage et au contact des outils employés dans les transformations successives du cuivre en rosettes, en verges et en fils. Et il ne faut pas s'imaginer que le fil ordinaire d'argent soit en de meilleures conditions; car l'argent, qui se trouve presque toujours en présence du fer pendant les opérations nécessaires à son extraction, ne se convertit en fil qu'à l'aide du marteau et des filières d'acier.'

Du Bois-Raymond conducts his researches in animal electricity is that here alluded to. The needle never comes to zero, but is drawn to it by a minute magnet. This defect may be completely removed. By making sure at the outset that the naked wire is diamagnetic, and by the substitution of clean white silk for green, the compensator may be dispensed with and a perfect instrument secured. It is never necessary to wait for the completion of the coil to test the quality of the wire.*

§ 2.

First Experiments on Absorption by Ordinary Methods.

Our present knowledge of the deportment of liquids and solids would lead to the inference that, if gases and vapours exercised any appreciable absorptive power on radiant heat, the absorption would make itself most manifest on heat emanating from an obscure source. But an experimental difficulty occurs at the outset in dealing with such heat. How must we close the chamber containing the gases through which the calorific rays are to be sent? Melloni found that a glass plate one-tenth of an inch in thickness intercepted all the rays emanating from a source of the temperature of boiling water, and fully 94 per cent. of the heat from a source of 400° Centigrade. Hence a tube closed with glass plates would be scarcely more suitable for the purpose now under consideration than if its ends were stopped by plates of metal.

Rock-salt immediately suggests itself as the proper substance; but to obtain plates of suitable size and transparency was exceedingly difficult. Indeed, had I been less efficiently seconded, the obstacles thus arising might have been insuperable. To the Trustees of the British Museum I am indebted for the material of one good plate of salt; to Mr. Harlin for another; while Mr. Lettsom, at the instance of Mr. Darker,† brought me a piece of salt from Germany from which two fair plates were

* Mr. Becker, to whose skill and intelligence I have been greatly indebted, furnished me with several specimens of wire of the same fineness as that used by Du Bois-Raymond, some covered with green silk and others with white. The former were invariably attracted, the latter invariably repelled. In all cases the *naked* wire was repelled.

† During the course of the inquiry I have often had occasion to avail myself of the assistance of this excellent mechanician.

taken. To Lady Murchison, Sir Emerson Tennant, Sir Philip Egerton, and Mr. Pattison my best thanks are also due for their friendly assistance.

The first experiments were made with a tube of tin polished inside, 4 feet long and 2·4 inches in diameter, the ends of which were furnished with brass appendages to receive the plates of rock-salt. Each plate was pressed firmly against a flange by a bayonet joint, being separated from the flange by a suitable washer. Various descriptions of leather washers were tried for this purpose and rejected. The substance finally chosen was vulcanised indiarubber very lightly smeared with a mixture of bees'-wax and spermaceti. A T-piece was attached to the tube, communicating on one side with a good air-pump, and on the other with the external air, or with a vessel containing the gas to be examined.

The tube being mounted horizontally, a Leslie's tube containing hot water was placed close to one of its ends, while an excellent thermo-electric pile, connected with its galvanometer, was presented to the other. The tube being exhausted, the calorific rays sent through it fell upon the pile, a permanent deflection of 30° being the consequence. The temperature of the water was in the first instance purposely so arranged as to produce this deflection.

Dry air was now admitted into the tube, while the needle of the galvanometer was observed with all possible care. Even by the aid of a magnifying lens I could not detect the slightest change of position. Oxygen, hydrogen, and nitrogen, subjected to the same test, gave the same negative result. The temperature of the water was subsequently lowered, so as to produce a deflection of 20° and 10° in succession, and then heightened till the deflection amounted to 40°, 50°, 60°, and 70°; but in no case did the admission of air, or any of the above gases, into the exhausted tube produce any sensible change in the position of the needle.

It is a well-known peculiarity of the galvanometer, that its higher and lower degrees represent different amounts of calorific action. In my instrument, for example, the quantity of heat necessary to move the needle from 70° to 71° is about twenty times that required to move it from 11° to 12°.* Now in the

* See *Remarks*, 1872, at the end of this memoir.

case of the small deflections above referred to the needle was, it is true, in a sensitive position; but then the total amount of heat passing through the tube was so inconsiderable that a small percentage of it, even if absorbed, might well escape detection. In the case of the large deflections, on the other hand, a very considerable abstraction of heat would be necessary to produce any sensible diminution of the deflection. Hence arose the thought of operating, if possible, with large quantities of heat, while the needle intended to reveal its absorption should continue to occupy its position of maximum delicacy.

§ 3.

Method of Compensation.

The first attempt at solving this problem was as follows:—My galvanometer is a differential one—the coil being composed of two wires wound side by side, so that a current can be sent through either of them independent of the other. The thermo-electric pile was placed at one end of the tin tube, and the ends of one of the galvanometer wires were connected with it. A copper ball heated to low redness being placed at the other end of the tube, the needle of the galvanometer was propelled to its stops at 90°. The ends of the second wire were now so attached to a second pile that when the latter was caused to approach the copper ball, the current excited passed through the coil in a direction opposed to the first one. Gradually, as the second pile was brought nearer to the source of heat, the needle descended from the stops, and when the two currents were equal the position of the needle was at zero.

Here, then, we had a powerful flux of heat through the tube; and if a column of gas four feet long exercised any sensible absorption, the needle was in the position best calculated to reveal it. In the first experiment made in this way, the neutralisation of one current by the other occurred when the tube was filled with air; on exhausting the tube, the needle started suddenly off in a direction which indicated that a *less* amount of heat passed through the partially exhausted tube than through the filled one. The needle, however, soon stopped, turned, descended quickly to zero, and passed on to the other side, where its deflection became permanent. The air employed in this experiment came direct from the laboratory, and the first impulsion of the needle was probably due to the aqueous vapour precipitated as a cloud by the sudden exhaustion of the tube; for when, previous to its admission, the air passed over chloride of calcium, or pumice-stone moistened with sulphuric acid, no such effect was observed. The needle moved steadily in one direction till its maximum deflection was attained, and this deflection showed that in all cases radiant heat was absorbed by the air within the tube.

These experiments were begun in the spring of 1859, and continued without intermission for seven weeks. The course of the inquiry during this whole period was an incessant struggle with experimental difficulties. Approximate results were easily obtainable; but I aimed at exact measurements, which could not be made with a varying source of heat like the copper ball. To obtain a high and steady source of heat I resorted to copper cubes containing fusible metal, or oil, but was not satisfied with their action. Finally, a lamp was constructed which poured a steady sheet of flame against a plate of copper; and, to keep the flame constant, a gas regulator specially constructed for me by Mr. Hulet was made use of. It was also arranged that the radiating plate should form one of the walls of a chamber which could be connected with the air-pump and exhausted, so that the heat emitted by the copper plate might cross a vacuum before entering the experimental tube. With this apparatus, during the summer of 1859, I approximately determined the absorption of nine gases and twenty vapours. The results would furnish materials for a long memoir; but increased experience and improved methods have enabled me to substitute for those results others of greater accuracy; I shall therefore pass over the work of these seven weeks without further allusion to it.

On September 9 of the present year (1860) the inquiry was resumed. For three weeks the heated plate of copper was my source of heat, but it was finally rejected on the score of insufficient constancy. The cube of hot oil was again resorted to, and with it I continued to work up to Monday, October 29. During these seven weeks, from eight to ten hours daily were devoted to experiments; but the results, though more accurate, must unhappily share the fate of those obtained in 1859. In fact, these fourteen weeks of labour constituted a period of discipline, during which a continued struggle was carried on against the difficulties of the subject and the defects of the locality in which the inquiry was conducted.

My reason for trying these high sources of heat was this: the absorptive power of some of the gases examined was so small that, to make it clearly evident, a powerful beam of radiant heat was essential. For other gases, and for *all* the vapours that had come under my notice, sources of lower temperature would have been not only sufficient, but far preferable. I was finally induced to resort to boiling water, which, though it gave greatly diminished effects, was capable of being preserved at so constant a temperature that deflections which, with the other sources, would be disturbed by errors of observation, became with it true quantitative measures of absorption.

§ 4.

Final Form of Apparatus.

The entire apparatus made use of in the experiments on absorption is figured on the Frontispiece. S S′ is the *experimental tube*, composed of brass, polished within, and connected, as shown in the figure, with the air-pump A A. At S and S′ are the plates of rock-salt which close the tube air-tight. The length from S to S′ is 4 feet. C is a cube, containing boiling water, in which is immersed the thermometer *t*. The cube is of cast-copper, and on one of its faces was a projecting ring, to which a brass tube of the same diameter as S S′, and capable of being connected air-tight with the latter, was carefully soldered. The face of the cube within the ring is the radiating plate, which is coated with lampblack. Thus between the cube C and the first plate of rock-salt there is *a front chamber* F, connected with the air-pump of the flexible tube D D, and capable of being exhausted independently of S S′. To prevent the heat of conduction from reaching the plate of rock-salt S, the tube F is caused to pass through a vessel V, being soldered to the latter where it enters it and issues from it. This vessel is supplied with a continuous flow of cold water through the influx tube *i i*, which dips to the bottom of the vessel; the water escapes through the efflux tube *e e*, and the continued circulation of the cold liquid completely intercepts the heat that would otherwise reach the plate S.

The cube C is heated by the gas-lamp L. P is the thermo-electric pile placed on its stand at the end of the experimental tube, and furnished with two conical reflectors, as shown in the figure. C′ is the *compensating cube*, used to neutralise by its radiation * the effect of the rays passing through S S′. The regulation of this neutralisation was an operation of some delicacy; to effect it the double screen H was connected with a winch and screw arrangement, by which it could be advanced or withdrawn through extremely minute spaces. For this most useful adjunct I am indebted to the kindness of my friend, Mr. Gassiot. N N is the galvanometer, with perfectly astatic needles and perfectly non-magnetic coil; it is connected with the pile P by the wires *w w*; Y Y is a system of six chloride of calcium tubes, each 32 inches long; R is a U-tube, containing fragments of pumice-stone, moistened with strong caustic potash; and Z is a second similar tube, containing fragments of pumice-stone wetted with strong sulphuric acid. When *drying* only was aimed at, the potash tube was suppressed. When, on the contrary, as in the case of atmospheric air, both moisture and carbonic acid were to be removed, the potash tube was included. G G is a holder from which the gas to be experimented with was sent through the drying tubes, and thence through the pipe *p p* into the experimental tube S S′. The appendage at M and the arrangement at O O may for the present be disregarded; I shall refer to them particularly by-and-by.

The mode of proceeding was as follows: The tube S S′ and the chamber F being exhausted as perfectly as possible, the connexion between them was intercepted by shutting off the cocks *m m′*. The rays from the interior blackened surface of the cube C passed first across the vacuum F, then through the plate of rock-salt S, traversed the experimental tube, crossed the second plate S′, and being concentrated by the anterior conical reflector, impinged upon the adjacent face of the pile P.

* It will be seen that in this arrangement I have abandoned the use of the differential galvanometer, and made the thermo-electric pile itself the differential instrument.

Meanwhile the rays from the hot cube C′ fell upon the opposite face of the pile, and the position of the galvanometer needle declared at once which source was predominant. A movement of the screen H back or forward with the hand sufficed to establish an approximate equality; but to make the radiations perfectly equal, and thus bring the needle exactly to 0°, the fine motion of the screw above referred to was necessary.

The needle being at 0°, the gas to be examined was admitted into the tube, passing, in the first place, through the drying apparatus. Any required quantity of the gas might be admitted; and here experiments on gases and vapours enjoy an advantage over those with liquids and solids—namely, the capability of changing the density at pleasure. When the required quantity of gas had been admitted the galvanometer was observed, and from the deflection of its needle the absorption was accurately determined.

The galvanometer was calibrated by the method recommended by Melloni ('Thermochrose,' p. 59), the precise value of its larger deflections being at once obtained by reference to a table. Up to the 30th degree, or thereabouts, the deflections may be regarded as the expression of the absorption; but beyond this the absorption equivalent to any deflection was obtained from the table of calibration.

§ 5.

ABSORPTION OF RADIANT HEAT.

First Results.—Action of Ozone and of Compound Gases on Radiant Heat.

The air of the laboratory, freed from its moisture and carbonic acid, and permitted to enter until the tube was filled, produced a deflection of about . . . 1°

Oxygen obtained from chlorate of potash and peroxide of manganese produced a deflection of about . . . 1°

One specimen of nitrogen, obtained from the decomposition of nitrate of potash, produced a deflection of about 1°

Hydrogen from zinc and sulphuric acid produced a deflection of about . . . . 1°

Hydrogen obtained from the electrolysis of water produced a deflection of about . . . . 1°

Oxygen obtained from the electrolysis of water, and sent through a series of eight bulbs containing a strong solution of iodide of potassium, produced a deflection of about . . . . 1°

In the last experiment the electrolytic oxygen was freed from its ozone. The iodide of potassium was afterwards suppressed, and the oxygen, plus its ozone, admitted into the tube; the deflection produced was . . . 4°

Hence the small quantity of ozone which accompanied the oxygen in this case trebled the absorption of the oxygen itself.*

I have repeated this experiment many times, employing different sources of heat. With sources of high temperature the difference between the ozone and the ordinary oxygen comes out very strikingly. By careful decomposition a much larger amount of ozone might be obtained, and a correspondingly large effect on radiant heat.

In obtaining the electrolytic oxygen two different vessels were made use of. To diminish the resistance of the acidulated water to the passage of the current, I placed in one vessel a pair of very large platinum plates, between which the current from a battery of ten Grove's cells was transmitted. The oxygen bubbles liberated were extremely minute, and the gas, on being sent through iodide of potassium, scarcely coloured the liquid; the characteristic odour of ozone was almost entirely absent, and there was little or no action upon radiant heat. In the second vessel smaller plates were used. The bubbles of oxygen were much larger, and did not come into such intimate contact with either the platinum or the water. The oxygen thus obtained showed the characteristic reactions of ozone; and with it the above result was obtained.

The total amount of heat transmitted through the tube in these experiments produced a deflection of . . . 71·5°

Taking as unit of heat the quantity necessary to cause the needle to move from 0° to 1°, the number of units expressed by the above deflection is 308

Hence the absorption by the above gases amounted to about 0·33 per cent.

* It will be shown subsequently that this result is in harmony with the supposition that ozone is a *compound* body. See Memoir II. §§ 17, 18, and 19.

I am unable at the present moment to range with certainty oxygen, hydrogen, nitrogen, and atmospheric air in the order of their absorptive powers, though several hundred experiments have been made with the view of doing so. The proper action of these gases is so small that the slightest foreign impurity gives one a predominance over the other. In preparing the gases the methods recommended in chemical treatises have been resorted to, but as yet only to discover the defects incidental to these methods. Augmented experience and the assistance of my friends will, I trust, enable me to solve this point by-and-by. An examination of the whole of the experiments induces me to regard hydrogen as the gas which exercises the lowest absorptive power.*

We have here the cases of minimum gaseous absorption. It will be interesting to place in juxtaposition with the above results some of those obtained with olefiant gas—the most highly absorbent permanent gas that I have hitherto examined. I select for this purpose an experiment made on November 21.

The needle being steady at zero in consequence of the equality of the actions on the opposite faces of the pile, the admission of olefiant gas gave a permanent deflection of . . . . 70·3°

The gas being completely removed, and the equilibrium re-established, a plate of polished metal was interposed between one of the faces of the pile and the source of heat adjacent. The total amount of heat passing through the exhausted tube was thus found to produce a deflection of . . . . 75°

Now a deflection of 70·3° is equivalent to 290 units, and a deflection of 75° is equivalent to 360 units; hence more than seven-ninths of the total heat, about 81 per cent., were cut off by the olefiant gas.

* The test to which these gases were subjected was far more severe than any previously applied, but the result was in practical accordance with the conviction that, at all events, in transparent gases, the absorption of radiant heat, if not absolutely insensible, was, at all events, beyond the reach of experiment. But early in 1859, coal-gas being at hand, I tried it, and found its deportment more like that of an adiathermic solid than that of a gas. A crowd of other gases was immediately added. In fact, this single experiment with coal-gas opened the door to all the researches recorded in this volume. [1872.]

The extraordinary energy with which the needle was deflected when the olefiant gas was admitted into the tube, was such as might occur had the plates of rock-salt become suddenly covered with something opaque. To test whether any such action occurred, I carefully polished a plate, and against it was projected for a considerable time a stream of the gas; there was no dimness produced. The plates of rock-salt, moreover, which were removed daily from the tube, usually appeared as bright when taken out as when they were put in.*

The gas in these experiments issued from its holder, where it had been in contact with cold water. To test whether it had so chilled the plates of rock-salt as to produce the effect, a similar holder was filled with atmospheric air and permitted it to attain the temperature of the water; but the action of the air was not thereby sensibly augmented.

In order to subject the gas to ocular examination, I had a glass tube constructed and connected with the air-pump. On permitting olefiant gas to enter it not the slightest dimness or opacity was observed. To remove the last trace of doubt as to the possible action of the gas on the plates of rock-salt, the tin tube referred to at page 12 was perforated at its centre and a cock inserted into it; the source of heat was at one end of the tube, and the thermo-electric pile at some distance from the other. The plates of salt were entirely abandoned, the tube being open at its ends and consequently full of air. On allowing the olefiant gas to stream for a second or two into the tube through the central cock, the needle flew off and struck against its stops. It was held steadily for a considerable time between 80° and 90°.

A slow current of air sent through the tube gradually removed the gas, and the needle returned accurately to zero.

The gas within the holder being under a pressure of about twelve inches of water, the cock attached to the cube was turned quickly on and off; the quantity of gas which entered the tube in this brief interval was sufficient to cause the needle to be driven to the stops, and steadily held between 60° and 70°.

The gas being again removed, the cock was turned once half round as quickly as possible. The needle was driven in the

* From the very beginning of the inquiry my attention was awake to the possibility of precipitation upon the plates of rock-salt. [1872.]

first instance through an arc of 60°, and was held permanently at 50°.

The quantity of gas which produced this last effect, on being admitted into a graduated tube, was found not to exceed one-sixth of a cubic inch in volume.

The tin tube was now taken away, and both sources of heat allowed to act from some distance on the thermo-electric pile. When the needle was at zero, olefiant gas was allowed to issue from a common Argand burner into the air between one of the sources of heat and the pile. The gas was invisible—nothing was seen in the air—but the needle immediately declared its presence, being driven through an arc of 41°. In the four experiments last described the source of heat was a cube of oil heated to 250° Centigrade, the compensation cube being filled with boiling water.*

Those who, like myself, have been taught to regard transparent gases as sensibly diathermanous, will probably share the astonishment with which I witnessed the foregoing effects. I was, indeed, slow to believe it possible that a body so constituted, and so transparent to light as olefiant gas, could be so densely opaque to any kind of rays; and, to secure myself against error, several hundred experiments were executed with this single substance. But the citing of them at greater length could not add to the conclusiveness of the proofs just furnished, that the case is one of true calorific absorption.†

§ 6.

Variations of Density.—Relation of Absorption to Quantity of Matter.

Having thus established in a general way the absorptive power of olefiant gas, the question arises, What is the relation which subsists between the density of the gas and the quantity of heat extinguished?

* With a cube containing boiling water this experiment has been since made visible to a large assembly.

† It is evident that the old mode of experiment might be applied to this gas. Indeed, several of the solids examined by Melloni are inferior to it in absorptive power. Had time permitted, I should have checked my results by experiments made in the usual way; this, however, will be done on a future occasion.

I sought at first to answer this question in the following way: An ordinary mercurial gauge was attached to the air-pump. The experimental tube being exhausted, and the needle of the galvanometer at zero, olefiant gas was admitted until it depressed the mercurial column 1 inch, the consequent deflection being noted. The gas was then admitted until a depression of 2 inches was observed, and thus the absorption effected by gas of 1, 2, 3, and more inches' pressure was determined. In the following table the first column contains the pressures in inches of mercury, the second the deflections, and the third the absorption equivalent to each deflection.

TABLE I.—*Olefiant Gas.*

Pressures in inches	Deflections °	Absorption per 100	Pressures in inches	Deflections °	Absorption per 100
1	56	90	7	61·4	182
2	58·2	123	8	61·7	186
3	59·3	142	9	62	190
4	60	157	10	62·2	192
5	60·5	168	20	66	227
6	61	177			

No definite relation between the density of the gas and its absorption is here exhibited. We see that an augmentation of the density *seven times* about *doubles* the amount of the absorption; while gas of 20 inches' pressure effects only 2½ times the absorption of gas under 1 inch of pressure.

But here the following reflections suggest themselves: It is evident that olefiant gas of 1 inch pressure, producing so large a deflection as 56°, must extinguish a large proportion of the rays which are capable of being absorbed by the gas, and hence the succeeding measures, having a less and less amount of heat to act upon, must produce a continually smaller effect. But supposing the quantity of gas first introduced to be so inconsiderable that the number of rays extinguished by it is a vanishing quantity compared with the total number capable of absorption, we might reasonably expect that in this case a double quantity of gas would produce a double effect, a treble quantity a treble effect, or, in general terms, that the absorption would, for a time, be proportional to the density.

To test this idea a portion of the apparatus, purposely omitted in the description already given, was made use of.

O O (see Frontispiece) is a graduated glass tube, the end of which dips into the basin of water B. The tube can be closed above by means of the stopcock *r*; *d d* is a tube containing fragments of chloride of calcium. The tube O O is first filled with water to the cock *r*; this water is then displaced by olefiant gas; and afterwards the tube S S′, and the entire space between the cock *r* and the experimental tube, is exhausted. The cock *n* being now closed and *r′* left open, the cock *r* at the top of the tube O O is carefully turned on and the gas permitted to enter the tube S S′ with extreme slowness. The water rises in O O, each of the smallest divisions of which represents a volume of $\frac{1}{50}$th of a cubic inch. Successive measures of this capacity were admitted into the tube, the absorption in each case being determined.

In the following table the first column expresses the quantity of gas admitted into the experimental tube, and the second the corresponding deflection, which, within the limits of the table, expresses the absorption; the third column contains the absorption, calculated on the supposition that it is proportional to the density.

TABLE II.—*Olefiant Gas.*

(Unit-measure $\frac{1}{50}$th of a cubic inch.)

Measures of Gas	Absorption per 100		Measures of Gas	Absorption per 100	
	Observed	Calculated		Observed	Calculated
1	2·2	2·2	9	19·8	19·8
2	4·5	4·4	10	22	22
3	6·6	6·6	11	24	24·2
4	8·8	8·8	12	25·4	26.4
5	11	11	13	29	28·6
6	12	13·2	14	30·2	29·8
7	14·8	15·4	15	33·5	33
8	16·8	17.6			

This table shows the correctness of the foregoing surmise, and proves that for small quantities of gas the absorption is exactly proportional to the density.

Let us pause for a moment to estimate the tenuity of the gas with which we have here operated. The length of the experimental tube is 48 inches, and its diameter 2·4 inches; its volume is therefore 218 cubic inches. Adding to this the contents of the cocks and other conduits leading to the tube, we may assume that each fiftieth of a cubic inch of the gas had to diffuse itself through a space of 220 cubic inches. The pressure, therefore, of a single measure of the gas thus diffused would be $\frac{1}{11000}$th of an atmosphere,—a tension capable of depressing the mercurial column connected with the pump $\frac{1}{367}$th of an inch, or about $\frac{1}{15}$th of a millimetre!

§ 7.

Action of Sulphuric-ether Vapour on Radiant Heat.

But the absorptive energy of olefiant gas, extraordinary as it is shown to be by the above experiments, is far exceeded by that of some of the vapours of volatile liquids. A glass flask was provided with a brass cap furnished with an interior thread, by means of which a stopcock could be screwed airtight to the flask. Sulphuric ether being placed in the latter, the space above the liquid and the liquid itself were completely freed of air by means of an air-pump. The flask, with its closed stopcock, being attached to the experimental tube; the latter was exhausted and the needle brought to zero. The cock was then turned on so that the ether-vapour slowly entered the experimental tube. An assistant observed the gauge of the air-pump, and when it had sunk an inch, the stopcock was promptly closed. The consequent galvanometric deflection was then noted; a second quantity of the vapour, sufficient to depress the gauge another inch, was then admitted, and in this way the absorptions of five successive measures, each possessing within the tube 1 inch of pressure, were determined.

In the following table the first column contains the pressures in inches, the second the deflection due to each, and the third the amount of heat absorbed, expressed in the units already referred to. For the purpose of comparison the corresponding absorptions of olefiant gas are placed in the fourth column.

Table III.—*Sulphuric Ether.*

Pressures in inches	Deflections	Absorption per 100	Corresponding Absorption by Olefiant Gas
1	64·8	214	90
2	70	282	123
3	72	315	142
4	73	330	154
5	73	330	163

For these pressures the absorption of radiant heat by the vapour of sulphuric ether is more than twice that of olefiant gas. We also observe that in the case of the vapour the successive absorptions approximate more quickly to equality. In fact, the absorption produced by 4 inches of the vapour is sensibly the same as that produced by 5.

But reflections similar to those already applied to olefiant gas are also applicable to ether. Supposing we make our unit-measure small enough, the number of rays first destroyed will vanish in comparison with the total number, and for a time the fact will probably manifest itself that the absorption is directly proportional to the density. To examine whether this is the case, a portion of the apparatus, omitted in the general description, was made use of. K is a small flask, with a brass cap, which is closely screwed to the stopcock c'. Between the cocks c' and c, the latter connected with the experimental tube, is the chamber M, the capacity of which was accurately determined. The flask K being partially filled with ether, the air above the liquid is removed. The stopcock c' being shut off and c turned on, the tube S S′ and the chamber M are exhausted. The cock c being now shut off, and c' turned on, the chamber M is filled with pure ether-vapour. By turning c' off and c on, this vapour is allowed to diffuse itself through the experimental tube; successive measures are thus introduced, and the effect produced by each is noted. Measures of various capacities were made use of, according to the requirements of the vapours examined.

In the first series of experiments made with this apparatus, I omitted to remove the air from the space above the liquid; each measure therefore sent in to the tube was a mixture of vapour and air. This diminished the effect of the vapour; but the proportionality, for small quantities, of density to absorption exhibits itself so decidedly as to induce me to record the observations. The first column, as usual, contains the measures of vapour, the second the observed absorption, and the third the calculated absorption. The galvanometric deflections are omitted, their values being contained in the second column. In fact, as far as the seventh observation, the absorptions are merely the record of the deflections.

Table IV.—*Mixture of Ether Vapour and Air.*

(Unit-measure $\frac{1}{50}$th of a cubic inch.)

Measures	Absorption per 100 Observed	Absorption per 100 Calculated	Measures	Absorption per 100 Observed	Absorption per 100 Calculated
1	4·5	4·5	21	82·8	95
2	9·2	9	22	84	99
3	13·5	13·5	23	87	104
4	18	18	24	88	108
5	22·8	23·5	25	90	113
6	27	27	26	93	117
7	31·8	31·5	27	94	122
8	36	36	28	95	126
9	39·7	40	29	98	131
10	45	45	30	100	135
20	81	90			

Up to the 10th measure we find that density and absorption augment in precisely the same ratio. While the former varies from 1 to 10, the latter varies from 4·5 to 45. At the 20th measure, however, a deviation from proportionality is apparent, and the divergence gradually augments from 20 to 30. In fact 20 measures tell upon the heat capable of being absorbed—the quantity quenched becoming so considerable that at length every additional measure encounters a materially enfeebled beam, and hence produces a diminished effect.

With ether vapour alone, the results recorded in the following table were obtained; and as I wished to know how far the pressure of the vapour might be diminished, the capacity of the unit-measure was reduced to $\frac{1}{100}$th of a cubic inch.

Table V.—*Sulphuric Ether.*

(Unit-measure $\frac{1}{100}$th of a cubic inch.)

Measures	Absorption per 100 Observed	Absorption per 100 Calculated	Measures	Absorption per 100 Observed	Absorption per 100 Calculated
1	5	4·6	17	65·5	77·2
2	10·3	9·2	18	68	83
4	19·2	18·4	19	70	87·4
5	24·5	23	20	72	92
6	29·5	27	21	73	96·7
7	34·5	32·2	22	73	101·2
8	38	36·8	23	73	105·8
9	44	41·4	24	77	110·4
10	46·2	46·2	25	78	115
11	50	50·6	26	78	119·6
12	52·8	55·2	27	80	124·2
13	55	59·8	28	80·5	128·8
14	57·2	64·4	29	81	133·4
15	59·4	69	30	81	138
16	62·5	73·6			

We here find the proportion between density and absorption sensibly preserved for the first eleven measures, after which the deviation gradually augments. Some specimens of ether have been examined which acted still more energetically on the thermal rays than that just referred to.

No doubt for smaller measures than $\frac{1}{100}$th of a cubic inch the above law holds still more rigidly true; and in a suitable locality it would be easy to determine with perfect accuracy $\frac{1}{10}$th of the absorption produced by our first measure; this would correspond to $\frac{1}{1000}$th of a cubic inch of vapour. But on entering the tube the vapour has only the tension due to the temperature of the laboratory, namely 12 inches. This would require to be multiplied by 2·5 to bring it up to that of the atmosphere. Hence the $\frac{1}{1000}$th of a cubic inch, the absorption of which has been affirmed capable of measurement, would, on being diffused through a tube possessing a capacity of 220 cubic inches, have a pressure of $\frac{1}{220} \times \frac{1}{2\cdot5} \times \frac{1}{1000} = \frac{1}{500{,}000}$th of an atmosphere!

§ 8.

Extension of Inquiry to other Vapours.

I have now to record the results obtained with thirteen other vapours. The method of experiment was in all cases the same as that employed in the case of ether, the only variable element being the size of the unit-measure. For with many substances no sensible effect could be obtained with the unit volume employed in the experiments last recorded. With bisulphide of carbon, for example, it was necessary to augment the unit-measure 50 times to render the measurements satisfactory.

TABLE VI.—*Bisulphide of Carbon.*

(Unit-measure ½ a cubic inch.)

Measures	Absorption per 100		Measures	Absorption per 100	
	Observed	Calculated		Observed	Calculated
1	2·2	2·2	11	16·2	24·2
2	4·9	4·4	12	16·8	26·4
3	6·5	6·6	13	17·5	28·6
4	8·8	8·8	14	18·2	30·8
5	10·7	11	15	19	33
6	12·5	13	16	20	35·2
7	13·8	15·4	17	20	37·4
8	14·5	17·6	18	20·2	39·6
9	15	19	19	21	41·8
10	15·6	22	20	21	44

As far as the sixth measure the absorption is proportional to the density; after which the effect of each successive measure diminishes. Comparing the absorption effected by a quantity of vapour which depressed the mercury column half an inch, with that effected by vapour of one inch pressure, the same deviation from proportionality is observed. Thus:—

By mercurial gauge.

Pressure	Absorption per 100
½ inch	14·8
1 inch	18·8

These numbers simply express the galvanometric deflections, which, as already stated, are strictly proportional to the absorption as far as 30° or thereabouts. Did the law of proportion hold good, the absorption due to 1 inch of tension ought of course to be 29·6 instead of 18·8.

Whether for equal volumes of the vapours at their maximum density, or for equal pressures as measured by the depression of the mercurial column, bisulphide of carbon exercises the lowest absorptive power of all the vapours hitherto examined. For very small quantities, a volume of sulphuric-ether vapour, at its maximum density in the unit-measure, and expanded thence into the experimental tube, absorbs 100 times the quantity of heat intercepted by an equal volume of bisulphide of carbon vapour at its maximum density. These substances mark the extreme limits of the scale, as far as my inquiries have hitherto proceeded. The action of every other vapour is less than that of sulphuric ether, and greater than that of bisulphide of carbon.

Remarks on the Explosion of Bisulphide-of-Carbon Vapour in the Air-pump Cylinders.

A very singular phenomenon was repeatedly observed during the experiments with bisulphide of carbon. After determining the absorption of the vapour, the tube was exhausted as perfectly as possible, the trace of vapour left behind being exceedingly minute. Dry air was then admitted to cleanse the tube. On again exhausting, after the first few strokes of the pump a jar was felt and a kind of explosion heard, while dense volumes of blue smoke immediately issued from the cylinders. The action was confined to these, and never propagated backwards into the experimental tube.

It is only with bisulphide of carbon that this effect has been observed. It may, I think, be explained in the following manner:—To open the valve of the piston, the gas beneath it must have a certain pressure, which, when suddenly produced, is sufficient to cause the combination of the constituents of the bisulphide of carbon with the oxygen of the air. Such a combination certainly takes place, for the odour of sulphurous acid is unmistakable amid the fumes.

To test this idea I tried the effect of compression in the air-syringe. A bit of tow or cotton wool moistened with bisulphide of carbon, and placed in the syringe, emitted a bright flash when the air was compressed. By blowing out the fumes with a glass tube, this experiment may be repeated twenty times with the same bit of cotton.

It is not necessary even to let the moistened pellet remain in the syringe. If the bit of tow or cotton be thrown into the syringe, and out again as quickly as it can be ejected, on compressing the air the luminous flash is seen. Pure oxygen produces a brighter flash than air. These facts are in harmony with the above explanation.

Continuation of Experiments on Vapours.

TABLE VII.—*Amylene.*

(Unit-measure $\frac{1}{10}$th of a cubic inch.)

Measures	Absorption per 100		Measures	Absorption per 100	
	Observed	Calculated		Observed	Calculated
1	3·4	4·3	6	26·5	25·8
2	8·4	8·6	7	30·6	30·1
3	12	12·9	8	35·3	34·4
4	16·5	17·2	9	39	38·7
5	21·6	21·5	10	44	43

For these quantities the absorption is proportional to the density, but for large quantities the usual deviation is observed as shown by the following observations:—

By mercurial gauge.

Pressure	Deflection	Absorption per 100
½ inch	60°	157
1 inch	65	216

Did the proportion hold good, the absorption for an inch of pressure ought to be 314, instead of 216.

TABLE VIII.—*Iodide of Ethyl.*

(Unit-measure $\frac{1}{10}$th of a cubic inch.)

Measures	Absorption per 100 Observed	Absorption per 100 Calculated	Measures	Absorption per 100 Observed	Absorption per 100 Calculated
1	5·4	5·1	6	31·8	30·6
2	10·3	10·2	7	35·6	35·9
3	16·8	15·3	8	40	40·8
4	22·2	20·4	9	44	45·9
5	26·6	25·5	10	47·5	51

By mercurial gauge.

Pressure	Deflection	Absorption per 100
½ inch	56°·3	94
1 inch	58·2	120

TABLE IX.—*Iodide of Methyl.*

(Unit-measure $\frac{1}{10}$th of a cubic inch.)

Measures	Absorption per 100 Observed	Absorption per 100 Calculated	Measures	Absorption per 100 Observed	Absorption per 100 Calculated
1	3·5	3·4	6	20·5	20·4
2	7	6·8	7	24	23·8
3	10·3	10·2	8	26·3	27·2
4	15	13·6	9	30	30·6
5	17·5	17	10	32·3	34

By mercurial gauge.

Pressure	Deflection	Absorption per 100
½ inch	48°·5	60
1 inch	56·5	96

TABLE X.—*Iodide of Amyl.*

(Unit-measure $\frac{1}{10}$th of a cubic inch.)

Measures	Absorption per 100 Observed	Absorption per 100 Calculated	Measures	Absorption per 100 Observed	Absorption per 100 Calculated
1	0·6	0·57	6	3·8	3·4
2	1	1·1	7	4·5	4
3	1·4	1·7	8	5	4·6
4	2	2·3	9	5	5·1
5	3	2·9	10	5·8	5·7

The deflections here are very small; the substance, however, possesses such feeble volatility that the pressure of a measure of its vapour, when diffused through the experimental tube, must be infinitesimal. With the specimen examined, it was not practicable to obtain a pressure sufficient to cause the mercury gauge to sink half-an-inch; hence no observations of this kind are recorded.

TABLE XI.—*Chloride of Amyl.*

(Unit-measure $\frac{1}{10}$th of a cubic inch.)

	Absorption per 100			Absorption per 100	
Measures	Observed	Calculated	Measures	Observed	Calculated
1	1·3	1·3	6	8·5	7·8
2	3	2·6	7	9	9·1
3	3·8	3·9	8	10·9	10·4
4	5·1	5·2	9	11·3	11·7
5	6·8	6·5	10	12·3	13

By mercurial gauge.

Pressure	Deflection	Absorption per 100
½ inch	59°	137
1 inch	not practicable.	

TABLE XII.—*Benzol.*

(Unit-measure $\frac{1}{10}$th of a cubic inch.)

	Absorption per 100			Absorption per 100	
Measures	Observed	Calculated	Measures	Observed	Calculated
1	4·5	4·5	11	47	49
2	9·5	9	12	49	54
3	14	13·5	13	51	58·5
4	18·5	18	14	54	63
5	22·5	22·5	15	56	67·5
6	27·5	27	16	59	72
7	31·6	31·5	17	63	76·5
8	35·5	36	18	67	81
9	39	40	19	69	85·5
10	44	45	20	72	90

Up to the 10th measure, or thereabouts, the proportion between density and absorption holds good, from which onwards the deviation from the law gradually augments.

By mercurial gauge.

Pressure	Deflection	Absorption per 100
½ inch	54°	78
1 inch	57	103

Table XIII.—*Methylic Alcohol.*

(Unit-measure $\frac{1}{10}$th of a cubic inch.)

Measures	Absorption per 100 Observed	Absorption per 100 Calculated	Measures	Absorption per 100 Observed	Absorption per 100 Calculated
1	10	10	6	53·5	60
2	20	20	7	59·2	70
3	30	30	8	71·5	80
4	40·5	40	9	78	90
5	49	50	10	84	100

By mercurial gauge.

Pressure	Deflection	Absorption per 100
½ inch	58°·8	133
1 inch	60·5	168

Table XIV.—*Formic Ether.*

(Unit-measure $\frac{1}{10}$th of a cubic inch.)

Measures	Absorption per 100 Observed	Absorption per 100 Calculated	Measures	Absorption per 100 Observed	Absorption per 100 Calculated
1	8	7·5	6	39·5	45
2	16	15	7	45	52·5
3	22·5	22·5	8	48	60
4	30	30	9	50·2	67·5
5	35·2	37·5	10	53·5	75

By mercurial gauge.

Pressure	Deflection	Absorption per 100
½ inch	58°·8	133
1 inch	62·5	193

Table XV.—*Propionate of Ethyl.*

(Unit-measure $\frac{1}{10}$th of a cubic inch.)

Measures	Absorption per 100 Observed	Absorption per 100 Calculated	Measures	Absorption per 100 Observed	Absorption per 100 Calculated
1	7	7	6	38·8	42
2	14	14	7	41	49
3	21·8	21	8	42·5	56
4	28·8	28	9	44·8	63
5	34·4	35	10	46·5	70

By mercurial gauge.

Pressure	Deflection	Absorption per 100
½ inch	60°·5	168
1 inch	not practicable.	

Table XVI.—*Chloroform.*

(Unit-measure $\frac{1}{10}$th of a cubic inch.)

Measures	Absorption per 100 Observed	Absorption per 100 Calculated	Measures	Absorption per 100 Observed	Absorption per 100 Calculated
1	4·5	4·5	6	27	27
2	9	9	7	31·2	31·5
3	13·8	13·5	8	35	36
4	18·2	18	9	39	40·5
5	22·3	22·5	10	40	45

Subsequent observations lead me to believe that the absorption by chloroform is a little higher than that given in the above table.

Table XVII.—*Alcohol.*

(Unit-measure $\frac{1}{2}$ a cubic inch.)

Measures	Absorption per 100 Observed	Absorption per 100 Calculated	Measures	Absorption per 100 Observed	Absorption per 100 Calculated
1	4	4	9	37·5	36
2	7·2	8	10	41·5	40
3	10·5	12	11	45·8	44
4	14	16	12	48	48
5	19	20	13	50·4	52
6	23	24	14	53·5	56
7	28·5	28	15	55·8	60
8	32	32			

By mercurial gauge.

Pressure	Deflection	Absorption per 100
$\frac{1}{2}$ inch	60°	175
1 inch	not practicable.	

The difference between the measurements when equal *pressures* and when equal *volumes* at the maximum density are made use of is here strikingly exhibited. In the case of alcohol, for example, a unit-measure of half a cubic inch was needed to obtain an effect about equal to that produced by benzol with a measure only $\frac{1}{10}$ of a cubic inch in capacity; and yet for a common pressure of 0·5 of an inch, alcohol cuts off precisely twice as much heat as benzol. There is also an enormous difference between alcohol and sulphuric ether when equal measures at the maximum density are compared; but to bring the alcohol and ether vapours up to a common pressure, the density of the former must be many times augmented. Hence it follows that when *equal pressures* of these two substances are compared, the difference between them diminishes considerably. Similar observations apply to many of the substances whose deportment is recorded in the foregoing tables; to the iodide and chloride of amyl, for example, and to the propionate of ethyl. Indeed it is not unlikely that with equal pressures the vapour of a perfectly pure specimen of the substance last mentioned would be found to possess a higher absorptive power than that of sulphuric ether itself.

§ 9.

Action of Chlorine.—Possible Influence of Vapours on the Interior Surface of the Experimental Tube.

It has been already stated that the experimental tube employed in these experiments was of brass, polished within for the purpose of augmenting by reflexion the calorific flux, and thus bringing into clearer light the action of the feebler gases and vapours. Wishing, however, to try the effect of chlorine, I admitted a quantity of the gas into the polished tube. The needle was deflected with prompt energy; but on pumping out,* it refused to return to zero. To cleanse the tube, dry air was introduced into it ten times in succession; but the needle pointed persistently to the 40th degree from zero. The cause was easily surmised: the chlorine had attacked the metal and partially destroyed its reflecting power; the absorption by the sides of the tube itself cutting off an amount of heat competent to produce the observed deflection. For subsequent experiments the interior of the tube had to be repolished.

Though no vapour previously examined had produced a permanent effect of this kind, it was necessary to be perfectly satisfied that this source of error had not vitiated the other experiments. To check the results, therefore, a length of 2 feet of the experimental tube was coated carefully on the inside with lampblack, and with it were determined the absorptions of all the vapours previously examined, at a common pressure of 0·3 of an inch. A general corroboration was all that was

* Dense dark fumes rose from the cylinders on this occasion. A similar effect was produced by sulphuretted hydrogen.

aimed at, and I am satisfied that the slight discrepancies which the measurements exhibit would disappear, or be accounted for, in a more careful examination.

In the following table the results obtained with the blackened and with the bright tubes are placed side by side, the pressure in the former being three-tenths, and in the latter five-tenths of an inch.

TABLE XVIII.

Vapour	Absorption per 100		Absorption in Blackened Tube proportional to
	Blackened Tube 0·3 pressure	Bright Tube 0·5 pressure	
Bisulphide of carbon	5	21	23
Iodide of methyl	15·8	60	71
Benzol	17·5	78	79
Chloroform	17·5	89	79
Iodide of ethyl	21·5	94	97
Wood-spirit	26·5	123	120
Methylic alcohol	29	133	131
Chloride of amyl	30	137	135
Amylene	31·8	157	143

The order of absorption is here shown to be the same in both tubes, the quantity absorbed in the bright tube being, in general, about 4½ times that absorbed in the black one. In the third column, indeed, I have placed the numbers contained in the first column multiplied by 4·5. These results completely dissipate the suspicion that the effects observed with the polished tube could be due to a change of the reflecting power of its inner surface by the contact of the vapours.

With the blackened tube the order of absorption of the following substances, commencing with the lowest, stood thus:—

Alcohol,
Sulphuric ether,
Formic ether,
Propionate of ethyl,

whereas with the bright tube they stood thus:—

Formic ether,
Alcohol,
Propionate of ethyl,
Sulphuric ether.

As already stated, these differences would in all probability disappear, or be accounted for on re-examination. Indeed very slight differences in the purity of the specimens used would be more than sufficient to produce the observed differences of absorption.*

§ 10.

Action of Permanent Gases on Radiant Heat.

The deportment of oxygen, nitrogen, hydrogen, atmospheric air, and olefiant gas has been already recorded. Besides these I have examined carbonic oxide, carbonic acid, sulphuretted hydrogen, and nitrous oxide. The action of these gases is so much feebler than that of any of the vapours referred to in the last section, that, in testing the relationship of absorption to density, the unit-measures used with the vapours were abandoned, the quantities of gas admitted being determined by the depression of the mercurial gauge.

TABLE XIX.—*Carbonic Oxide.*

Pressure in inches	Observed	Calculated
0·5	2·5	2·5
1	5·6	5
1·5	8	7·5
2	10	10
2·5	12	12·5
3	15	15
3·5	17·5	17·5

Up to a pressure of 3½ inches the absorption by carbonic oxide is proportional to the density of the gas. But this proportion does not obtain with large quantities of the gas, as shown by the following table:—

Pressure in inches	Deflection	Absorption per 100
	°	
5	18	18
10	32·5	32·5
15	41	45

* In illustration of this I may state, that of two specimens of methylic alcohol given me by two of my chemical friends, the one gave an absorption of 84 and the other of 203. The one had been purified with great care, but the other was not pure. Both specimens, however, went under the common name of methylic alcohol. I have had a special apparatus constructed with a view to examine the influence of ozone on the interior of the experimental tube.

TABLE XX.—*Carbonic Acid.*

Pressure in inches	Observed	Calculated
0·5	5	3·6
1	7·5	7
1·5	10·5	10·5
2	14	14
2·5	17·8	17·5
3	21·8	21
3·5	24·5	24·5

Here we have the proportion exhibited, but not so with larger quantities.

Pressure in inches	Deflection	Absorption per 100
5	25°	25
10	36	36
15	42·5	48

TABLE XXI.—*Sulphuretted Hydrogen.*

Pressure in inches	Absorption per 100		Pressure in inches	Absorption per 100	
	Observed	Calculated		Observed	Calculated
0·5	7·8	6	3	34·5	36
1	12·5	12	3·5	36	42
1·5	18	18	4	36·5	48
2	24	24	4·5	38	54
2·5	30	30	5	40	60

The proportion here holds good up to a pressure of 2·5 inches, when the deviation from it commences and gradually augments.

Though these measurements were made with all possible care, I should like to repeat them. Dense fumes issued from the cylinders of the air-pump on exhausting the tube of this gas, and I am not at present able to state with confidence that a trace of such in a very diffuse form within the tube did not interfere with the purity of the results.

TABLE XXII.—*Nitrous Oxide.*

Pressure in inches	Absorption per 100		Pressure in inches	Absorption per 100	
	Observed	Calculated		Observed	Calculated
0·5	14·5	14·5	3	45	87
1	23·5	29	3·5	47·7	101·5
1·5	30	43·5	4	49	116
2	35·5	58	4·5	51·5	130·5
2·5	41	71·5	5	54	145

Here the divergence from proportionality is manifest from the commencement.

Remarks on the Experiments of Dr. Franz.

I promised at the beginning of this memoir to allude to the results of Dr. Franz: this shall now be done. With a tube 3 feet long and blackened within, an absorption of 3·54 per cent. by atmospheric air was observed in his experiments. In my experiments, however, with a tube 4 feet long and polished within, which makes the distance traversed by the reflected rays far more than 4 feet, the absorption is only one-ninth of the above amount. In the experiments of Dr. Franz, moreover, carbonic acid appears as a feebler absorber than oxygen. According to my experiments, for small quantities the absorptive power of carbonic acid is about 150 times that of oxygen; and at the atmospheric pressure, carbonic acid probably absorbs nearly 100 times as much as oxygen.

The differences between Dr. Franz and me admit of the following simple explanation. His source of heat was an argand lamp, and the ends of his experimental tube were stopped with plates of glass. Now Melloni has shown that fully 61 per cent. of the heat rays emanating from a Locatelli lamp are absorbed by a plate of glass one-tenth of an inch in thickness. Hence the greater portion of the rays issuing from the lamp of Dr. Franz was expended in heating the two glass ends of his experimental tube. These ends became secondary sources of heat which radiated against his pile. On admitting cool air into the tube, the partial withdrawal by conduction and convection of the heat of the glass plates produced an effect exactly the same as that of true absorption. By allowing the air in my experimental tube to come into contact with the radiating plate, I have often obtained a deflection of twenty or thirty degrees,—the effect being due to the cooling of the plate, and not to absorption. It is also certain that, had I, like Dr. Franz, used heat from a luminous source, my small absorption of 0·4 per cent. would have been considerably diminished.

§ 11.

Action of Aqueous Vapour.—Possible Effect of an Atmospheric Envelope on the Temperature of a Planet.

I have now to refer briefly to a point of considerable interest —the effect, namely, of our atmosphere on solar and terrestrial heat. In examining the separate effects of the air, carbonic acid, and aqueous vapour of the atmosphere, on the 20th of last November, the following results were obtained:—

Air sent through the caustic-potash tube and through the drying-tubes produced an absorption of about . . 1

Air direct from the laboratory, containing therefore its carbonic acid and aqueous vapour, produced an absorption of 15

Deducting the effect of the gaseous acid, it was found that the quantity of aqueous vapour diffused through the atmosphere on the day in question, produced an absorption at least equal to thirteen times that of the atmosphere itself.

It is my intention to repeat and extend these experiments on a future occasion*; but even at present conclusions of great importance may be drawn from them. It is exceedingly probable that the absorption of the solar rays by the atmosphere, as established by M. Pouillet, is mainly due to the watery vapour contained in the air. The vast difference between the temperature of the sun at midday and in the evening, is also probably due in the main to that comparatively shallow stratum of aqueous vapour which lies close to the earth. At noon the depth of it pierced by the sunbeams is very small; in the evening very great in comparison.

The intense heat of the sun's direct rays on high mountains is not, I believe, due to his beams having to penetrate only a small depth of air, but to the comparative absence of aqueous vapour at those great elevations.†

But this vapour, which exercises such a destructive action on the obscure rays is comparatively transparent to the rays of light. Hence the differential action of the heat coming from the sun to the earth, and that radiated from the earth into space, is vastly augmented by the aqueous vapour of the atmosphere.

De Saussure, Fourier, M. Pouillet, and Mr. Hopkins regard this interception of the terrestrial rays as exercising a most

* The peculiarities of the locality in which this experiment was made render its repetition under other circumstances necessary.

† As proved by the observations of Welsh and Hooker, reasoned out by Strachey. [1872.]

important influence on climate. Now if, as the above experiments indicate, the chief influence be exercised by the aqueous vapour, every variation of this constituent must produce a change of climate. Similar remarks would apply to the carbonic acid diffused through the air, while an almost inappreciable admixture of any of the stronger hydrocarbon vapours would powerfully hold back the terrestrial rays and produce corresponding climatic changes. It is not, therefore, necessary to assume alterations in the density and height of the atmosphere to account for different amounts of heat being preserved to the earth at different times; a slight change in the variable constituents of the atmosphere would suffice. Such changes in fact may have produced all the mutations of climate which the researches of geologists reveal. However this may be, the facts above cited remain firm; they constitute true causes, the *extent* alone of the operation remaining doubtful.*

The measurements recorded in the foregoing pages constitute only a small fraction of those actually made; but they fulfil the object of the present portion of the inquiry. They establish the existence of enormous differences among colourless gases and vapours as to their action upon radiant heat; and they also show that when the quantities are sufficiently small, the absorption in the case of each particular vapour is exactly proportional

The experiments, moreover, furnish us with purer cases of *molecular action* than have been hitherto attained in researches of this nature. In both solids and liquids the cohesion of the particles is implicated; they mutually control and limit each other. A certain action, over and above that which belongs to them separately, comes into play and embarrasses our conceptions. But in the cases above recorded the molecules are perfectly free, and we fix upon them individually the effects which the experiments exhibit; thus the mind's eye is directed more firmly than ever on those distinctive physical qualities whereby a ray of heat is stopped by one molecule and unimpeded by another.

* On this point, see Section 23, Mem. II.

§ 12.

RADIATION OF HEAT BY GASES.

Reciprocal Experiments on Radiation and Absorption.

It is known that the quantity of light emitted by a flame depends chiefly on the incandescence of solid matter; the brightness of an ignited jet of ordinary gas, for example, being chiefly due to the solid particles of carbon liberated in the flame.*

Melloni found the radiation of heat from his alcohol lamp to be greatly augmented by plunging a spiral of platinum wire into the flame. He also proved that a bundle of wire placed in the current of hot air ascending from an argand chimney gave a copious radiation, while when the wire was withdrawn no trace of radiant heat could be detected by his apparatus. He concluded from this experiment that air possesses the power of radiation in so feeble a degree that our best thermoscopic instruments fail to detect this power.† These are the only experiments hitherto published upon this subject, and they are negative.

I have now to record some affirmative ones. The pile, furnished with its conical reflector, was placed upon a stand, with a screen of polished tin in front of it. An alcohol lamp was placed behind the screen, so that its flame was entirely hidden by the latter; on rising above the screen, the gaseous column radiated its heat against the pile and produced a considerable deflection. The same effect was produced when a candle or an ordinary jet of gas was substituted for the alcohol lamp.

The heated products of combustion acted on the pile in the above experiments, but the radiation from ordinary undried air was easily demonstrated by placing a heated iron spatula or metal sphere behind the screen. A deflection was thus obtained which, when the spatula was raised to a red heat, amounted to more than sixty degrees. No radiation from the spatula to the pile, was here possible, and no portion of the heated air itself approached the pile, so as to communicate its warmth by contact to the latter.

* By a suitable arrangement the particles may be liberated *red-hot,* instead of white-hot, each individual particle describing a *line* of red light. [1872.]

† *La Thermochrose,* p. 94. Further: 'No fact is yet known which directly proves the emissive power of pure and transparent elastic fluids.'—Taylor's *Scientific Memoirs,* vol. v. p. 551.

My next care was to examine whether different gases possessed different powers of radiation; and for this purpose the following arrangement was devised. P (fig. 1) represents the thermo-electric pile with its two conical reflectors; S is a double screen of polished tin; A is an argand burner consisting of two concentric rings perforated with orifices for the escape of the gas; C is a heated copper ball; the tube *t t* leads to a gas-holder containing the gas to be examined. When the ball C is placed on the argand burner, it of course heats the air in contact with it; an ascending current is established, which acts on the

Fig. 1.

pile as in the experiments last described. It was found necessary to neutralize this radiation, and for this purpose a large Leslie's cube L, filled with water a few degrees above the temperature of the air, was allowed to act on the opposite face of the pile.

When the needle was thus brought to zero, the cock of the gas-holder was turned on; the gas passed through the burner, came into contact with the ball, and ascended afterwards in a heated column in front of the pile. The galvanometer was now observed, and the limit of the arc through which its needle was urged was noted. It is needless to remark that the ball was entirely hidden by the screen from the thermo-electric pile, and that, even were this not the case, the mode of compensation adopted would still give us the pure action of the gas.

The results of the experiments are given in the following table, the figure appended to the name of each gas marking the number of degrees through which the radiation from the latter urged the needle of the galvanometer*:—

	Degrees
Air	0
Oxygen	0
Nitrogen	0
Hydrogen	0
Carbonic oxide	12
Carbonic acid	18
Nitrous oxide	29
Olefiant gas	53

The radiation from air, it will be remembered, was neutralized by the large Leslie's cube, and hence the 0° attached to it merely denotes that the propulsion of air from the gas-holder through the argand burner did not augment the effect. Oxygen, hydrogen, and nitrogen, sent in a similar manner over the ball, were equally ineffective. The other gases, however, not only exhibit a marked action, but also marked differences of action. Their radiative powers follow precisely the same order as their powers of absorption. In fact, the deflections actually produced by their respective absorptions at a common pressure of 5 inches are as follows:—

Air	A fraction of a degree
Oxygen	" "
Nitrogen	" "
Hydrogen	" "
Carbonic oxide	18°
Carbonic acid	25°
Nitrous oxide	44°
Olefiant gas	61°

It would be easy to give these experiments a more elegant form, and to arrive at greater accuracy, which I intend to do on a future occasion; but my object now is simply to establish the general order of the radiative powers of these gases, as contrasted with their powers of absorption.

* I have also rendered these experiments on radiation visible to a large assembly. They may be readily introduced in lectures on radiant heat.

§ 13.

The Varnishing of Polished Metal Surfaces by Gases.

When the polished metallic face of a Leslie's cube is turned towards a thermo-electric pile, the effect produced is inconsiderable, but it is greatly augmented when a coat of varnish is laid upon the polished surface. Now a film of gas may be employed instead of the coat of varnish. Such a cube, containing boiling water, had a polished silver face turned towards the pile, and its effect on the galvanometer neutralized in the usual manner. The needle being at 0°, a film of olefiant gas, issuing from a narrow slit, was caused to pass over the metal. The consequent radiation produced a deflection of 45°. When the gas was cut off, the needle returned accurately to 0°.

Reciprocally, absorption by a coating film of gas may be shown by filling a cube with cold water, but not so cold as to produce the precipitation of the aqueous vapour of the air, and allowing the pile to radiate against it. A gilt copper ball, cooled in a freezing mixture, being placed in front of the pile, its effect was neutralized by presenting a beaker containing a little iced water to the face opposite. A film of olefiant gas was sent over the ball, but the consequent deflection proved that the absorption by the ball, instead of being greater, was less than before. On examination, the ball was found coated with a crust of ice, which is one of the best absorbers of radiant heat. The olefiant gas, being warmer than the ice, partially neutralized its action. When, however, the temperature of the ball was only a few degrees lower than that of the atmosphere, and its surface quite dry, the film of gas was found, like a film of varnish, to augment the absorption.

§ 14.

First Observation of the Radiation of a Vapour heated dynamically.

A remarkable effect, which contributed at first to the complexity of the experiments, can now be explained. Conceive the experimental tube exhausted and the needle at zero; conceive a small quantity of alcohol or ether vapour admitted; it cuts off a portion of the heat from one source, and the opposite source triumphs. Let the consequent deflection be 45°. If dry air be now admitted till the tube is filled, its normal effect of course would be slightly to augment the absorption and make the deflection greater. But the following action is really observed:—When the air first enters, the needle, instead of ascending, descends; it falls to 26°, as if a portion of the heat originally cut off had been restored. At 26°, however, the needle stops, turns, moves quickly upwards, and takes up a permanent position a little higher than 45°. Let the tube now be exhausted, the withdrawal of the mixed air and vapour ought of course to restore the equilibrium with which we started; but the following effects are observed:—When the exhaustion commences, the needle moves upwards from 45° to 54°; it then halts, turns, and descends speedily to 0°, where it permanently remains.

After many attempts to account for the anomaly, I proceeded thus:—A thermo-electric couple was soldered to the external surface of the experimental tube, and its ends connected with a galvanometer. When air was admitted, a deflection was produced, showing that the air, on entering the vacuum, was heated. On exhausting, the needle was also deflected, showing that the interior of the tube was chilled. These are indeed known effects; but I was anxious to make myself perfectly sure of them. I subsequently had the tube perforated and delicate thermometers screwed into it airtight. On filling the tube the thermometric columns rose, on exhausting it they sank, the range between the maximum and minimum amounting in the case of air to 5° Fahr.

Hence the following explanation of the above singular effects. The absorptive power of the ether or alcohol vapour is very great, and its radiative power is equally so. The heat generated by the air on its entrance is communicated to the vapour, which thus becomes a temporary source of radiant heat, and diminishes the deflection produced in the first instance by its presence. The reverse occurs when the tube is exhausted; the vapour is then chilled, its great absorptive action on the heat radiated from the adjacent face of the pile comes into play, the original absorption being apparently augmented. In both cases, however, the action is transient; the vapour soon loses the heat communicated to it, and soon gains the heat which it has lost, and matters then take their normal course.*

§ 15.

On the Physical Connexion of Radiation, Absorption, and Conduction.

Notwithstanding the great accessions of late years to our knowledge of the nature of heat, we are as yet, I believe, quite gnorant of the atomic conditions on which radiation, absorption, and conduction depend. What are the specific qualities which cause one body to radiate copiously and another feebly? Why, on theoretic grounds, must the equivalence of radiation and absorption exist? Why should a highly diathermanous body, as shown by Mr. Balfour Stewart, be a bad radiator, and an athermanous body a good radiator? How is heat conducted? and what is the strict physical meaning of good conduction and bad conduction? Why should good conductors be, in general, bad radiators, and bad conductors good radiators? These, and other questions, referring to facts more or less established, have still to receive their complete answers. It was less with a hope of furnishing such than of shadowing forth the possibility of uniting these various effects by a common bond, that I submitted the following reflexions to the notice of the Royal Society.

In the experiments recorded in the foregoing pages, we have dealt with *free* atoms, both simple and compound, and it has been found that in all cases absorption takes place. The meaning of this, according to the dynamical theory of heat, is that no atom is capable of existing in vibrating æther without accepting a portion of its motion. We may, if we wish, imagine a certain roughness of the surface of the atoms which enables the æther to *bite* them and carry the atom along with it. But no matter what the quality may be which enables any atom to accept motion from the agitated æther, the same quality must enable it to impart motion to still æther when it is plunged in the latter and agitated. It is only necessary to imagine a body immersed in water to see that this must be the case.

* Under the head Dynamic Radiation and Absorption, this result is fully developed in subsequent memoirs.

There is a polarity here as rigid as that of magnetism. From the existence of absorption, we may on theoretic grounds infallibly infer a capacity for radiation; from the existence of radiation, we may with equal certainty infer a capacity for absorption; and each of them must be regarded as the measure of the other.*

This reasoning, founded simply on the mechanical relations of the æther and the atoms immersed in it, is completely verified by experiment. Great differences have been shown to exist among gases as to their powers of absorption, and precisely similar differences as regards their powers of radiation. But what specific property is it which makes one free molecule a strong absorber, while another offers scarcely any impediment to the passage of radiant heat? I think the experiments throw some light upon this question. If we inspect the results above recorded, we shall find that the *elementary* gases—hydrogen, oxygen, nitrogen—and the *mixture* atmospheric air, possess absorptive and radiative powers beyond comparison less than those of the *compound* gases. Uniting the atomic theory with the conception of an æther, this result appears to be exactly what ought to be expected. Taking Dalton's idea of an elementary body as a single sphere, and supposing such a sphere to be set in motion in still æther, or placed without motion in moving æther, the communication of motion by the atom in the first instance, and the acceptance of it in the second, must be less than when a number of such atoms are grouped together and move as a system. Thus we see that hydrogen and nitrogen, which, when *mixed* together, produce a small effect, when *chemically united* to form ammonia, produce an enormous effect. Thus oxygen and hydrogen, which, when mixed in their electrolytic proportions, show a scarcely sensible action, when chemically combined to form aqueous vapour exert a powerful action. So also with oxygen and nitrogen, which, when mixed, as in our atmosphere, both absorb and radiate feebly, when united to form oscillating systems, as in nitrous oxide, have their powers vastly augmented. Pure atmospheric air, of 5 inches' mercury pressure, does not effect an absorption equivalent

* This was written long before Kirchhoff's admirable papers on the relation of emission to absorption were known to me. The vibrating period is not here taken into account, but simply the *amount* of molecular motion capable of being received or imparted.

to more than one-fifth of a degree, while nitrous oxide of the same pressure effects an absorption equivalent to fifty-one degrees. Hence the absorption by nitrous oxide at this pressure is about 250 times that of air.

No fact in chemistry carries the same conviction to my mind, that air is a *mixture* and not a *compound*, as that just cited.

In like manner, the absorption by carbonic oxide of 5 inches pressure is nearly 100 times that of oxygen alone; the absorption by carbonic acid is about 150 times that of oxygen; while the absorption by olefiant gas of this pressure is 1,000 times that of its constituent hydrogen. Even the enormous action last mentioned is surpassed by the vapours of many of the volatile liquids, in which the atomic groups are known to attain their highest degree of complexity.

But, besides molecular complexity, another important consideration remains. All the gases and vapours hitherto mentioned are transparent to light; that is to say, the waves of the visible spectrum pass among them without sensible absorption. Hence it is plain that their absorptive power depends on the *periodicity* of the undulations which strike them. At this point the present inquiry connects itself with the experiments of Nièpce,* the observations of Foucault,† the theoretic notions of Euler, Ångström,‡ Stokes, and Thomson, and those splendid researches of Kirchhoff and Bunsen which so immeasurably extend our experimental range. By Kirchhoff it has been conclusively shown that every atom absorbs in a special degree those waves which are synchronous with its own periods of vibration. Now, besides presenting broader sides to the æther, the association of simple atoms to form groups must, as a general rule, render their motions through the æther more sluggish, and tend to bring the periods of oscillation into isochronism with the slow undulations of obscure heat, thus enabling the molecules to absorb more effectually such rays as have been made use of in our experiments.

Let me here state briefly the grounds which induce me to conclude that an agreement in period alone is not sufficient to cause powerful absorption and radiation—that in addition to

* Referred to by Ångström. See below.

† *Annales de Chimie*, 1860, vol. lviii. p. 476.

‡ Poggendorff's *Annalen*, vol. xciv. p. 41. *Philosophical Magazine*, vol. ix. p. 327.

this the molecules must be so constituted as to furnish *points d'appui* to the æther. The heat of contact is accepted with extreme freedom by rock-salt, but a plate of the substance once heated requires a great length of time to cool. This surprised me when I first noticed it. But the effect is explained by the experiments of Mr. Balfour Stewart, by whom it has been proved that the radiative power of heated rock-salt is extremely feeble. Periodicity can have no influence here, for the æther is capable of accepting and transmitting impulses of all periods; and the fact that rock-salt requires more time to cool than alum simply proves that the molecules of the former glide through the æther with comparatively small resistance, and thus continue moving for a long time; while those of the latter speedily communicate to it the motion which we call radiant heat. This power of gliding through still æther possessed by the rock-salt molecules, must of course enable the moving æther to glide round them, and no coincidence of period could, I think, make such a body a powerful absorber.*

Many chemists, I believe, are disposed to reject the idea of an atom, and to adhere to that of equivalent proportions merely. They figure the act of combination as a kind of interpenetration of one substance by another. But this is a mere masking of the fundamental phenomenon. The value of the atomic theory consists in its furnishing the physical explanation of the law of equivalents: assuming the one, the other follows; † and assuming the act of chemical union as Dalton figured it, we see that it blends harmoniously with the perfectly independent conception of an æther, and enables us to reduce the phenomena of radiation and absorption to the simplest mechanical principles.

Considerations similar to the above may, I think, be applied to the phenomena of *conduction*. In the 'Philosophical Magazine' for August 1853, I have described an instrument used in examining the transmission of heat through cubes of wood and other substances. When engaged with this instrument, I

* With reference to the question of vibrating period and molecular form, see *Fragments of Science*, earlier editions, p. 210 *et seq.*, or *Radiation*, p. 52 *et seq.*

† For the further treatment of this subject, see *Fragments of Science*, earlier editions, pp. 135 and 136. See also *Faraday as a Discoverer*, cheap edition, p. 146 *et seq.*

had also cubes of various crystals prepared, and determined approximately their powers of conduction. With one exception, it was found that the conductivity augmented with the diathermancy. The exception was that a cube of very perfect rock-crystal conducted heat slightly better than a cube of rock-salt. The latter, however, had a very high conductive power; in fact rock-salt, calcareous spar, glass, selenite, and alum stood in these experiments, and as regards conductivity, exactly in their order of diathermancy in the experiments of Melloni. Considerations have been already adduced which show that the molecules of rock-salt glide with facility through the æther; but the ease of motion which these molecules enjoy must facilitate their mutual collision. Their motion, instead of being expended on the æther between them, and communicated by it to the external æther, is in great part transferred directly from particle to particle, or in other words, is freely conducted. When a molecule of alum, on the contrary, oscillates, it produces a swell in the intervening æther, which swell is in part transmitted, not to the molecules, but to the general æther of space, and thus lost as regards conduction. This lateral waste prevents the motion from penetrating the alum to any great extent, and the substance is what we call a bad conductor.*

Such considerations as these could hardly occur without carrying the mind to the kindred question of electric conduction; but the speculations have been pursued sufficiently far for the present, and they must now abide the judgment of those competent to decide whether they are the mere emanations of fancy, or a fair application of principles which are acknowledged to be secure.

The present paper, I may remark, embraces only the first section of these researches.

* In the above considerations regarding conduction, I have limited myself to the illustration furnished by two compound bodies; but the elementary atoms also differ among themselves as regards their powers of accepting motion from the æther and of communicating motion to it. I should infer, for example, that the atoms of platinum encounter more resistance in moving through the æther than the atoms of silver; simply because platinum is a worse conductor than silver.

Supplementary Remarks, 1872.

For the use of the younger student here follow three extracts from my book on 'Heat, a Mode of Motion.'

1. Note on the construction of the thermo-electric pile.
2. Note on the construction of the galvanometer.
3. Remarks on the different values of galvanometric degrees.

Also Melloni's account of the mode of calibrating a galvanometer; in other words, of reducing all its degrees, high and low, to a common value, extracted from 'La Thermochrose.'

1. *Note on the Construction of the Thermo-electric Pile.*

Let A B (fig. 2) be a bar of antimony, and B C a bar of bismuth, soldered together at B. Let the free ends A and C be united by a wire, A D C. On warming the place of junction, B, an electric current is generated, the direction of which is from bismuth to antimony (or against the alphabet), across the junction, and from antimony to bismuth (or with the alphabet), through the connecting wire, A D C. The arrows indicate the direction of the current.

Fig. 2.

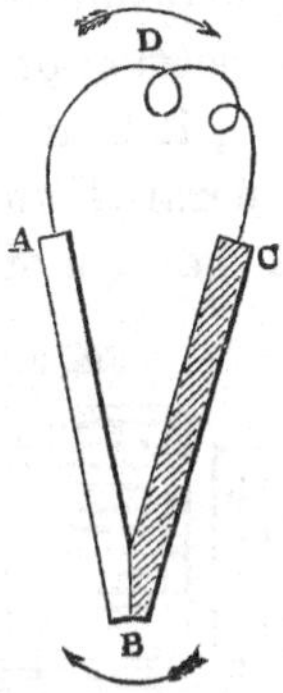

If the junction B be *chilled*, a current is generated opposed in direction to the former. The figure represents what is called a thermo-electric pair or couple.

By the union of several thermo-electric pairs a more powerful current can be generated than that obtained from a single pair. Fig. 3 (next page), for example, represents such an arrangement, in which the shaded bars are supposed to be all of bismuth, and the unshaded ones of antimony. On warming all the junctions, B, B, &c., a current is generated in each, and the sum of these currents, all of which flow in the same direction, produces a stronger resultant current than that obtained from a single pair.

The V formed by each pair need not be so wide as it is shown in fig. 3; it may be contracted without prejudice to the couple. And if it is desired to pack several pairs into a small compass, each separate couple may be arranged as in fig. 4, where

Fig. 3.

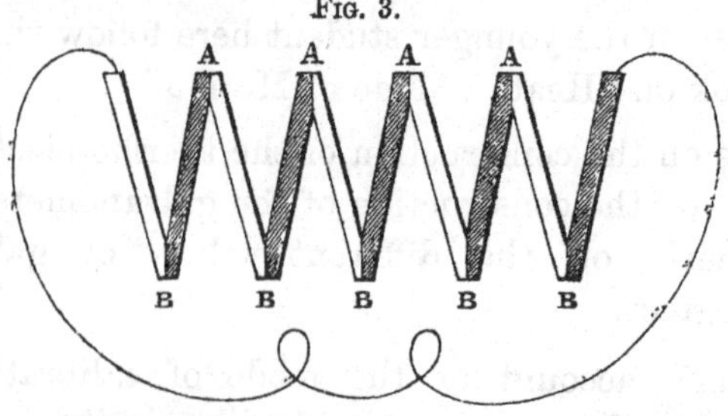

the black lines represent small bismuth bars, and the white ones small bars of antimony. They are soldered together at the ends, and throughout their length are usually separated by strips of paper merely. A collection of pairs thus compactly set together constitutes a thermo-electric pile, a drawing of which is given in fig. 5.

Fig. 4.

The current produced by heat being always from bismuth to antimony across the heated junction, a moment's inspection of fig. 3 will show that when any one of the junctions A, A, is heated, a current is excited opposed in direction to that generated when the heat is applied to the junctions B, B. Hence in the case of the thermo-electric pile, the effect of heat falling upon its two opposite faces is to produce currents in opposite directions. If the temperature of the two faces be alike, they neutralise each other, no matter how highly they may be heated absolutely; but if one of them be warmer than the other, a current is produced. The current is thus due to a *difference* of temperature between the two faces of the pile, and within certain limits the strength of the current is exactly proportional to this difference.

Fig. 5.

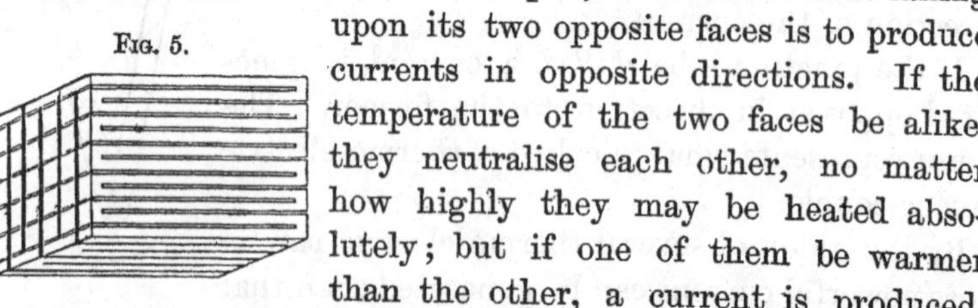

From the junction of almost any other two metals, thermo-electric currents may be obtained, but they are most readily generated by the union of bismuth and antimony.*

* The discovery of thermo-electricity is due to Thomas Seebeck, Professor in the University of Berlin. Nobili constructed the first thermo-electric pile; but in Melloni's hands it became an instrument so important as to supersede all others in researches on radiant heat.

2. *Note on the Construction of the Galvanometer.*

The existence and direction of an electric current are shown by its action upon a freely suspended magnetic needle.

But such a needle is held in the magnetic meridian by the magnetic force of the earth. Hence, to move a single needle, the current must overcome the magnetic force of the earth.

Very feeble currents are incompetent to do this in a sufficiently sensible degree. The following two expedients are, therefore, combined to render sensible the action of such currents:—

The wire through which the current flows is coiled so as to surround the needle several times; the needle must swing freely within the coil. The action of the single current is thus multiplied.

The second device is to neutralise the directive force of the earth, without prejudice to the magnetism of the needle. This is accomplished by using two needles instead of one, attaching them to a common vertical stem, and bringing their opposite poles over each other, the north end of the one needle and the south end of the other being thus turned in the same direction. The double needle is represented in fig. 6.

Fig. 6.

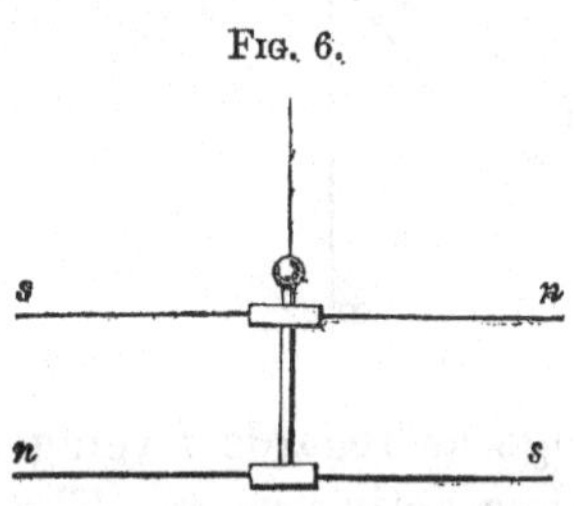

It must be so arranged that one of the needles shall be within the coil through which the current flows, while the other needle swings freely above the coil, the vertical connecting-piece passing through an appropriate slit in the coil. Were both the needles within the coil, the same current would urge them in opposite directions, and thus one needle would neutralise the other. But when one is within and the other without, the current urges both needles in the same direction.

The way to prepare such a pair of needles is this. Magnetise both of them to saturation; then suspend them in a vessel, or under a shade, to protect them from air-currents. The system will probably set in the magnetic meridian, one needle being in almost all cases stronger than the other. Weaken the stronger needle carefully by the touch of a second smaller magnet. When the needles are precisely equal in strength, they will set at *right angles to the magnetic meridian.*

It might be supposed that when the needles are equal in strength, the directive force of the earth would be completely annulled, that the double needle would be perfectly *astatic,* and perfectly neutral as regards direction; obeying simply the torsion of its suspending fibre. This would be the case if the magnetic axes of both needles could be caused to lie with mathematical accuracy in the same vertical plane. In practice this is almost impossible; the axes always cross each other. Let *n s*, *n′ s′* (fig. 7) represent the axes of two needles thus crossing, the magnetic meridian being parallel to M E; let the pole *n* be drawn by the earth's attractive force in the direction *n m*; the pole *s′* being urged by the repulsion of the earth in a precisely opposite direction. When the poles *n* and *s′* are of exactly equal strength, it is manifest that the force acting on the pole *s′*, in the case here supposed, would have the advantage as regards leverage, and would therefore overcome the force acting on *n*. The crossed needles would therefore turn away still further from the magnetic meridian, and a little reflexion will show that they cannot come to rest until the line which bisects the angle enclosed by the needles is at right angles to the magnetic meridian.

Fig. 7.

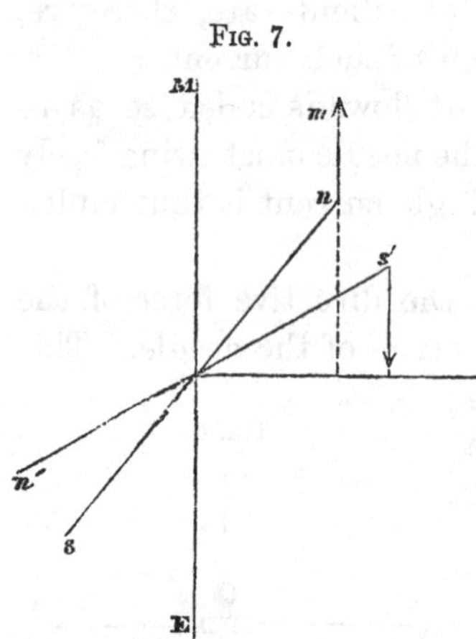

This indeed is the test of perfect equality as regards the magnetism of the needles; but in bringing them to this state of perfection we have often to pass through various stages of obliquity to the magnetic meridian. In these cases the superior strength of one needle is compensated by an advantage, as regards leverage, possessed by the other. By a happy accident a touch is sometimes sufficient to make the needles perfectly equal; but many hours are often expended in securing this result. It is only of course in very delicate experiments that this perfect equality is needed; but in such experiments it is essential.

3. *Remarks on the different Values of Galvanometric Degrees.*

The needle being at zero, let us suppose a quantity of heat to fall upon the pile, sufficient to produce a deflection of one degree. Suppose the quantity to be afterwards augmented, so as to produce deflections of two degrees, three degrees, four degrees, five degrees; then the amounts of heat which produce these deflections stand to each other in the ratios of 1 : 2 : 3 : 4 : 5 : the heat which produces a deflection of 5° being exactly five times that which produces a deflection of 1°. But this proportionality exists only so long as the deflections do not exceed a certain magnitude. For, as the needle is drawn more and more aside from zero, the current acts upon it at an ever-augmenting disadvantage. The case is illustrated by a sailor working a capstan; he always applies his strength at right angles to the lever, for, if he applied it obliquely, only a portion of that strength would be effective in turning the capstan round. And in the case of our electric current, when the needle is very oblique to the current's direction, only a portion of its force is effective in moving the needle. Thus it happens, that though the quantity of heat may be, and, in our case, *is*, accurately expressed by the strength of the current which it excites, still the larger deflections, inasmuch as they do not give us the action of the whole current, but only of a part of it, cannot be a true measure of the amount of heat falling upon the pile.

The galvanometer here employed is so constructed that the angles of deflection up to 30° or thereabouts, are proportional to the quantities of heat; the quantity necessary to move the needle from 29° to 30° is sensibly the same as that required to move it from 0° to 1°. But beyond 30° the proportionality ceases. The quantity of heat required to move the needle from 40° to 41° is three times that necessary to move it from 0° to 1°; to deflect it from 50° to 51° requires five times the heat necessary to move it from 0° to 1°; to deflect it from 60° to 61° requires about seven times the heat necessary to move it from 0° to 1°; to deflect it from 70° to 71° requires eleven times, while to move it from 80° to 81° requires more than fifty times the heat necessary to move it from 0° to 1°. Thus, the higher we go, the greater is the quantity of heat represented by a degree of deflection; the reason being, that the force which then moves the needle is only a fraction of the force of the current really circulating in the wire, and hence represents only a fraction of the heat falling upon the pile.

4. *Calibration of the Galvanometer.*

The following method of calibrating the galvanometer is recommended by Melloni as leaving nothing to be desired in regard to facility, promptness, and precision. His own statement of the method, translated from 'La Thermochrose,' p. 59, is as follows:—

Two small vessels, V V, are half-filled with mercury, and connected separately, by two short wires, with the extremities G G of the galvanometer. The vessels and wires thus disposed make no change in the action of the instrument; the thermo-electric current being freely transmitted, as before, from the pile to the galvanometer. But if, by means of a wire F, a communication be established between the two vessels, part of the current will pass through this wire and return to the pile. The quantity of electricity circulating in the galvanometer will be thus diminished, and with it the deflection of the needle.

Fig. 8.

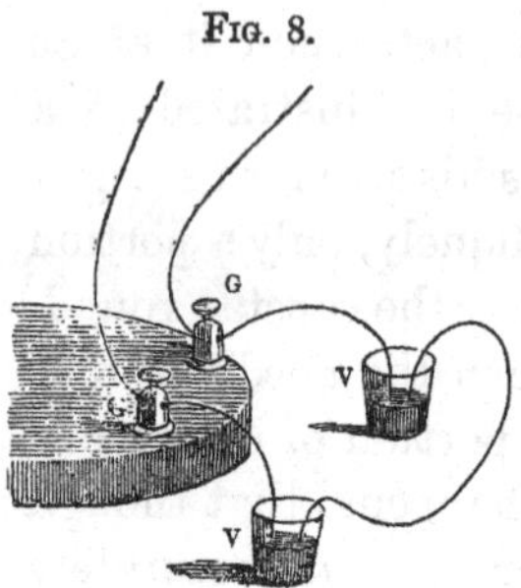

Suppose, then, that by this artifice we have reduced the galvanometric deviation to its fourth or fifth part; in other words, supposing that the needle, being at 10 or 12 degrees, under the action of a constant source of heat, placed at a fixed distance from the pile, descends to 2 or 3 degrees, when a portion of the current is diverted by the external wire; I say, that by causing the source to act from various distances, and observing in each case the *total* deflection, and the *reduced* deflection, we have all the data necessary to determine the ratio of the deflections of the needle, to the forces which produce these deflections.

To render the exposition clearer, and to furnish, at the same time, an example of the mode of operation, I will take the numbers relating to the application of the method to one of my thermo-multipliers.

The external circuit being interrupted, and the source of heat being sufficiently distant from the pile to give a deflection not exceeding 5 degrees of the galvanometer, let the wire be placed from v to v; the needle falls to 1·5°. The connection between the two vessels being again interrupted, let the source be brought near enough to obtain successively the deflections:—

5°, 10°, 15°, 20°, 25°, 30°, 35°, 40°, 45°.

Interposing after each the same wire between v and v, we obtain the following numbers:—

1·5°, 3°, 4·5·°, 6·3°, 8·4°, 11·2°, 15·3°, 22·4°, 29·7°.

Assuming the force necessary to cause the needle to describe each of the first degrees of the galvanometer to be equal to unity, we have the number 5 as the expression of the force corresponding to the first observation. The other forces are easily obtained by the proportions:—

$$1·5 : 5 = a : x = \frac{5}{1·5} a = 3·333.*$$

where a represents the deflection when the exterior circuit is closed. We thus obtain

5, 10, 15·2, 21, 28, 37·3

for the forces, corresponding to the deflections,

5°, 10°, 15°, 20°, 25°, 30°.

In this instrument, therefore, the forces are sensibly proportional to the arcs, up to nearly 15 degrees. Beyond this, the proportionality ceases, and the divergence augments as the arcs increase in size.

The forces belonging to the intermediate degrees are obtained with great ease, either by calculation or by graphical construction, which latter is sufficiently accurate for these determinations.

By these means we find,

Degrees . . .	13°,	14°,	15°,	16°,	17°,	18°,	19°,	20°,	21°.
Forces	13,	14·1,	15·2,	16·3,	17·4,	18·6,	19·8,	21,	22·3.
Differences . . .		1·1,	1·1,	1·1,	1·1,	1·2,	1·2,	1·2,	1·3.

Degrees . . .	22°,	23°,	24°,	25°,	26°,	27°,	28°,	29°,	30°.
Forces . . .	23·5,	24·9,	26·4,	28,	29·7,	31·5,	33·4,	35·3,	37·3.
Differences . . .		1·4,	1·5,	1·6,	1·7,	1·8,	1·9,	1·9,	2.

* That is to say, one reduced current is to the total current to which it corresponds as any other reduced current is to its corresponding total current.

In this table we do not take into account any of the degrees preceding the 13th, because the force corresponding to each of them possesses the same value as the deflection.

The forces corresponding to the first 30 degrees being known, nothing is easier than to determine the values of the forces corresponding to 35, 40, 45 degrees, and upwards.

The reduced deflections of these three arcs are,

15·3°, 22·4°, 29·7°.

Let us consider them separately, commencing with the first. In the first place, then, 15 degrees, according to our calculation, are equal to 15·2; we obtain the value of the decimal 0·3 by multiplying this fraction by the difference 1·1 which exists between the 15th and 16th degrees; for we have evidently the proportion

$$1 : 1{\cdot}1 = 0{\cdot}3 : x = 0{\cdot}3.$$

The value of the reduced deflection corresponding to the 35th degree, will not, therefore, be 15·3°, but 15·2° + 0·3° = 15·5°. By similar considerations we find 23·5·° + 0·6° = 24·1°, instead of 22·4°, and 36·7° instead of 29·7° for the reduced deflections of 40 and 45 degrees.

It now only remains to calculate the forces belonging to these three deflections, 15·5°, 24·1°, and 36·7°, by means of the expression 3·333 a; this gives us—

the forces, 51·7, 80·3, 122·3;
for the degrees, 35, 40, 45.

Comparing these numbers with those of the preceding table, we see that the sensitiveness of our galvanometer diminishes considerably when we use deflections greater than 30 degrees.

HISTORIC REMARKS ON MEMOIR I.

It will lessen the labours of those who may hereafter wish to refer to this subject if I here indicate the exact relationship, in point of time, of the foregoing investigation to the labours of a distinguished philosopher, now unhappily removed from us, who worked, as he stated, and as I believe, independently of me, on a portion of the area covered by these researches.

Prompted by the beautiful experiment of Grove, in which a white-hot platinum wire plunged into hydrogen gas is immediately chilled to darkness, the late Professor Magnus entered, in 1860, on a skilful and ingenious experimental inquiry regarding the conduction of heat by gases. The most noteworthy result deduced by its author from this investigation is that hydrogen conducts heat like a metal, and that the quenching of the glowing wire in Grove's experiment is a consequence of conduction.

A preliminary notice (vorläufige Mittheilung) of this investigation was communicated to the Academy of Sciences in Berlin on July 30, 1860, being published in the 'Monatsberichte der Akademie' for that date.

Though the transmission of radiant heat through gases is not alluded to, nor even the term 'radiant heat' employed throughout this notice, still its subject and results naturally suggested the possible action of radiant heat. Accordingly we find Professor Magnus following up his first inquiry by another 'On the Transmission of Radiant Heat through the Gases,' which was communicated to the Academy of Sciences on February 7, 1861, and published in the 'Monatsberichte' for that date.

The two investigations here referred to appear together *in extenso* in Poggendorff's 'Annalen' for April 1861, and I had the pleasure of recommending them for translation in the 'Philosophical Magazine,' where they appeared in July and August 1861.

Professor Magnus's memoir contains the following historic references:—'As far as I know, Dr. Franz's are the only experiments which have hitherto been published on the diathermancy of gases. These, which moreover refer only to atmospheric air, hydrogen, and carbonic acid, could not be sufficient for the present purpose, because an argand lamp was used as a source of heat. But it was not merely possible, but probable, that the transmission of thermal rays would differ with their source. If, therefore, the experiments were to be conclusive, the transmission must be investigated for rays proceeding from the same source of heat, that of boiling water.'

The paragraph preceding this one refers to 1860 as 'last year,' hence it is to be inferred that this was written in 1861.

In the next paragraph but one of his paper Professor Magnus makes the following passing reference to the relationship of his labours to mine:—'These experiments were already ended when I saw in the "Proceedings of the Royal Society" that Dr. Tyndall in London is occupied with an investigation on the transmission of heat through gases. As Dr. Tyndall, whose investigation has been thus far only announced, has subjected his gases to experiment in tubes closed by plates of rock-salt, I believe the following results to be entirely independent of the investigation of Mr. Tyndall.'

These two paragraphs occurring so nearly together, and yet so different, render it probable that the paper from which they are taken was written at different times, and that it was not until its later stages that Professor Magnus became aware of my researches.

In the early part of 1861 he wrote to me to inquire about the nature and course of my investigation, and I immediately sent him a sketch of my work. This was the origin of the following letter, which, to ensure perfect accuracy, I asked my friend Dr. Debus to translate for me soon after I received it. It clearly indicates the view then entertained by Professor Magnus of our respective labours:—

Berlin: March 17, 1861.

My dear Tyndall,—My apprehensions have indeed been realised. Both of us have been working on the same subject. I reproach myself for having been led into a mistake by my consideration for you. In October of last year [*i.e.* October 1860], on occasion of the fiftieth anniversary of our University, and at a time when my experiments on the passage of radiant heat, using boiling water as a source of heat, had long been finished, Clausius told me that you were engaged with researches on the passage of radiant heat through gases. I confess the announcement of this investigation in the 'Proceedings' had escaped my notice. I was doubtful if I should or should not write to you and communicate my results. I considered the case with Poggendorff, and it appeared to me that a communication of my results to you, who had been engaged on the subject for a whole year, would be a claim for priority. Also, as your notice appeared already in January, 1860,* I supposed you had nearly or quite finished the investigation. I perceive now the better course would have been to have written to you, because Poggendorff tells me that your experiments were only made in November,† whilst the first part of my investigation, mentioned above, was already finished in July, when I was lecturing on the conduction of heat by gases. I believe you are sufficiently acquainted with me and my solicitude, to know that I should not have spoken on a conduction of gases if I had not convinced myself that the difference in the rise of temperature in the same was not caused by a difference in the transmission of radiant heat. It was necessary for this reason to examine this transmission for the heat of boiling water which was used for conduction. The striking results obtained by me as well as by you made me desirous to examine, before publishing these results, how the gases would comport themselves with a source of heat of a high temperature, and this subject has occupied me during last autumn. It is now the third time that it has been my peculiar fate to have been working simultaneously with another person on the same subject, and also publishing the results at the

* This ought to be May 1859, instead of January 1860. The mistake may, perhaps, be thus accounted for. The 'Philosophical Magazine' usually gives a short account of the communications made to the Royal Society, and the number of that Journal for January 1860 does contain a reprint of the communication made by me to the Royal Society eight months previously. Professor Magnus must have overlooked the heading of this section of the 'Philosophical Magazine,' which runs thus:—'May 26, 1859. Sir B. C. Brodie in the Chair. The following communications were read.' Then follows mine among others.

† I do not understand this.—[J. T.]

same time. The only satisfaction I derive from this circumstance is, that I had to compete with the most eminent experimentalists, such as Regnault and yourself. It has happened and cannot be altered. If we had more frequently corresponded it would not have occurred. Without doubt we both shall pursue the subject; let us therefore in future communicate to each other the special line of inquiry which each of us intends to follow, in order that we may not clash again. As soon as our papers are published, such an understanding can be more easily established than at present. You will have learnt from the February number of Poggendorff that my experiments do not extend to vapours, and the investigation of this part of the subject, excepting aqueous vapour, belongs exclusively to you. On the other hand, you have not examined ammonia, probably because it would have attacked your tube. This gas intercepts the rays of heat in a far higher degree than olefiant gas. On the radiation from gases no experiments have been made by me, and the interesting discoveries on this point are due to you alone. I intend to examine next the influence of reflecting plane surfaces, but before this I have to finish another investigation on the conducting power of gases for E., which also yields most remarkable results. Your statement in the 'Comptes-rendus,' that moist air has a 15-times greater absorbing power for radiant heat than dry air, is not intelligible to me, except on condition that your air was cloudy or perhaps that a layer of moisture had settled on your rock-salt plates. For I was not able to notice a difference or perceptible difference between dry air and such which was saturated with vapour of water at 17° C.

Heartily yours,
G. Magnus.

To this letter I immediately sent the following reply:—

My dear Magnus,—Though science is sure to be the gainer from our working independently at the same subject, it is not quite agreeable to find one's own personal claims interfering with those of a valued friend. My simple course, however, is to lay before you a brief statement of the circumstances attending the investigation of the mutual action of gases and radiant heat, feeling assured that you will draw just conclusions from the statement.

It is several years since the thought of examining the absorption of radiant heat by gases first occurred to me, but until the early part of 1859 I was so much occupied with other matters that I could not turn my attention specially to it. In 1859 I reflected a good deal on the question, made many experiments, devised new methods, and, after some weeks of successful work, communicated in a note to the Royal Society the mode of experiment and the general character of the results. This note, which was written with the express intention of setting my mind at rest as to the possible claims of other investigators, is published in the 'Proceedings' of the Society for May 26, 1859. I saw clearly that the subject would occupy me for several years. A new and vast field had in fact been laid open to experiment, and having by thought and labour fairly mastered it and obtained definite results, I felt at liberty to cultivate it calmly and quietly without fear of being interfered with by anybody. In the note referred to I state distinctly that *the investigation is in progress, a full account of it at a future day being promised to the Royal Society.*

I continued to work; and on June 10, 1859, gave a Friday evening discourse on the subject before the members of the Royal Institution. The Prince

Consort was in the chair; and by throwing the image of the dial of my galvanometer upon a screen, I succeeded in making the experiments on the radiation and absorption of heat by gases and vapours visible to an assembly of many hundred persons. A report of the Lecture is published in the 'Proceedings of the Royal Institution' for June 10, 1859.

You tell me that you had overlooked the notice in the 'Proceedings of the Royal Society,' and that Clausius was the first to inform you in October 1860 of my being engaged in experiments on the radiation of heat through gases. I, on the other hand, did not neglect to give my results due publicity. They were communicated to De la Rive in a letter published in the 'Bibliothèque universelle' for July 1859, vol. v. page 232. My lecture of the 10th of June was also reproduced in 'Cosmos,' vol. xv. page 321, where you will find the following words:—'Ces recherches de M. Tyndall sont encore à leur début, il les continuera avec ardeur; déjà il a perfectionné sa mode d'expérimentation. Il emploie un galvanomètre d'un seul fil, avec deux sources de chaleur agissant sur les deux faces d'une même pile,' &c. You will also find the subject referred to in the 'Nuovo Cimento,' vol. x. p. 196. Also in the 'Comptes-rendus,' and in other journals. When I was in Paris in 1859, my experiments formed a common topic of conversation among scientific men. Foucault, Jamin, Moigno, Wertheim and others, spoke to me about them. In fact they aroused an amount of attention far greater than I had any reason to expect.

In London it was known to everybody that I was engaged on this question; Hofmann, Williamson, and Frankland can all call to mind my obtaining volatile liquids from them in 1859. Faraday also recollects how often I have expressed to him my satisfaction that a new field had been opened which I might cultivate at my leisure without fear of interfering with the claims of anybody. In 1859 I had measured the absorption of oxygen, hydrogen, nitrogen, air, carbonic oxide, carbonic acid, nitrous oxide, coal gas, and olefiant gas; and I had also determined the absorption produced by the vapours of the following substances:—bisulphide and bichloride of carbon, sulphuric ether, cyanide of ethyl, chloroform, benzol, methylic alcohol, amylene, iodide of ethyl, acetate of ethyl, propionate of ethyl, formate of ethyl, iodide of amyl, chloride of amyl, ethylic ether, ethyl amylic ether, amylic alcohol, absolute alcohol, and others. *In fact in* 1859 I *had the materials for an elaborate memoir on this subject already in my hands.*

You say that I had not examined ammonia even in 1860. The fact, however is that early in 1859 I had examined it and established its great absorptive power. If you will cast your eye on my letter to De la Rive published in the 'Bibliothèque universelle' for July 1859, you will see that ammonia did not escape me.*

Two circumstances offered a direct hindrance to the working out of my

* The substances mentioned in my letter to M. De la Rive are ranged in the order of their absorptive power, which augments from the elementary gases through carbonic oxide, carbonic acid, nitrous oxide, olefiant gas, coal gas, up to ammonia.

So also at page 47 of the foregoing memoir, which at the date (March 17) of Professor Magnus's letter had been already more than two months in the possession of the Royal Society (handed in January 10), the following passage occurs:—'Thus we see that hydrogen and nitrogen, which when *mixed* together produce a small effect, when *chemically united* to form ammonia, *produce an enormous effect.*' But again there is some ground for the results in 1859: the first was the desire to get the 'Glaciers of the Alps' off my hands; the second was a sudden call to undertake the duties of Professor at the School of Mines.* I resumed my experiments in 1860, and you will find a general statement of my results at pp. 244-5 of the 'Glaciers of the Alps,' a copy of which I had the pleasure of forwarding to Mrs. Magnus. Referring to the preface to that work, you will see that it was published in June 1860; and the preface was written several months subsequently to the pages above referred to. In fact the experimental notes now before me would satisfy you that in July 1859 I had obtained, in a substantially correct form, almost all the results of which you first heard in October 1860, and which you suppose I did not obtain until the November of that year.

Believe me ever yours faithfully,

JOHN TYNDALL.

This small history, then, may be thus summed up:—

On May 26, 1859, I communicated to the Royal Society a preliminary notice of an investigation 'On the Transmission of Radiant Heat through Gaseous Bodies;' and on the 10th of June following I delivered before the Royal Institution a discourse 'On the Transmission of Heat of different qualities through Gases of different kinds,' illustrating the subject by experiments.

The former communication was published in the 'Proceedings of the Royal Society,' the latter in the 'Proceedings of the Royal Institution' for the respective dates referred to. Soon afterwards notices and descriptions of the investigation appeared in various other journals both English and Continental.

On July 30, 1860, or fourteen months after my first communication to the Royal Society, Professor Magnus communicated to the Academy of Sciences in Berlin a preliminary notice of an investigation on the *Conduction* of Heat by Gases. In this notice there is no mention of either radiation or absorption.

After various and long-continued efforts to exalt the power and delicacy of my apparatus, to vary the source of heat, to bring clearly into view the action of the feeblest substances, and to confer by repetition quantitative certainty on the results, I communicated to the Royal Society on January 10, 1861, the foregoing memoir 'On the Absorption and Radiation of Heat by Gases and Vapours, and on the Physical Connexion of Absorption, Radiation, and Conduction.'

My reasons for thus postponing publication are glanced at in pp. 14 and 15 of the foregoing memoir, where it is stated that during the summer of 1859 I had determined approximately the absorption of nine gases and of twenty vapours, having then in my possession the materials for a long memoir. Seven weeks of labour in 1859, and seven other weeks in 1860, are practically ignored through my reluctance to publish results which, though approximately correct, did not satisfy my desire for accuracy.

On February 7, 1861, or about a year and three-quarters after my first, and about a month subsequent to my second communication to the Royal Society, Professor Magnus read before the Academy of Sciences in Berlin a memoir

misapprehension of Professor Magnus; because though I had established the *position* of ammonia in the list of absorbing gases, the apparatus I employed did not warrant me in assigning to it a *numerical value*. This I hastened to do in the very next inquiry.

* I wished also to make my apparatus as perfect as possible. Indeed, having already secured myself by the wide publicity referred to, I was in no hurry to urge on my researches.

entitled 'Ueber den Durchgang der Wärmestrahlen durch die Gase.' It is stated to form the second part of an investigation the first part of which had been communicated to the Academy on the 30th of July previously.

Until I received the foregoing letter, dated from Berlin, March 17, I had no notion that Professor Magnus or any other investigator was engaged on this question. As stated by him in the letter just referred to, he had confined himself to the question of absorption, and he operated upon gases only. He did not embrace radiation nor the action of vapours in his inquiry.

I unreservedly accept his statement that he broke ground in this field independently of me, and that he was not aware until October 1860 that I had been engaged upon the subject.

Were it correct, as supposed by Professor Magnus, that my experiments were not executed prior to November 1860, I would willingly, on the mere strength of his assurance that his had been completed in July 1860, concede to him the priority which in his letter he seems to claim. I would not for a moment insist upon the circumstance that I had preceded him by a month in the publication of my investigation; but the fact is that results of far greater variety and number than those obtained by him were already in my possession in July 1859. In a subsequent paper samples of these results shall be produced. (See Memoir XII. p. 405.)

II.

ON THE ABSORPTION AND RADIATION OF HEAT BY GASEOUS MATTER.

ANALYSIS OF MEMOIR II.

THE description of the instruments employed in the last investigation is here recapitulated, and the modifications of them which that inquiry suggested are described. The experiments on chlorine, ozone, and aqueous vapour are followed up. Chlorine is found to be far feebler as an absorbent than many light and transparent gases, while both ozone and aqueous vapour are found to exercise a far more powerful absorbent action than ordinary oxygen. Experiments on the human breath follow. Chlorine is then compared with hydrochloric acid and bromine with hydrobromic acid, the lighter and more transparent compound gas being proved in each case more destructive of the heat-rays than the elementary one.

The experiments with gases are extended, the gases being now enclosed in a *glass* experimental tube, and the source of heat changed from a cube of boiling water to a plate of copper, against which a constant sheet of flame is permitted to play. As in the last investigation, the elementary gases begin the list, as the lowest absorbers, while ammonia concludes the list, as the highest absorber.

At the pressure of the atmosphere, ammonia is found to quench more than one thousand times the quantity of radiant heat intercepted by dry air.

Smaller volumes of the gases are then compared together, the difference between the compound and elementary gases being thereby far more impressively brought out. At a common pressure of one inch of mercury, for example, the absorption of ammonia is proved to be over five thousand times, that of olefiant gas over six thousand times, while that of sulphurous acid is six thousand four hundred and eighty times the absorption of dry atmospheric air. In this table chlorine and bromine, notwithstanding their colour and density, show themselves to be more diathermic than any of the transparent compound gases examined.

Experiments on lampblack are next described; and the fact that this substance exercises an elective absorption, acting differently upon different kinds of heat, is demonstrated.

The experiments with vapours described in the last investigation are then followed up, and the extraordinary energy of absorption made manifest in Memoir I. is brought out with still greater emphasis. Some of the stronger vapours at a pressure of $\frac{1}{10}$th of an inch, or $\frac{1}{300}$th of an atmosphere, are proved to exert 600 times the absorbent power of 30 inches of atmospheric air; and probably more than 180,000 times the power of air reduced to the same pressure as the vapour.

It is pointed out that one gas or vapour may start, at a small pressure, with a lower power of absorption than another, and that when the pressure is gradually augmented, the one may overtake, and even transcend, the other. Further experiments on this point are contemplated.

Following up an observation made in the last inquiry, several sections of this memoir are devoted to the *dynamic radiation* and the *dynamic absorption* of gases and vapours. An illustration or two will suffice to show the origin and character of these experiments.

Alcohol vapour, under a pressure of half an inch of mercury, was admitted into the experimental tube: the consequent deflection was 72°.

Dry air was permitted to enter by my assistant, while I, expecting only a slight augmentation of the deflection, was looking through the telescope at the needle. Instead of advancing, it fell to 0°, and swung to 25° on the opposite side of zero.

A small modicum of ether vapour was admitted into the experimental tube: the deflection was 30°. On the entrance of dry air the needle fell to 0°, and swung to 60° on the other side.

After the first moment of surprise and perplexity caused by this unexpected result, the explanation suggested itself that, as air is dynamically warmed on entering a vacuum, the reversal of the deflection might be due to the heat radiated from the vapour warmed by the air. If this were true, no external source of heat would be necessary to the production of the effect; but, on the contrary, an entirely novel method of determining radiant power would be available. This proved to be the case. The source of heat was abolished, and the dynamic radiation of vapours was determined by allowing a measured quantity of each to enter the experimental tube, and sending after it an atmosphere of air. These determinations harmonised perfectly with those obtained by the totally different methods described in the last memoir.

Dynamic absorption was determined by chilling the mixed air and vapour by partial exhaustion, and permitting the pile to radiate its heat into the chilled vapour. Here, also, the results agree with those obtained by causing the vapours to intercept the heat from an external source on its way to the pile.

The dynamic radiation and absorption of gases were also determined, and a similar agreement found to exist between them and former results.

It is then shown that, by simply reasoning upon the physical conditions involved, we arrive at the conclusion that when a vapour exists under a small pressure in a long experimental tube, its dynamic radiation may exceed that of a gas which entirely fills the tube; while when both columns are rendered sufficiently short, the radiation of the gas may far exceed that of the vapour. Experiment is shown to completely confirm this reasoning.

It is worth noting that in the case of the sulphuric-ether vapour above referred to, which produced a deflection of 30°, a powerful flow of heat from a source of 212° Fahr. was at the time passing through the experimental tube to the pile. On the entrance of the air the vapour was raised five or six degrees in temperature, and this moderate amount of heating enabled the vapour not only to neutralise but to far overmatch, as a radiator, the selfsame vapour as an absorber of the heat emitted from the source.

A section of the memoir is devoted to the action of odours upon radiant heat.

A section is also devoted to the action of electrolytic oxygen, comprising its ozone. It is proved that the action on radiant heat augments as the size of the electrode diminishes, the action varying in the experiments recorded from 20 to 136. The harmony of this result with the totally different experiments of De la Rive and Meidinger is pointed out. The question whether ozone is a *form of oxygen* or a *peroxide of hydrogen* is subjected to an experimental test, the conclusion from which is, that ozone is produced by the packing together of oxygen atoms into molecular groups—that heat dissolves the bond of union between the atoms, and converts them into ordinary oxygen.

Two sections are devoted to a comparison of the experiments on air, oxygen, hydrogen, and aqueous vapour with those of Professor Magnus. This distinguished investigator, without any knowledge of what had been previously accomplished, had taken up a portion of this inquiry; and found the action of air, oxygen, and hydrogen to be much greater, and that of aqueous vapour much less, than my experiments make them. This whole question will be subsequently subjected to analysis.

Experiments on the reduction of the temperature of a source by bringing air into contact with it are described. The action of an atmospheric envelope is again adverted to, and an experiment which offers a hope of determining the temperature of space is indicated. The memoir winds up by showing how the internal friction of air is affected by variations of density; the velocity of the atoms, caused by the same difference of temperature, being almost doubled when the density of the air is reduced to one-half.

II.

FURTHER RESEARCHES ON THE ABSORPTION AND RADIATION OF HEAT BY GASEOUS MATTER.*

§ 1.

Recapitulation.

THE apparatus made use of in this inquiry is the same in principle as that employed in my last investigation.† It grew up in the following way:—A wide tube was prepared for the gases through which radiant heat was to be transmitted; but it was necessary to close the ends of this tube by a substance pervious to all kinds of heat, obscure as well as luminous. Rock-salt fulfils this condition; and accordingly plates of the substance an inch in thickness, so as to be able to endure considerable pressure, were resorted to. In the earliest experiments a cube of boiling water was placed before one end of this tube, and a thermo-electric pile connected with a galvanometer at the other; it was found that if the needle pointed to any particular degree when the tube was exhausted, it pointed to the same degree when the tube was filled with air. By this mode of testing, the presence of dry air, oxygen, nitrogen, or hydrogen had no sensible influence on the radiant heat passing through the tube.

In some of these trials the needle stood at 80°, in some at 20°, and in others at intermediate positions. I, however, reasoned thus:—The quantity of heat which produces the deflection of 20° is exceedingly small, and hence a minute fraction of this quantity, even if absorbed, might well escape detection. On the other hand, the quantity of heat which produces the deflection of 80° is large, but then it would require a large

* Received by the Royal Society, January 9, and read before the Society, January 30, 1862.

† *Philosophical Transactions*, 1861, and *Phil. Mag.* vol. xxii. p. 169.

absorption to move the needle even half a degree in this position. A deflection of 20° is represented by the number 20, but a deflection of 80° is represented by the number 710. While pointing to 80, therefore, an absorption capable of producing a deflection equal to 15 or 20 of the lower degrees of the galvanometer, would hardly produce a sensible motion of the needle. The problem then was, to work with a copious radiation, and at the same time to preserve the needle in a position where it would be sensitive to the slightest fluctuations in the absolute amount of heat falling upon the pile.

This problem was first solved by the employment of a differential galvanometer, and afterwards by converting the thermo-pile into a differential thermometer. Its second face was exposed, and a second source of heat was placed in front of that face. A moveable screen was interposed between the two, by the motion of which the same amount of heat could be caused to fall upon the posterior surface of the pile as that received from the experimental tube by its anterior surface. When this was effected, no matter how high the previous deflection might be, it was completely neutralised, and the needle descended to zero.

The experimental tube being exhausted and the equilibrium established, it was immediately destroyed by the entrance of any gas, capable of absorbing even an extremely small fraction of the radiant heat. The second source predominated, and a deflection followed which, when properly reduced, became a strict measure of the absorption.

But in these early experiments my radiating source stood at some distance from the anterior end of the tube, and the heat, previously to entering the latter, had to cross a space of air. This air and its possible sifting influence I wished to abolish, so as to allow the calorific rays to enter the gas with all the qualities which they possessed at the moment of emission. I first thought of soldering the end of the experimental tube direct to the radiating surface, thus allowing the gas to come into direct contact with the source. But it immediately occurred to me that the introduction of cool gas into the tube would lower the temperature of the source, and that it could never be known how far the indication of the galvanometer under such circumstances could be regarded as a true effect of absorption.

An independent tube, 8 inches long, and of the same diameter as the experimental tube, was therefore soldered on to the radiating plate. By a screw joint, the free end of this tube was connected air-tight with the experimental tube. Thus a chamber, from which the air could be removed, was introduced between the first plate of salt and the radiating surface, which was thereby withdrawn from all possible convective action. The radiant heat also entered the tube unchanged in quality save the infinitesimal change due to its passage through the diathermic salt.

§ 2.

New Apparatus.

I will ask permission to refer in this memoir to the Plate facing page 64: a verbal reference will in most cases be sufficient to indicate the changes recently introduced. S S′, it will be remembered, represented the experimental tube, which was first made of brass polished within. Such a tube could not be used for gases or vapours capable of attacking brass; and though this difficulty was combated by blackening the tube within, I could never feel at ease regarding the action of the gases upon the blackening substance. Many gases, moreover, present great difficulties on account of their affinity for atmospheric moisture. Hydrobromic and hydrochloric acid, for example, form dense fumes in the air; and however carefully they might have been dried, I should have been reluctant to base any inference on their deportment without actually having them under my eyes during experiment.

The brass tube, then, which stretched from S to S′ in the figure referred to is now replaced by one of glass, 2 feet 9 inches long, and 2·4 inches in diameter. The source of heat in my last-published inquiry was the cube of hot water C; but glass being far inferior to brass in reflecting power, I could not with this source bring out with the desired force the vast differences existing between various kinds of gaseous matter. A hood of plate copper (mentioned at p. 14), was therefore employed. It was united by brazing to a tube 8 inches long, destined to form the vacuous chamber in front of the first plate of rock-salt. To heat the copper plate, a lamp constructed on the principle of

Bunsen's burner was made use of. The gas passed upwards by four hollow columns, each perforated for the admission of air. The mixture of air and gas issued into a chamber shaped like the frustrum of a cone, and over this chamber was placed a shade of thin sheet-iron, the top of which was narrowed to a slit one-eighth of an inch wide and 2 inches long. From this slit the mixture of gas and air issued, and formed upon ignition a sheet of flame. This was caused to glide along the back of the copper plate before referred to, which was thereby raised to a temperature of about 270° C.

To preserve this source constant was one of the greatest difficulties of the investigation; for the slightest agitation of the surrounding air, or the least flickering of the flame itself, was sufficient to disturb the steadiness of the galvanometer and to render experiments in the most delicate cases impossible. The flame was therefore surrounded by screens of pasteboard, these being again encompassed by towels, through the meshes of which the flame was fed; a suitable chimney also thickly covered, produce a gentle draught and carried off the products of combustion; the rhythmic jumping of the flame which sets in so readily was destroyed by screens of wire-gauze. The 'compensating cube' C′, the double screen H, and the thermo-electric pile P remain as before. They are exposed in the figure, but during the experiments they were surrounded by a close hoarding, all the chinks of which were stuffed with tow, so as to protect the cube and pile from the disturbing action of the air. The vacuous front chamber passed as before through a vessel V in which a current of cold water, constantly renewed, was caused to circulate. Six weeks' practice was required to master all the difficulties of this portion of the apparatus.

§ 3.

Preliminary Efforts and Precautions.—Chlorine, Ozone, and Aqueous Vapour.

On the 16th, 17th, and 18th of June, 1861, I experimented on chlorine and on ozone, and satisfied myself that as an absorber of radiant heat, chlorine was far outstripped by many light colourless gases, and that ozone had a power of absorption very much greater than common oxygen.

The work was resumed on the 12th of September, and my first care was to examine whether the experiments on moist and dry air described in Memoir I. stood the test of repetition.* Professor Magnus had found that the presence of aqueous vapour in air had no influence on the absorption of radiant heat; while according to my experiments dry air exercised only a small fraction of the absorptive energy of saturated air. I commenced my researches in September with a few experiments on this subject.

Half an atmosphere of the undried air of the laboratory, admitted directly into the experimental tube, cut off an amount of heat which produced a deflection of 30 degrees.

The drying-apparatus at this time consisted of a U-tube filled with fragments of pumice-stone wetted with sulphuric acid. Associated with this was a similar tube filled with like fragments, but moistened with caustic potash in solution to remove the carbonic acid of the air.

The air of the laboratory, passed through both of these tubes in succession, till a pressure of 15 inches was attained, gave a deflection of 26 degrees.

This result surprised me, showing, as it seemed to do, a very close agreement between dry and moist air. On examining the drying-tubes, however, it was found that, by a mistake of arrangement, the air had entered the sulphuric-acid tube first, and passed straight from the potash into the experimental tube, thus partially reloading itself with moisture after it had been dried by the sulphuric acid.

On causing the air to pass first through the potash, the deflection fell to less than 5 degrees. Hence this experiment showed the absorption due to the moisture and carbonic acid of the air to be more than six times as great as that of the atmosphere itself. It will presently be shown that this difference falls far short of the truth.

The potash and sulphuric acid were now abandoned, the air being dried by passing it through a U-tube filled with fragments of chloride of calcium, which had lain in the tube for some months. The observed deflection was 40 degrees; that is to say, 10 degrees more than that produced by the *undried* air.

This result, and many others of a similar nature, were due to

* See Sections 20 to 22, Memoir II., and the whole of Memoir III.

the imperfection of the chloride of calcium. Chemists, I think, ought to be very cautious in the use of this substance as a drying agent. When freshly fused it answers well for this purpose, but when old it yields an impalpable powder, which proved in the highest degree perplexing to me in my first experiments. It is generally found, I believe, that a drying-tube of sulphuric acid gains more in weight than one of chloride of calcium, and from this it has been inferred that the quantity of moisture taken up by the former is greater than that taken up by the latter. The difference, however, may really be due to the mechanical carrying away of a portion of the chloride by the current of air.

On the 13th of September these experiments were resumed. The dry air then gave a deflection of less than 2 degrees; while the air direct from the laboratory caused, in one experiment, the needle to move from 20 degrees on one side of zero to 28 on the other. In a second experiment the undried air caused the needle to move from 18° on one side of zero to 32° on the other.

Experiments made on the 17th entirely corroborated this result. Three successive trials with the undried air of the laboratory yielded the deflections 29°, 31°, and 30° respectively, while the deflection produced by the dried air *was less than a single degree*. On this day, therefore, the action of the aqueous vapour of the air was at least thirty times that of the air itself.

Almost every week-day during the last four months, experiments similar to the above have been executed, and in no case have I observed a deviation from the result that the absorptive action of the aqueous vapour of the air is great in comparison with that of the air itself. Further on, this subject will receive additional illustration.

As my mastery over the apparatus, and my acquaintance with the precautions necessary in delicate cases, increased, the absorption by air, and by the transparent elementary gases generally, diminished more and more. I was induced to abandon the use of pumice-stone as well as chloride of calcium, and to construct the drying-apparatus in the following way. The internal portion of a massive block of pure glass was pounded to small fragments in a mortar; these were boiled in pure nitric acid, and afterwards washed several times with distilled water, so as to remove all trace of the acid. They were then dried, afterwards moistened with pure sulphuric acid, and introduced by means of a funnel into a U-tube. The funnel was necessary to preserve the neck of the tube from all contact with the acid, the least action of which upon the corks employed to close the tube being sufficient to entirely vitiate the experiments. At the top of each arm of the U-tube fragments of dry glass were placed, upon which any accidental dust or particles from the cork or sealing-wax might fall.

Similar precautions were taken with the caustic-potash tube. Pure white marble was pounded to fragments and subjected to the action of a dilute acid, which removed the outer surface. The fragments were afterwards washed in distilled water and dried, then moistened with pure caustic potash, and introduced into the U-tube in the manner already described. It was sometimes necessary to perform this operation daily, and never on any occasion have I used tubes to dry a feeble gas which had been previously used to dry a powerful one.

§ 4.

First Experiments on the Human Breath.—Chlorine and Hydrochloric Acid.—Bromine and Hydrobromic Acid.

In the present communication many subjects will be touched upon which for want of time I have been unable to develop. The following is an example of these. Choosing a day of suitable temperature and moisture—a day on which the human breath shows no signs of precipitation—the action of the substances expired from the lungs may be determined. By breathing directly into the experimental tube, the action produced by the sum of the products of respiration might be accurately measured; by breathing through the sulphuric-acid tube, the moisture of the breath would be withdrawn, and the difference between the action then observed and the former action would give that of the carbonic acid. In this way the products of respiration might be estimated singly, and the influence of various kinds of food and drink, or of physical exertion, on the respiration might be investigated in a manner hitherto unthought of.

I have to record the following experiments only in connexion with this point. Placing a suitable tube between my lips, and filling my lungs with air, a stopcock which was interposed between me and the experimental tube was partially opened, and through it I breathed slowly into the tube until the mercury gauge of the pump was depressed 15 inches. I had, at the time, two assistants, C. A. and R. C., and they subsequently breathed into the experimental tube the same quantity as myself. In the following table the absorption produced by the breath of each is stated; the initials J. T. are my own :—

Action of the Products of Respiration on Radiant Heat.

Initials of person's name	Absorption per 100	Initials of person's name	Absorption per 100
J. T.	62	J. T.	59
J. T.	62	R. C.	63
R. C.	66	C. A.	62
R. C.	68	J. T.	60·5
J. T. again	59		

The absorption of dry air on the day that these results were obtained was found to be 1. *The same dry air inhaled, underwent a chemical change which augmented its absorptive energy at least* 60 *times.* This is given as a minor limit,; it is unnecessary to say how much I regard it as falling short of the truth.

The day afterwards the following results were obtained, the same amount as before being exhaled :—

Initials	Absorption per 100
J. T.	56
R. C.	62
J. T.	56
R. C.	59

In all cases R. C., who is the smallest and least robust man of the three, appeared to have the advantage. I will only add a few results obtained on the 6th of October, the quantity of air expired on the occasion depressing the mercurial column 5 inches :—

Initials		Absorption per 100
J. T.		33·5
R. C.		35
R. C.	After half a glass of Trinity Audit Ale	41
	Again	35
	After a teaspoonful of brandy	35
	After chewing and swallowing a small quantity of onion	40

After taking the ale and brandy my assistant washed his mouth and gargled his throat several times with cold water. I give these results merely as illustrative of one of the numerous applications of the apparatus. In all the experiments the tube remained perfectly transparent throughout, and, on pumping out, the needle in each case returned accurately to zero.

In my last paper the fact was brought prominently forward that the elementary bodies then examined were far less hostile to the passage of the longer heat undulations than the compound ones. I was desirous this year to extend the experiments to one or two of the coloured gases and vapours, and on the 20th of September resumed the experiments on chlorine. This gas is highly coloured, and of a specific gravity of 2·45; one of its compounds, hydrochloric acid, is quite transparent, and of a specific gravity of only 1·26. Does the act of combination with hydrogen, which renders chlorine gas more transparent to light, render it also more transparent to heat?

Chlorine prepared from hydrochloric acid and peroxide of manganese, and dried by passing it through sulphuric acid, was admitted into the tube till it depressed the mercury gauge 21 inches; the consequent absorption was expressed by the number 44.

Hydrochloric acid was admitted till it depressed the gauge 19 inches; the absorption was 68.

The following results were afterwards obtained:—

	Absorption per 100
Chlorine 15 inches	32
Chlorine 14 inches	30
Chlorine 14 inches	30
Hydrochloric acid 14 inches	47
Chlorine again	30
Hydrochloric acid	56

In all cases the effect of the compound gas was found to exceed that of the elementary one; so that *the chemical change which renders chlorine more transparent to light renders it more opaque to obscure heat.*

I may remark that a subsidiary gauge was here used, so as to prevent the chlorine from entering the air-pump.

Great care is required in experiments on hydrochloric acid, and great care was bestowed on the above. Previous to the introduction of the gas the experimental tube was filled with perfectly dry air, so as to leave a perfectly dry residue on exhaustion. The gas was allowed to stream through the drying-tube until all traces of air were expelled both from it and the retort; then a joint was suddenly broken, and the retort was connected with the experimental tube. The gas thus passed directly from the retort through the drying apparatus into the experimental tube. It was difficult to avoid sending in with the gas a few particles of moisture; but these, if such existed, appeared to be dissipated by the dynamic heating of the gas on entering the tube, and kept in a gaseous state by the flux of heat passing through it. At all events the closest scrutiny could detect no trace of mist or turbidity within the tube; it was perfectly transparent throughout. The chlorine, on the contrary, was intensely coloured.

Many experiments were made with chlorine which had been collected over water, but something (what I know not yet) which materially augmented its absorption appeared to be in all cases carried along with the gas from the water into the tube.

These experiments were made in the early part of the present inquiry, and before I had become aware of all the peculiarities of my apparatus. Subsequent efforts reduced in some degree the absorption both of chlorine and hydrochloric acid. Very careful experiments made on the 29th of October gave the following absorptions for these two gases, at a pressure of 30 inches:—

Chlorine	39
Hydrochloric acid	53

After each experiment the chlorine and hydrochloric acid were removed from the experimental tube in the following manner:—A cock and connecting-piece were attached to one end of the experimental tube, and from them a length of india-rubber tubing led to the flue of the laboratory stove. A gas-holder of air was put in connexion with the other end of the experimental tube, a system of drying-tubes intervening between the latter and the holder. By water-pressure a stream of dry air was forced gently through the experimental tube into the flue, and in this way the gases, which if pumped out would have injured the pistons, were speedily removed. As the dry air replaced the gases, the needle gradually descended to zero, its arrival there being indicative of the complete displacement of the gas. The perfect dryness of the air thus made use of was beautifully proved. Had the air contained moisture, it would instantly on its mixture with the hydrochloric acid have rendered the medium within the tube turbid; and however slight this turbidity might be, and however invisible to the eye, the galvanometer would have revealed it. But there was no movement in an upward direction; the needle gradually sunk from the moment the air entered.

Bromine vapour and hydrobromic acid furnish another illustration of the influence of chemical union on the absorption of radiant heat. The opacity of the former to light is far greater than that of chlorine, while the two compounds are equally transparent. The density of bromine vapour, moreover, is 5·54, whereas that of hydrobromic acid is only 2·75. The difficulty of operating with this acid is at least equal to that attendant on hydrochloric acid; and several successive days were spent in endeavouring to arrive at safe conclusions in connexion with this subject. Bromine dried with phosphoric acid was introduced into a flask furnished with a screw-cap, by which it was attached to the experimental tube. By turning a stop-cock, the pure vapour was allowed slowly to enter until the mercury column was depressed two inches. From more than twenty experiments made with this substance, I should infer that the absorption of the quantity mentioned does not exceed 11, while the absorption of hydrobromic acid *of the same pressure* amounts to 30.

The hydrobromic acid was prepared by the action of glacial phosphoric acid (for a pure specimen of which I have to thank Dr. Frankland) on bromide of potassium. If the above figures represent the truth (and I have spared no pains to arrive at a right conclusion), we have here a most striking instance of *transparency to light and opacity to obscure heat being promoted by the self-same chemical act.**

* A layer of liquid bromine, sufficiently opaque to intercept the entire luminous rays of a gas-flame, is highly diathermanous to its obscure rays. An opaque solution of iodine in bisulphide of carbon behaves similarly. The details of these experiments shall be published in due time: they were publicly shown in my lectures many months ago.—June 13, 1862.

§ 5.

New Experiments on Gases.

In the following table are given the absorptions of a number of gases at a common pressure of one atmosphere, as determined with the new apparatus :—

TABLE I.

Name	Absorption per 100	Name	Absorption per 100
Air	1	Carbonic acid	90
Oxygen	1	Nitrous oxide	355
Nitrogen	1	Sulphuretted hydrogen	390
Hydrogen	1	Marsh-gas	403
Chlorine	39	Sulphurous acid	710
Hydrochloric acid	62	Olefiant gas	970
Carbonic oxide	90	Ammonia	1195

Air, oxygen, nitrogen, and hydrogen are all set down as equal to unity in the above table. I do not mean thereby to affirm that there are no differences between these gases, but that the most powerful and delicate tests hitherto applied have failed to establish a difference in a satisfactory manner. It is not improbable that the action of these gases may turn out to be even less than it is here found to be; for who can say that the best-constructed drying-apparatus is really perfect? Besides, stop-cocks must be greased, and hence may contribute an infinitesimal impurity to the air passing through them. It is not even certain that monohydrated sulphuric acid may not deliver a modicum of vapour to the current of air passing through it. At all events, if any further advance should be made in the purification of the gases, it will certainly only tend to augment the enormous differences exhibited in the above table.

Ammonia, of the pressure mentioned, stands highest in the above list as regards absorptive energy. I believe a length of less than 3 feet of this gas, which to the vision is as transparent within the tube as the vacuum itself, to be *perfectly black* to the rays emanating from the source here made use of. When the gas was in the tube, the interposition of a double metallic screen between the pile and source augmented the deflection very slightly. But it will be shown, further on, that the ammonia in this experiment could not exhibit its full energy of absorption, and that in the length indicated it is in all probability absolutely impervious to the heat issuing from our source.

It would be a mere affectation of accuracy to try to deal with smaller quantities of the first four substances mentioned in the table than those here examined. Still, if such small quantities could be directly measured, the action of air, oxygen, hydrogen, and nitrogen, in comparison with that of the other substances at the same pressure, would doubtless be greatly reduced. With the energetic gases the rays are most copiously quenched by the portions which first enter the tube, the portions which enter last producing in many cases an infinitesimal effect. Now it has been shown in the last paper that, for very small absorptions, the effect is sensibly proportional to the quantity of gas present; and this would seem to justify the assumption that for 1 inch of pressure the absorption of air, oxygen, nitrogen, and hydrogen would be $\frac{1}{30}$th of the absorption at 30 inches. In the case of each of the other gases I have measured directly the absorption of a quantity corresponding to a single inch of pressure. Assuming the proportionality just referred to, and again calling the effect of air unity (the unit, however, being only $\frac{1}{30}$th of that in the last table), the following are the relative absorptions:—

TABLE II.

Gas	Absorption	Gas	Absorption
Air	1	Carbonic oxide	750
Oxygen	1	Nitric oxide	1590
Nitrogen	1	Nitrous oxide	1860
Hydrogen	1	Sulphide of hydrogen	2100
Chlorine	60	Ammonia	5460
Bromine	160	Olefiant gas	6030
Hydrobromic acid	1005	Sulphurous acid	6480

Here we have the extraordinary result, that, for pressures of 1 inch of mercury, *the absorption of ammonia is over five thousand times, the absorption of olefiant gas over six thousand times, while the absorption of sulphurous acid is six thousand four hundred and eighty times that of air.*

It is impossible not to be struck by the position of chlorine and bromine in this table. They are elements, and, notwithstanding their colour and density, they take rank after the transparent elementary gases. The perfectly transparent olefiant gas absorbs more than one hundred times the amount absorbed by chlorine, and nearly forty times the quantity absorbed by the intensely brown vapour of bromine. I cannot think this fact insignificant. Hitherto chemists have spoken to us of elements, and we have helped ourselves to conceptions regarding them and their compounds in the only way possible to our mental constitution. But our conceptions remained purely subjective, nor were we acquainted with any physical trait which would in any degree justify these conceptions. Here, however, we seem to touch the ultimate particles of matter. Starting from the idea that a gas absorbs such vibrations as are isochronous with its own, in all cases the compound gas reveals itself to the mind's eye with its molecules swinging more slowly than the atoms of which it is composed, *when uncombined.* The absorption of the longer undulations proves the general coincidence in period with those undulations. We, as it were, load the atom by the act of chemical union, and thereby render its vibrations more sluggish, that is to say, more fit to synchronise with the slowly recurrent waves of obscure heat.

In the foregoing table the absorption of nitric oxide is given as 1590, which is less than that of nitrous oxide, though the molecule of the former contains a greater number of atoms than that of the latter. It will be noticed *that those gases which on combining suffer no condensation are less energetic absorbers than those which suffer a reduction of volume.* Whether this rule is universal I am as yet unable to say.

It is very difficult to operate with nitric oxide; the affinity of the gas for oxygen is so enormous that the slightest trace of this substance gives rise to the brown fumes of nitrous acid. On first sending this gas into the experimental tube, 1 inch of it gave an absorption of 2040; but the needle slowly went up afterwards, until it finally indicated an absorption of 5100. On looking across the tube at this time, the brown hue of nitrous acid was discernible.

In a second experiment the vacuum was made as perfect as possible; and, on allowing nitric oxide to enter, the absorption was found to be 1860, but the needle soon afterwards declared an absorption of 3060, the brown fumes appearing as before.

On filling the experimental tube with nitrogen, then exhausting, and allowing nitric oxide to enter, the absorption of 1 inch of the gas was 1680. On filling the experimental tube previously with hydrogen the absorption was 1590, which is that given in the table. On letting in a mixture of air and nitric oxide till the tube was filled, the action last mentioned was augmented nearly twentyfold. Nitrous acid is therefore an extremely energetic gas. *The difference between it and bromine is enormous, even when their colours are undistinguishable.*

A close inspection of Melloni's Table* reveals, I think, the tendency of solid bodies also to become more transparent to heat as their composition becomes more simple. After rock-salt itself, comes the element sulphur, and after it fluor-spar. But the case of lampblack will here occur to many, as the most powerful absorber and radiator yet discovered. No doubt the grouping of the atoms of an elementary substance may make it tantamount to a compound, and no doubt this is actually the case with lampblack; another eminent example of this kind is ozone. Leslie, however, found water to be a better radiator than lampblack, and Wells found several substances which were more capable of being chilled by nocturnal radiation. On reflexion, moreover, the following considerations arise. The lampblack of commerce and the soot of a lamp or candle—that is to say, the substances hitherto employed in experiments on radiant heat—are copiously mixed with hydrocarbons, which are the most powerful absorbers and radiators in Nature. It might fairly be questioned whether the reputed experiments with lampblack really dealt with lampblack at all. But even the impure substance is to some extent transparent to radiant heat.

§ 6.

Radiation through Black Glass and Lampblack.

I have plates of black glass, rendered so by the solution of carbon in the glass while in a state of fusion, which, though impervious to the rays of the most intense electric-light, allow of a copious transmission of obscure heat. Melloni's beautiful experiments on glass of this character are well known. Another of Melloni's experiments which I have recently verified is the following. A plate of transparent rock-salt was

* *La Thermochrose*, p. 164.

placed over a smoky camphine lamp, soot being deposited on its surface until it intercepted every ray of a brilliant jet of gas. The smoked plate was placed between a source of heat of a temperature of 100° C. and a thermo-electric pile, a polished screen being placed between the salt and the source of heat. As long as the screen remained, the needle of the galvanometer connected with the pile stood at zero; but the moment the screen was removed the needle promptly advanced, showing the instantaneous transmission across the layer of soot of a portion of the heat incident upon the salt. The actual numbers obtained in this experiment are these:—The deflection produced by the heat transmitted through the soot was 52°; which is equal to 90 units. The deflection produced when the layer of soot had been carefully removed, so as to leave both surfaces of the salt smooth and transparent, was 71°, which is equal to 300 units. The quantity transmitted through the soot is therefore to the total quantity as

$$90 : 300,$$

or as

$$30 : 100;$$

that is to say, the lampblack, which was perfectly opaque to the light of a gas-jet, was transparent to fully 30 per cent. of the incident heat. On consulting Melloni's Table, I was gratified to find that he made the transmission by a plate similarly prepared 27 per cent.; while a layer so opaque that it cut off the beams of the sun itself transmitted 23 per cent. of the rays emitted by a source heated to 100° C.

§ 7.

Selective Absorption by Lampblack.

At page 93 of 'La Thermochrose,' Melloni examines the absorption of this substance for all sorts of rays, and by a series of ingenious experiments, and reasonings remarkable for their clearness and precision, he arrives at the conclusion that lampblack absorbs with the same intensity all descriptions of radiant heat.* At page 284, however, he cites and discusses with the

* 'Donc, le noir de fumée absorbe avec la même intensité toute sorte de rayonnements calorifiques' (p. 101).

same precision a series of experiments made with smoked rock-salt, in which he shows that the same layer of lampblack transmits 8 per cent. of the rays from a lamp of Locatelli, 10 per cent. of those of incandescent platinum, 18 per cent. of those from copper heated to 400° C., and fully 23 per cent. of those emitted by a source of 100° C. Now a transmission of 8 per cent. implies an absorption of 92; while transmissions of 10, 18, and 23 per cent. imply absorptions of 90, 82, and 77. But that the self-same layer of lampblack absorbs 77 per cent. of the rays from one source and 92 per cent. of the rays from another, is at variance with the statement that lampblack absorbs heat from all sources with the same intensity. Suppose the surface of a thermo-electric pile to be coated by a layer of lampblack of the same thickness as that which coated Melloni's plate of salt; 23 per cent. of the rays from a source of 100° C. would go right through such a layer and impinge upon the metal face of the pile; and if the latter were a good reflector, the heat incident upon it would be in great part retransmitted through the lampblack and lost to the instrument. For a source of 100° C., this loss would be many times greater than for a Locatelli lamp. Possibly, however, Melloni may have meant simply to assert that *for practical purposes* the absorption at the face of his pile might be considered to be the same for all kinds of heat.*

§ 8.

New Experiments on Vapours.—Further Proof of the Influence of Chemical Combination on the Absorption of Radiant Heat.

I have now to record some new experiments on the action of *vapours* upon radiant heat. A number of glass flasks were prepared, of the shape and size of common test-tubes, each of which was furnished with a brass cap carefully cemented on to it. By means of this it could be attached to a stopcock, and thus connected with the experimental tube. The mode of operation was this:—The liquid was introduced into the flask

* The sun, through the floating carbon of the London atmosphere, sometimes presents a most instructive appearance. Entirely shorn of his rays and of perfectly uniform brightness, his colour at times is as red as blood. This is doubtless in part due to the comparative transparency of the floating carbon for the longer undulations.

by means of a small glass funnel. A stopcock was then attached to the flask and connected with a second air-pump, which was always kept at hand. The air above the liquid was removed, and the air dissolved in it was allowed to bubble away, until nothing remained but the pure liquid below and the pure vapour above it. The stopcock was then shut off, and the flask united to the experimental tube. The exhaustion of the tube being complete, and the needle of the galvanometer at zero, the cock attached to the flask was turned on and the mercury-gauge carefully observed at the same time. No bubbling of the liquid was in any case permitted. The vapour entered silently and without the slightest commotion; and when the mercurial column was depressed to the extent required, the vapour was promptly intercepted.

The energy with which the needle moves the moment a strong vapour enters is so extraordinary that, lest the shock against them should derange the magnetism of the astatic pair, I removed the stops which arrested the swing of the needle at 90°. It often swung far beyond a quadrant; and after it had come finally and permanently to rest, its position was observed in the following manner:—The dial of the galvanometer being well illuminated, a looking-glass was placed behind the instrument, at such an angle that when looked at horizontally the image of the dial was clearly seen. This image was observed by an excellent telescope fixed at a distance of 11 feet from the galvanometer. Attached to the needle, and in continuation of it, was a bit of glass fibre of extreme fineness, blackened with Indian ink. This index ranged over the graduated circle, and by means of it a very small fraction of a degree could be easily read off. The expedient of observing from a distance was resorted to, because it was found that my approach to the galvanometer, perhaps through the diamagnetic action of my own body, had a sensible effect upon the needle, which, I believe, surpasses in delicacy any hitherto employed.

The *permanent* deflection of the needle was noted in all these experiments, and the value of the deflection, expressed in terms of one of the lower degrees of the galvanometer, was obtained from a table of calibration. To spare unnecessary labour I omit the deflections in the following table, and give the absorptions only produced by the vapours, at 0·1, 0·5, and 1·0 inch of pressure.

TABLE III.

Name of substance	Pressures 0·1 inch	0·5 inch	1·0 inch
Bisulphide of carbon	15	47	62
Iodide of methyl	35	147	242
Benzol	66	182	267
Chloroform	85	182	236
Iodide of ethyl	158	290	390
Methylic alcohol	109	390	590
Amylene	182	535	823
Alcohol	325	622	
Sulphuric ether	300	710	870
Formic ether	480	870	1075
Acetic ether	590	980	1195
Propionate of ethyl	596	970	
Boracic ether	620		

Let us compare some of the results of this table of transparent vapours with the action of the highly coloured vapour of bromine. The absorption of bromine vapour at 1 inch pressure is about 6, and at 0·1 of an inch pressure would probably not exceed 1; hence at 0·1 of an inch pressure, bisulphide of carbon exerts probably 15 times the absorbent power of bromine; but bisulphide of carbon is the feeblest of the compound vapours hitherto discovered. The strongest of these, boracic ether, has, according to the above estimate, and at the pressure stated, *more than* 600 *times the absorbing energy of the strongly coloured bromine.*

The whole of the numbers in the above table are referred to atmospheric air as unity; 0·1 of an inch of bisulphide-of-carbon vapour, for example, absorbs 15 times as much as a whole atmosphere of air. Let us compare, for an instant, the action of boracic ether with that of air. We arrive at an approximate comparison in this way. The absorption of the tenth of an inch of boracic ether is something more than that of a whole inch of methylic alcohol; by diminishing the quantity of methylic alcohol to one-tenth, we reduce its absorption from 590 to 109. The absorption of one-tenth of an inch of boracic ether is 620°; suppose its absorption to diminish with diminished quantity in the proportion of methylic alcohol, we should then have for 0·01 of an inch of boracic ether an absorption of 111; that is to say, for $\frac{1}{3000}$th of an atmosphere of boracic ether, we should have an action 111 times that of a whole atmosphere of oxygen, nitrogen, hydrogen, or atmospheric air.

With the transparent elementary gases it is impossible to measure directly the absorption of 0·1 of an inch; but assuming, as before, that up to an absorption of 1 the effect is proportional to the quantity of gas present, the absorption of each of the elementary gases, at a pressure of 0·1 of an inch, would be about 0·0033; hence the absorption of boracic ether of 0·1 of an inch pressure is to that of air at the same pressure as

$$620 : 0{\cdot}0033,$$

which would give to the ether an energy 186,000 *times that of air.*

I have already spoken of the blackness of ammonia at 30 inches pressure. Referring to Table I., its absorption is found to be 1195. In the last table the vapour of acetic ether, under only one-thirtieth of the pressure of the ammonia, produces apparently the same effect; its absorption is also 1195. Such facts give one entirely new ideas of the capabilities of matter; and our wonder will not be diminished by the results to be recorded further on.

§ 9.

Superior Action at one Pressure does not prove Superiority at all Pressures.

With both gases and vapours we find that it does not follow that a gas which produces a larger effect than another at one pressure should surpass it at all other pressures. Some gases start from a lower level than others, but finally attain an equal, or even a greater elevation. If their absorptions were represented by curves plotted from the same datum-line, these curves would in some cases approach, and in some cases, cross each other. At a pressure of 1 inch, for example, carbonic acid has more than double the absorptive power of carbonic oxide, whereas at a pressure of 30 inches they are equal; indeed some of my experiments show carbonic oxide to have the advantage. On the 22nd of October, for example, the deflection produced by 2 inches of carbonic oxide was found to be 15°, while that of 2 inches of carbonic acid was 38°. The two gases at a pressure of 30 inches gave these results :—

	Degrees
Carbonic oxide	52
Carbonic acid	51·5

And again, on the 4th of November I obtained the following relative effects :—

	Pressures	
	1·2 inch	24 inches
	°	°
Carbonic oxide	12	57
Carbonic acid	37	54

The same remarks apply to vapours. Methylic alcohol, for example, starts at a lower level than the iodide of ethyl, but ascends more quickly, and finally reaches a much higher elevation. The same observation applies to benzol and the iodide of ethyl, in comparison with chloroform.

§ 10.

DYNAMIC RADIATION AND ABSORPTION.

A class of facts are now to be referred to which surprised and perplexed me when I first observed them. As an illustration, I will take the case of alcohol vapour. A quantity of this substance, sufficient to depress the mercury gauge 0·5 of an inch, produced an absorption which caused a deflection of 72° of the galvanometer needle.

While the needle pointed to this high figure, and previously to pumping out the vapour, dry air was allowed to stream into the experimental tube, and I happened while it entered to observe the effect upon the galvanometer. The needle, to my astonishment, sank speedily to zero, and went to 25° at the opposite side. The entry of the almost neutral air here not only abolished the absorption previously observed, but left a considerable balance in favour of the face of the pile turned towards the source. A repetition of the experiment brought the needle down to zero, and sent it to 38° on the opposite side. In like manner a very small quantity of the vapour of sulphuric ether produced a deflection of 30°; on allowing dry air to fill the tube the needle descended speedily to zero, and swung to 60° at the opposite side.

These results both perplexed and distressed me, imagining

as I did, on first observing them, that I had been dealing throughout with an effect totally different from absorption. I thought, indeed, that the vapours had deposited themselves in opaque films on the plates of rock-salt, and that the dry air on entering had cleared these films away, and allowed the heat from the source free transmission.

But a moment's reflexion dissipated this supposition. The clearing away of such a film could at best but restore the state of things existing prior to its formation. It might be conceived of as bringing the needle again to 0°; but it could not possibly produce the negative deflection, which, in the case of ether-vapour, amounted to the vast amplitude of 60°. Nevertheless I dismounted the tube, and subjected the plates of salt to a searching examination. No such deposition as that above surmised took place. The salt remained perfectly transparent while in contact with the vapour.

Some of the experiments recorded in the Bakerian Lecture for 1860 had taught me that the dynamic heating of the air when it entered the exhausted tube was sufficient to produce a very sensible radiation on the part of any powerful vapour contained within the tube, though I was slow to believe that the enormous effect now under consideration could be thus accounted for. My first care was to determine the difference of temperature within the experimental tube at the end furthest from the source of heat, and the air without. I then examined, by an extremely sensitive thermometer, the increase of temperature produced by the admission of dry air into the tube, and the decrease of temperature consequent on pumping out, and found the former to be a considerable fraction of the total heat transmitted from the source. Could it be that the heat thus imparted to the alcohol and ether vapours, and radiated by them against the adjacent face of the pile, was more than sufficient to make good the loss by absorption? The *experimentum crucis* at once suggested itself. If the effects observed were due to the dynamic heating of the air, we ought to obtain them even when the sources of heat are entirely abolished. We should thus arrive at the solution of the novel, and at first sight utterly paradoxical problem, dealt with in the next section.

§ 11.

To determine the Radiation and Absorption of Gases and Vapours without any Source of Heat external to the Gaseous Body itself.

I.—*Vapours.*

For the sake of brevity, I will call the heating of a gas on its admission into a vacuum, the *dynamic heating* of the gas; and the chilling accompanying its pumping out, *dynamic chilling.* It would also contribute to brevity if I were allowed to call the radiation and absorption of the gaseous body, consequent on such heating and chilling, *dynamic radiation* and *dynamic absorption,* though the terms are not unobjectionable.

Both the source of heat and the compensating cube were dispensed with, and the thermo-electric pile was presented to the end of the cold experimental tube. By a little management, the slight inequality of radiation against both faces of the pile, arising from differences of temperature in the various parts of the laboratory, was obliterated, and the needle of the galvanometer was brought to 0°.

The vapours were admitted in the manner already described, until a pressure of 0·5 of an inch was obtained. The air was then allowed to enter through a drying-apparatus by an orifice of a constant magnitude. Two stopcocks, in fact, were introduced between the drying-tube and the experimental tube; one of these was kept partially turned on, and formed a gauge for the admission of the air. When the tube was to be exhausted, the second stopcock was turned quite off. When the tube was to be filled, this stopcock was turned full on; but the *gauge-cock* was never touched during the entire series of experiments.

Before, however, the mode of experiment was thus strictly arranged, a few preliminary trials gave me the following results:—

Nitrous oxide on entering caused the needle to swing in a direction which indicated the heating of the gas; the limit of its excursion was 28°, after which it slowly sank to 0°.

The pump was now worked; the propulsion of the first portions of the gas from the tube was so much work done by the residue. That residue became consequently chilled; into it the adjacent face of the pile poured its heat, and a swing of the needle on the negative side of 0° was the consequence. The limit of the excursion was 20°.

Olefiant gas, operated on in the same manner, produced on entering the tube a swing of 67°, showing radiation; and on pumping out, a swing of 41°, showing absorption. After the pumping out of the gas, and without introducing a fresh quantity, dry air was again admitted; the swing produced by the dynamic radiation of the residue of the gas (0·2 of an inch in pressure) was 59°. On pumping out *very quickly*, the dynamic absorption produced a deflection of nearly 40°.

A little of the vapour of sulphuric ether was admitted into the tube; on the admission of dry air afterwards, the needle swung from 0° to 61°; on pumping out, the needle ran up to 40° on the opposite side.

These and other experiments, which gratified me exceedingly, showed that, without resorting to any source of heat external to the gaseous body itself, its radiation and absorption might be determined with extreme accuracy, and the reciprocity of both phenomena rendered strikingly clear. In fact, at this very time I had been devising an elaborate apparatus for the purpose of examining the radiation of gases and vapours, with a view to comparison with their absorption; but no such apparatus would have given me results equal in accuracy to those placed within reach by the discovery here referred to.

The following table is the record of a series of experiments on dynamic radiation and absorption. The vapour in each case was admitted till the mercury column fell half an inch, and dry air was admitted afterwards.

TABLE IV.—*Dynamic Radiation and Absorption of Vapours.*

	Deflections Radiation	Absorption per 100
	°	°
Bisulphide of carbon	14	6
Iodide of methyl	19·5	8
Benzol	30	14
Iodide of ethyl	34	15·5
Methylic alcohol	36	
Chloride of amyl	41	23
Amylene	48	
Alcohol	50	27·5
Sulphuric ether	64	34
Formic ether	68·5	38
Acetic ether	70	43

The paradox already referred to is here solved, and an explanation given of the extraordinary effect observed in the case of the alcohol and ether vapours when dry air entered the experimental tube. Dynamic radiation, moreover, and dynamic absorption go hand in hand; and if we compare both with Table III. (middle column), we shall find the order of the substances precisely the same, although one set of results are obtained with a source of heat external to the gaseous body, and the other with a source of heat and cold within the body itself. Were sufficient time at my disposal, this subject could be developed with advantage. The measurements just recorded constitute my first regular series; and, no doubt, augmented experience will enable me to attain more perfect results.

Half an inch of my most energetically acting vapour—namely, boracic ether—could not well be obtained; but one-tenth of an inch admitted into the tube and dynamically heated and chilled, gave—

Radiation	Absorption per 100
56°	28°

§ 12.

Attempted Estimate of Quantity of Radiant Vapour.

Seeing the astonishing energy with which some of these vapours absorb and radiate heat, it may be asked how far the quantity of vapour may be reduced before its action becomes insensible. At present I will not venture to answer this question fully; certainly we should be dealing at least with millionths of our smallest weights. But I will here give a detailed account of one experiment, the result of which can hardly fail to excite surprise. The experimental tube being exhausted, one-tenth of an inch of boracic-ether vapour was admitted into it: the barometer stood at 30 inches at the time; hence the pressure of the vapour within the tube was $\frac{1}{300}$th of an atmosphere.

Dynamically heated by dry air, the radiation of this vapour produced a deflection of 56°.

The tube was then exhausted to 0·2 of an inch, and the quantity of vapour reduced thereby to $\frac{1}{150}$th part of its first amount; the needle was allowed to come to zero, and the residue of vapour was dynamically heated as before: its radiation produced a deflection of 42°.

The pump was again worked till a vacuum of 0·2 of an inch was obtained, this residue containing of course $\frac{1}{150}$th of the quantity of ether present in the last. On dynamically heating this residue, its radiation produced a deflection of 20°.*

Two additional exhaustions, succeeded by dynamic heating, gave the deflections 14° and 10° respectively.

Tabulating the results so as to place each deflection beside the vapour-pressure which produces it, we have the following view of the experiment:—

TABLE V.—*Dynamic Radiation of Boracic Ether.*

Pressure in parts of an atmosphere	Deflection °
$\frac{1}{300}$th	56
$\frac{1}{150} \times \frac{1}{300} = \frac{1}{45000}$th	42
$\frac{1}{150} \times \frac{1}{150} \times \frac{1}{300} = \frac{1}{6750000}$th	20
$\frac{1}{150} \times \frac{1}{150} \times \frac{1}{150} \times \frac{1}{300} = \frac{1}{1012500000}$th	14

The air itself, slightly warming the apparatus near the pile, produces a feeble radiation, amounting to 6° or 7°. I have purposely excluded the deflection 10°, in order to show that the effect was still diminishing when the experiment ended, the constant effect due to the air itself being not yet attained. Two 0s are thus excluded from the denominator of the fraction which might fairly have appeared in it. The above result is, however, sufficiently extraordinary, showing as it does that the radiation of an amount of vapour possessing in the experimental tube a pressure of less than the thousand-millionth of an atmosphere is perfectly measurable. It will also be borne in mind that the temperature imparted to this infinitesimal quantity of matter could not be high.†

These experiments, which I intend to develop on a future occasion, seem to give us new ideas as to the nature and capa-

* This is less than the truth—my assistant having executed three or four strokes of the pump inadvertently while the dry air was not shut off, removing thereby a considerable proportion of the vapour which ought to be present at this stage of the experiment.

† I should like to repeat these experiments on boracic ether for this reason: the liquid, when it evaporates in moist air, leaves a solid residue of boracic acid, which may be seen round the stopper of the bottle containing the liquid; and, though it is not probable that any such residue was formed in the foregoing experiments, I should like to re-examine the vapour with special reference to this point.

bilities of matter. A platinum wire heated to whiteness in a vacuum by an electric current, becomes comparatively cold within a second after the current has been interrupted; yet that wire, while ignited, was the repository of an immense amount of mechanical energy. What has become of this? It has been conveyed away by a substance so attenuated that its very existence must for ever remain a hypothesis. But here is matter that we can weigh, measure, taste, and smell, proved to be reducible to a tenuity which, though expressible by numbers, defies the imagination to conceive it. Still we see it competent to arrest and originate quantities of energy which in comparison with its own mass must be almost infinite, a small fraction of this energy causing the double needle of the galvanometer to swing through considerable arcs. When we find common ponderable matter producing these effects, we have less difficulty in investing the luminiferous æther with those mechanical properties which have long excited the interest and wonder of those who have dwelt on the mechanical conceptions involved in the undulatory theory of light.

§ 13.

II.—*Gases.*

In the foregoing experiments dry air was used to warm the vapours, but similar differences ought to be exhibited by gases when heated by their own dynamic action. That this is the case the following experiments show:—

TABLE VI.—*Dynamic Radiation of Gases.*

Name	Radiation
	°
Air	7
Oxygen	7
Nitrogen	7
Hydrogen	7
Carbonic oxide	19
Carbonic acid	21
Nitrous oxide	31
Olefiant gas	63

These results are in accordance with those recorded in Table I., p. 80.

The following two gases were used in irregular quantities, but the energy of their radiation is thereby established beyond a doubt. They were admitted into the experimental tube from a large bolthead, until a common pressure was established between the gas in the tube and the gas in the bolthead.

	Radiation	Absorption per 100
Ammonia 15 in. pressure	56·5°	33·5°
Sulphurous acid 16 in. pressure . . .	45	24

§ 14.

Influence of Length and Density of Radiating Column.

Let us reflect for an instant on the condition of our tube, containing its ½-inch of vapour, at the moment when the latter has been heated by the entrance of the air. The rays from the molecules at the end of the tube most distant from the pile have to cross a space of nearly 3 feet before they reach the adjacent end, this space being filled with molecules similar in all respects to the radiating ones. Hence absorption to a comparatively great extent must occur; and indeed we can imagine the tube so long that its frontal portion should furnish a vapour screen absolutely opaque to the radiation of its hinder portion. Now comparing ether-vapour with olefiant gas, it is, I think, evident that the radiant points of the attenuated vapour, which depresses the mercury column only 0·5 of an inch, are further apart than those of the gas which depresses the column 30 inches. Consequently there is a wider door open for the radiation of the distant ether particles towards the pile than for the distant particles of olefiant gas. The length of the whole column, in fact, might be more or less available for the radiation of the vapour, and a part of it only available for the gas. Cut off this useless portion from the gas column, and we do not injure its efficacy; but cut off a similar length from the vapour column, and we may materially diminish its effect. Speaking generally, in reducing the column of ether and that of gas by the same amount, the diminution of radiation will be more sensibly felt where the radiant points are furthest asunder. Reasoning thus, it becomes evident that in a long tube the vapour may excel the gas in its amount of radiation, while in a short tube the gas may excel the vapour. Let us now test this reasoning by experiment.

The dynamic radiation of the following four substances has been tabulated thus:—

	Degrees
Sulphuric ether	64
Formic ether	68·5
Acetic ether	70
Olefiant gas	63

The action of olefiant gas is therefore smallest when the length of the radiating column is 2 feet 9 inches.

With a tube 3 inches long, or one-eleventh of the former length, precisely similar experiments yielded the following results:—

	Degrees
Sulphuric ether	11
Formic ether	12
Acetic ether	15
Olefiant gas	39

The verification of the above theoretic reasoning is here complete. It is proved that *in a long tube the dynamic radiation of the vapour exceeds that of olefiant gas, while in a short tube the dynamic radiation of the gas far exceeds that of the vapour.*

§ 15.

Laplace's Correction for the Velocity of Sound.—Remarks on the Radiant Power of Molecules and Atoms.

Some years ago a discussion was carried on between Professors Challis and Stokes on Laplace's correction for the velocity of sound in air. Professor Challis contending that Laplace had no right to his correction, inasmuch as the heat developed in the condensations of the waves of sound would be instantly wasted by radiation. Experiments, he argued, conducted in confined vessels furnish no ground for conclusions regarding what occurs in the atmosphere, where the heat developed has an indefinite space to lose itself in. Now, our experimental tube, though mechanically closed, is thermally open; by employing the rock-salt plate, indefinite extension, as regards the radiation of heat, is secured in one direction, and the means also exist of measuring the flux of this heat. What is true for one direction would of course be true for all, so that the apparatus will inform us of what occurs in the open atmosphere. The fact, then, is that, with the most powerfully radiating gases hitherto examined, the radiation continues a very sensible time, while the heat acquired by air, on entering the tube, is often a source of inconvenience on account of the inability of the air to disperse its heat by radiation. The question seems therefore experimentally decided in favour of Laplace and his supporter.

This lack of radiating power on the part of air, and of the elementary gases generally, is very noteworthy. The dynamically-warmed air is the proximate source of the heat imparted to the vapours in our experiments on dynamic radiation. It is related to those vapours precisely as a hot plate of polished metal to the coat of varnish which makes it a radiator. Without the intermediation of a second body neither the air nor the metal (both of them elements or mixtures of elements) is competent to impart motion to the luminiferous æther. The atoms possess the motion of heat, but they cannot communicate it to the light-medium, save in the scantiest degree. We have here a definite mechanical result of chemical union, which, if the theory of an æther be true, is as certain as any conclusion of mathematics, and which would hardly be rendered more certain if the physical vision were so sharpened as to be able to see the oscillating atom and the medium in which it swings. I write thus definitely lest it should be imagined that we are dealing in vague conjectures. The connexion of chemical and mechanical phenomena here established must, I think, be pregnant of results.

Further, if, as all the facts declare, radiation and absorption are complementary acts, a giving and taking of motion, united by a bond of strict proportionality, then it may be affirmed that no coincidence in period between the vibrations of a radiating body and those of oxygen, hydrogen, or air could make any one of these substances a good absorber. They are physically incapacitated from communicating motion, and hence in an equal degree incapacitated from accepting motion. *The form of the atom, therefore, or some other attribute than its period of oscillation, must enter into the question of absorption.* The neutrality of the elementary gases in the foregoing experiments does not arise from the accident that a source of heat was chosen whose periods did not synchronize with those of the gas; for however both might synchronize, the gas would still be a bad absorber. Even when the motion of heat which their own absorbent power does not enable them to take up is mechanically imparted to the atoms, or is communicated to them by contact, elementary bodies expend it but sparingly upon the luminiferous æther, which accepts all vibrations alike.*

§ 16.

Action of Odours upon Radiant Heat.

Scents and effluvia generally have long excited the attention of observant men. They have formed favourite illustrations of the divisibility of matter. Several chapters in the works of the celebrated Robert Boyle are devoted to this subject, and philosophers in all countries have speculated more or less upon the extraordinary tenuity of the matter which is competent to produce sensible effects upon the olfactory nerves. We have here, of course, materials for a wide inquiry, which it is quite out of my power to undertake at present. I think, however, that the apparatus thus far made use of enables us to deal with the question in a manner hitherto unattainable.

The leaves and flowers of a number of dry aromatic plants,† obtained from Covent Garden, were stuffed into glass tubes 18 inches long and a quarter of an inch in diameter. A current of dry air was first sent through the tubes for some minutes. They were then connected with the exhausted experimental tube, with its sources of heat arranged as already described. Dry air was then passed over the scented herbs until the experimental tube was filled. The consequent deflection was noted, and from it the absorbent action of the odorous substance was deduced.

The odour of thyme thus treated intercepted thirty-three times the quantity of heat stopped by the air in which it was diffused.

* I can hardly imagine the bands in the spectra of metallic compounds to be produced by the vibration of the compound atom. All my experiments show the vast influence of chemical union on the rate of oscillation; the metal itself and the compound of that metal could hardly, in my opinion, oscillate alike. Hence, the fact that the lines, say, of sodium burnt in air, or vaporized by the electric spark, are the same as those of chloride of sodium, proves, in my opinion, that decomposition has occurred when the bright and constant spectral bands are seen.

† I mean 'dry' in the common acceptation of the term. They were not green, but withered; doubtless, strictly speaking, they contained aqueous vapour.

Peppermint intercepted thirty-four times that quantity.
Spearmint intercepted thirty-eight times the same amount.
Lavender produced thirty-two times the action of the air.
Wormwood forty-one times the action of the air.

A number of perfumes, obtained from Mr. Atkinson, of Bond Street, were examined in the following manner. Small squares of dried bibulous paper, all of the same size, were rolled into cylinders about 2 inches in length; each of these was moistened by an aromatic oil, and introduced into a glass tube between the drying-apparatus and the experimental tube. The latter being first exhausted, was afterwards filled by a current of dry air which had passed over the scented paper. Calling the action of the air which formed the vehicle of the perfumes 1, the following absorptions were observed in the respective cases:—

TABLE VII.

Name of Perfume	Absorption per 100	Name of Perfume	Absorption per 100
Pachouli	30	Lavender	60
Sandal Wood	32	Lemon	65
Geranium	33	Portugal	67
Oil of Cloves	33·5	Thyme	68
Otto of Roses	36·5	Rosemary	74
Bergamot	44	Oil of Laurel	80
Neroli	47	Cassia	109

In comparison with the air which carried them into the tube, the weight of these odours must be almost infinitely small. Still we find that the least energetic in the list produces thirty times the effect of the air, while the most energetic produces 109 times the same effect. Would it be absurd to entertain the notion that, as regards the absorption of radiant heat, the perfume of a flower-bed may be more efficacious than the entire oxygen and nitrogen of the atmosphere above it?

After each scent had been introduced, a stream of dry air was admitted at one end of the tube, while the pump was worked in connexion with the other. The perfume was thus cleared out until the needle returned to 0°. This was often a long operation, the odours clung with such tenacity to the apparatus. Even after the zero point had been attained in the case of a strong perfume, a few minutes' rest of the pump sufficed to bring the scent from its hiding-places in the crevices and cocks of the apparatus, and almost to restore the original deflection.

The quantity of those residues must be left to the imagination to conceive. If they were multiplied by billions they probably would not reach the density of ordinary air.

Fearing that the more active perfumes might possibly prejudice the deportment of the more feeble ones which succeeded them, I made a series of experiments with the following essences, and obtained these results:—

	Absorption per 100
Camomile Flowers	87
Spikenard	355
Aniseed	372

Immediately afterwards the experiment with bergamot was repeated, and its action found to be exactly the same as that recorded in the table.

In experiments on musk different results were obtained at different times. On the 16th of October some fresh musk from the perfumer's, placed in a small glass tube, had dry air carried over it into the experimental tube. The first experiment gave an absorption of 74, the air which carried the perfume being unity. A second experiment, in which the air was admitted more quickly, gave the absorption 72.

It would be idle to speculate upon the quantity of matter which produced this result. The stories regarding the unwasting character of this substance are well known; suffice it to say that a quantity of its odour carried into the tube by a current of air of a minute's duration, produced an effect seventy-two times that of the air which carried it. Long-continued pumping failed to cleanse the tube and passages of the musk. It cannot be volatile, for an amount of ether-vapour which produces a far greater action is speedily cleared away, while the cocks and connecting pieces of the air-pump had to be boiled in a solution of soda before they were fit for use after the experiments with this substance.

Two perfectly concurrent experiments with ordinary cinnamon, in which fragments of the substance were placed in a tube and had dry air passed over them, gave an absorption of 53.

Several kinds of tea, treated in the same manner, produced absorptions which varied between 20 and 28.

In the teas, cinnamon, musk, and the odorous plants already referred to, dry air had been passed over them for some time before they were examined. Still a small amount of aqueous vapour may have entered with the odours, and thus rendered the results to some extent of a mixed character.

§ 17.

Action of Ozone upon Radiant Heat.

In my last memoir the action of ozone was briefly alluded to. The experiments were executed with a brass tube polished within, and I was desirous of repeating them with a tube which could not be attacked by this extraordinary substance. Experiments with the glass tube, performed on the 16th, 17th, and 18th of last July, satisfied me that the power of ozone as an absorber of radiant heat had not been over-estimated.

For the purpose of lessening the resistance to the passage of the current through the decomposing liquid, large electrodes were used in the first experiments. The oxygen thus obtained differed but little from ordinary oxygen.

For my recent experiments I had three decomposing-vessels constructed: the first (No. 1) had platinum plates of about four square inches of surface, which were rolled up to economize space; the plates of the second (No. 2) had two square inches of surface; while those of the third (No. 3) had only a square inch of surface each. Numerous experiments with these cells gave the following constant results:—

Electrolytic Oxygen.

From plates	Absorption per 100
No. 1	20
No. 2	34
No. 3	47

The absorption by ordinary oxygen being unity.

A series of experiments executed on the following day gave these results:—

No. 1	21
No. 2	36
No. 3	47

Here the influence of the size of the electrodes is unmistakable.

A portion of the plates of No. 2 was then cut away so as to make them smaller than those of No. 3. The oxygen obtained with these plates gave an absorption of

65,

thus exceeding No. 3 considerably. The plates of No. 3 were now reduced so as to make them the smallest of all; the oxygen which they delivered gave an absorption of

85.

Fearing the development of heat with these smallest plates, and knowing heat to be very destructive of ozone; I surrounded the apparatus by a mixture of pounded ice and salt. The absorption rose immediately to

136.

Had we not been prepared, by the results already recorded, for the effect of minute quantities of matter on radiant heat, we could not fail to be struck with astonishment on finding a quantity of ozone, which would elude all attempts on the part of the chemist to determine its amount, producing an effect so stupendous in comparison with that of common oxygen. I have, moreover, strong reason to believe that the effect of the ozone is here understated.

§ 18.

Experiments of De la Rive and Meidinger.

All the results here recorded had been for some time obtained, when, turning to De la Rive's excellent treatise on electricity, I there found the experiments of M. Meidinger on ozone referred to. I had never previously heard any allusion made to this investigation, and was gratified to find it the record of a very interesting piece of work.

M. Meidinger commences by showing the absence of agreement between theory and experiment in the decomposition of water, the difference showing itself very decidedly in a deficiency of oxygen *when the current was strong.* On heating his electrolyte, he found that this difference disappeared, the proper quantity of oxygen being always liberated. He at once surmised that the defect of oxygen might be due to the formation of ozone; but in what way was still to be determined. If it were due to the greater density of ozone in the tube which received the oxygen, the destruction of this substance by heat would restore the true volume. Strong heating, however, which destroyed the ozone, produced no alteration of volume. Hence M. Meidinger concluded that the observed defect was not due to the ozone mixed with the oxygen itself. He finally concluded, and justified his conclusion by satisfactory experiments, that the loss of oxygen was caused by the formation of peroxide of hydrogen, which being dissolved in the liquid was withdrawn from the electrolytic gas. He was further led to experiment with electrodes of different sizes, and found the loss of oxygen to be more considerable with a small electrode than with a large one; whence he inferred that the formation of ozone was facilitated by *augmenting the density of the current at the place where the electrode and electrolyte meet.* Nothing could be more different than the methods independently pursued by M. Meidinger and me in arriving at the same conclusion; and though no doubt of the accuracy of my experiments existed in my mind, it was pleasant to find them supported in such a remarkable and unexpected way. Since the perusal of M. Meidinger's paper I have repeated his experiments with the decomposition-cells above described, and have found that those which yielded the greatest absorption also show the greatest deficiency in the amount of oxygen liberated.*

§ 19.

On the Constitution of Ozone.

The quantities of ozone brought to bear in the foregoing experiments must be perfectly unmeasurable by ordinary means. No elementary gas that I have examined behaves at all like ozone. Its action is like that of olefiant gas, or boracic-ether vapour; bulk for bulk it might indeed transcend either. If it be oxygen, *it must be oxygen packed into groups of atoms, which encounter vast resistance in moving through the æther.* Two views of its constitution are entertained; the one regarding it as a form of oxygen, the other as a compound of hydrogen. I sought to decide the question in the following way:—Heat destroys ozone. If it were oxygen only, heat would convert it into the common gas; if it were the hydrogen compound, heat would convert it into oxygen *plus* aqueous vapour. The gas alone admitted into the experimental tube would give the neutral action of oxygen, but the gas *plus* aqueous vapour would give a sensibly greater action. The dry electrolytic gas

* I have recently learned that M. de la Rive himself was the first to observe the influence of the size of the electrodes on the development of ozone.

was first caused to pass through a glass tube heated to redness, and thence directly into the experimental tube. The experiment was repeated with a drying apparatus introduced between the heated tube and the experimental tube. The result is, that hitherto I have not been able to establish with certainty a difference between the two cases. If, therefore, the act of heating developed aqueous vapour, I can only say that the most powerful experimental tests fail to prove its presence. For the present, therefore, I hold *that ozone is produced by the packing of the atoms of elementary oxygen into oscillating groups*—that heat dissolves the bond of union, and allows the atoms to swing singly, thus disqualifying them for either intercepting or generating the motion which in combination they are competent to intercept and generate.

§ 20.

Action of Aqueous Vapour upon Radiant Heat.—Experiments of Magnus.

Since these researches were commenced, an eminent experimenter has been led by his own inquiries in another field to enter upon the investigation of gaseous diathermancy. On the 7th of February of the present year (1861), Professor Magnus communicated to the Academy of Sciences in Berlin a memoir 'On the Transmission of Heat through Gases.'* The published notices of my experiments, commencing in May 1859, had escaped his attention, and his work is therefore to be regarded as independent of mine. Considering the very different methods which we have pursued, the general agreement between us must be regarded as remarkable.

The starting-point of Professor Magnus's investigation was the interesting experiment of Mr. Grove, in which a platinum wire heated to whiteness by an electric current is suddenly cooled when plunged into hydrogen. This action, which we have hitherto been disposed to attribute to the mobility of hydrogen, and its consequent high convective power, Professor Magnus holds to be an effect of conduction; and this belief induced him to examine the conductibility of gases generally.

* Poggendorff's *Annalen*, reprinted in *Philosophical Magazine*, S. 4. vol. xxii. p. 85.

The mode of experiment which he adopted led him, not, in my opinion, to the establishment of gaseous conductivity at all, but to results substantially the same as some of those that I had previously obtained. In fact the very experiments devised to show conductivity proved in a very striking manner the existence of athermancy, or opacity to radiant heat, in the case of a considerable number of gases.

The apparatus of Professor Magnus consisted of two glass vessels, one much larger than the other, with their bottoms fused together. The larger one being turned upside down, the smaller one stood upright on the top of it. The mouth of the larger vessel was ground down, so that it could be placed like an ordinary receiver on the plate of an air-pump and exhausted, while through proper cocks different gases could be afterwards admitted into it.

To the plate of the air-pump on which the vessel was placed was attached a thermo-electric pile with wires leading from it, through the plate, to a galvanometer; the axis of the pile was vertical, one face of it being turned downwards, and the opposite face turned upwards towards the common surface of the two vessels which had been fused together.

Water was placed in the uppermost vessel, and caused to boil by conducting hot steam through it. Its bottom, which formed the top of the lower vessel, was thus heated to a temperature of 100° C., and it formed the source of heat made use of in the experiments.

Here, therefore, Professor Magnus had a radiating surface of glass—a good radiator—kept at a constant temperature by the hot water above it. At a distance from this surface, and turned towards it, was the thermo-electric pile, defended from the radiation of the surface, or exposed to it, at pleasure, by the action of a moveable screen. The entire space between the pile and the radiating surface could either be rendered a vacuum, offering no resistance to the passage of the calorific rays, or it could be filled by a gas the diathermancy of which was to be examined.

The concurrence of the experiments made with this apparatus and some of those previously made with mine is, as I have stated, remarkable. Some differences, however, exist between my friend and myself, a few remarks on which will not be without their use to those who may afterwards enter upon this extensive field of inquiry.

Experimenting in the ordinary way with his thermo-electric pile—that is to say, using one of its faces only—Professor Magnus finds that air and oxygen respectively intercept more than 11 per cent. of the heat emanating from his source of heat, while hydrogen cuts off more than 14 per cent. I, on the contrary, with the most powerful and delicate means I could employ, failed to establish, by experiments made in the ordinary manner, any action whatever on the part of these gases.* In fact it was their neutrality that drove me to devise the principle of compensation, described in the last memoir and briefly referred to at the commencement of this one. I was so particular in the experiments which led to the above negative result, that if the absorption amounted to one-tenth of that found by Professor Magnus it could not have escaped me. Nor is it likely that, if such an action existed, Melloni could have concluded that the absorption of a column of air fifteen times the length of that employed by Professor Magnus was absolutely insensible.

In the account of experiments published in Memoir I., where the source of heat was also 100° C., and the powerful method of compensation was employed, the absorption of air, oxygen, and hydrogen is set down at about 0·33 per cent., which is for air and oxygen thirty times, and for hydrogen over forty times less than that found by Professor Magnus.

In fixing the above figure for the absorption of these gases, I protected myself by assigning the superior limit of the effect, but I was morally certain at the time that by the improvement of the apparatus in power and delicacy, the effect would be made less. In the present inquiry, accordingly, the absorption was found to be under 0·1 per cent., which in the case of oxygen is less than $\frac{1}{100}$th, and in the case of hydrogen less than $\frac{1}{140}$th of the effect obtained by Professor Magnus with a tube less than half the length of mine. Making every allowance for the difference between our two sources of heat, the discrepancy between us is still enormous. In fact my conclusion is that these gases are practical vacua to radiant heat, and that the mixture of oxygen and nitrogen which constitutes the body of our atmosphere is the same.

* Page 12.

While, however, in the case of the elementary gases the discrepancy between Professor Magnus and myself consists in a defect on my part, or in an excess on his, with the powerful compound gases I obtain a considerably stronger action than he does. Thus with olefiant gas his absorption amounts to less than 54 per cent., whereas in mine it amounts to more than 72. This last result, however, is only what might be expected, inasmuch as the length of gas traversed by the radiant heat was in the one case a little under 15 inches, and in the other 33.

Professor Magnus has further published an account of experiments in which his source of heat was a powerful gas-flame, surrounded by a glass cylinder, and provided with a polished parabolic mirror to reflect and concentrate the rays. In this case the gases were enclosed in a glass tube 1 metre long and 35 millimetres in diameter, the two ends of which were stopped with plates of glass 4 millimetres thick.

Two series of experiments were executed with this tube, in one of which the interior surface was covered with black paper, and in the other uncovered. The former method had been previously pursued by Dr. Franz; and the result obtained by Professor Magnus in the case of atmospheric air and oxygen closely agrees with that obtained by Dr. Franz for the same gases. Professor Magnus makes the absorption in the case of the blackened tube about 2½, and Dr. Franz about 3 per cent., for air and oxygen.

In the case of the unblackened tube, however, the absorption was found to be much more considerable. Here air and oxygen quenched each 14·75 per cent., while hydrogen intercepted 16·23 per cent. of the total radiation. This great difference between the unblackened and the blackened tube is ascribed by Professor Magnus to a change of quality on the part of the heat, produced by its reflexion at the interior glass surface.

One of my motives for introducing a glass tube into the present inquiry was to enable myself to investigate the question raised by this surmise of Professor Magnus. I have failed, however, to obtain his result. My naked glass tube, which is nearly of the same length as his, gives me an action which is more than 140 times less than his in the case of air and oxygen, and more than 160 times less than his in the case of hydrogen. Our sources of heat are, it is true, different, but the disadvantage is on my side; for assuredly the rays from a gas-jet are less affected by the transparent elementary gases than those from an obscure source. Were the time at my disposal, I would repeat the experiments with a flame; but this, I regret to say, is out of my power at present.

Another difference between Professor Magnus and myself has reference to the influence of aqueous vapour. With both the gas-flame and the boiling water as sources of heat, he finds the effect of dry air to be precisely the same as that of air which he has allowed to pass in minute bubbles through water, and thus saturated with aqueous vapour.

I was engaged in experiments on this substance when my other duties compelled me to close this inquiry for a time. It may, however, be safely affirmed that not only is the action of aqueous vapour on radiant heat measurable, but that *this action may be made use of as a measure of atmospheric moisture, the tube used in my experiments being thus converted into a hygrometer of surpassing delicacy.* Unhappily, as in other cases touched upon in this memoir, I have been unable to give this subject the desired development; but the results obtained are nevertheless interesting.

On a great number of occasions the air sent directly from the laboratory into the experimental tube was compared with the same air after it had been passed through a drying-apparatus. Calling the action of the dry air unity, or supposing it rather to oscillate about unity (for the temperature of the source of heat varied a little from day to day), on the following days the annexed absorptions were observed with the undried air of the laboratory:—

Absorptions by undried air.

October 23rd	63	November 1st	50
October 24th	62	November 4th	58
October 29th	65	November 8th	49
October 31st	56	November 12th	62

Nearly $\frac{9}{10}$ths of the above effects are due to aqueous vapour; which, therefore, in some instances *exerted nearly sixty times the action of the air in which it was diffused.*

The experiments executed on aqueous vapour have been very

numerous and varied. Differing, as I did, from so cautious and able an experimenter as Professor Magnus, I spared no pains to secure myself against error. Air moistened in various ways, sometimes by allowing small bubbles of it to ascend through water, sometimes dividing it by sending it through the pores of common cane immersed in water, has been experimented with. Between the drying apparatus and the experimental tube tubes have been introduced containing fragments of glass moistened with water, and the air allowed to pass over them; in all such cases large effects were obtained, the absorption being usually *more than eighty times that of dried air*. Fragments of unwetted glass, which had been merely exposed to the air of the laboratory, had dry air led over them into the experimental tube; the absorption was fifteen times that of dried air.* A roll of bibulous paper, taken from one of the drawers of the laboratory, and to all appearance perfectly dry, was enclosed in a glass tube, and dry air carried between its leaves. The experiment was made five times in succession with the same paper, and the following absorptions were observed:—

	Absorption per 100
No. 1	72
No. 2	62
No. 3	62
No. 4	47
No. 5	47

In fact, the action of aqueous vapour is exactly such as might be expected from the vapour of a liquid which Melloni found to be the most powerful absorber of radiant heat of all that he had examined.

§ 21.

Night-Moisture on the Interior Surface of Experimental Tube. —Abandonment of Rock-salt Plates.

Every morning, on commencing my experiments, I had an interesting example of the power of glass to gather a film of aqueous moisture on its surface. The air of the laboratory being removed from the experimental tube, on allowing dry air to enter for the first time, the needle would move from 0° to 50°. On

* These experiments have a direct bearing on the subsequent ones of Professor Magnus. [1872.]

pumping out it would return to 0°, and on letting in dry air a second time it would swing almost to 40°. Repeated exhaustions caused this action to sink almost to nothing. These results were entirely due to the vapour collected during the night in an invisible film on the inner surface of the tube, which was removed by the air on entering, and diffused through the tube. When the dry air entered at the end of the tube nearest the source of heat, on the first and second admissions, and sometimes even on a third, the vapour carried from the warm end to the cold end was precipitated as a mist upon the surface of the glass for a distance sometimes of nearly a foot. The mistiness always disappeared on pumping out. It is needless to remark that facts of this character, of which many could be cited, were not calculated to promote incautiousness on my part. I saw very clearly how easy it was to fall into the gravest errors, and took due precautions to prevent myself from doing so.

Knowing that a solution of salt was almost as opaque to radiant heat as water itself, I was careful to examine whether the effects observed with aqueous vapour might not be due to the precipitation of the vapour on the rock-salt surfaces. The substance is well known to be very hygroscopic; and during the last three years the knowledge of this fact has rendered me careful to remove the polished plates every evening from the apparatus, and to keep them in perfectly dry air. Still, when it is remembered that the air on entering the tube is raised in temperature and thus enabled to maintain a greater amount of vapour, and that the tube and plates of rock-salt form the channel for a flux of heat from the radiating source, the likelihood of precipitation occurring will seem but small. On examining the plates, moreover, after the undried air of the laboratory had been experimented with, no trace of precipitated moisture was observed upon their surfaces.

But, to place the matter beyond all doubt, I abolished the plates of rock-salt altogether, and operated thus:—An india-rubber bag B (Fig. 9) was filled with air, and to its nozzle a T-piece, with the cocks Q Q′, was attached. The cock Q′ was connected with two tubes, U′ U′, each of which was filled with fragments of glass moistened with distilled water. The cock Q was connected with the tubes U U, each of which was filled with fragments of glass moistened by sulphuric acid. The other ends of these two series of tubes were connected with the cocks O O′; and from the T-piece between these cocks a tube led to the end E′ of the open experimental tube T. The cock A at the other end

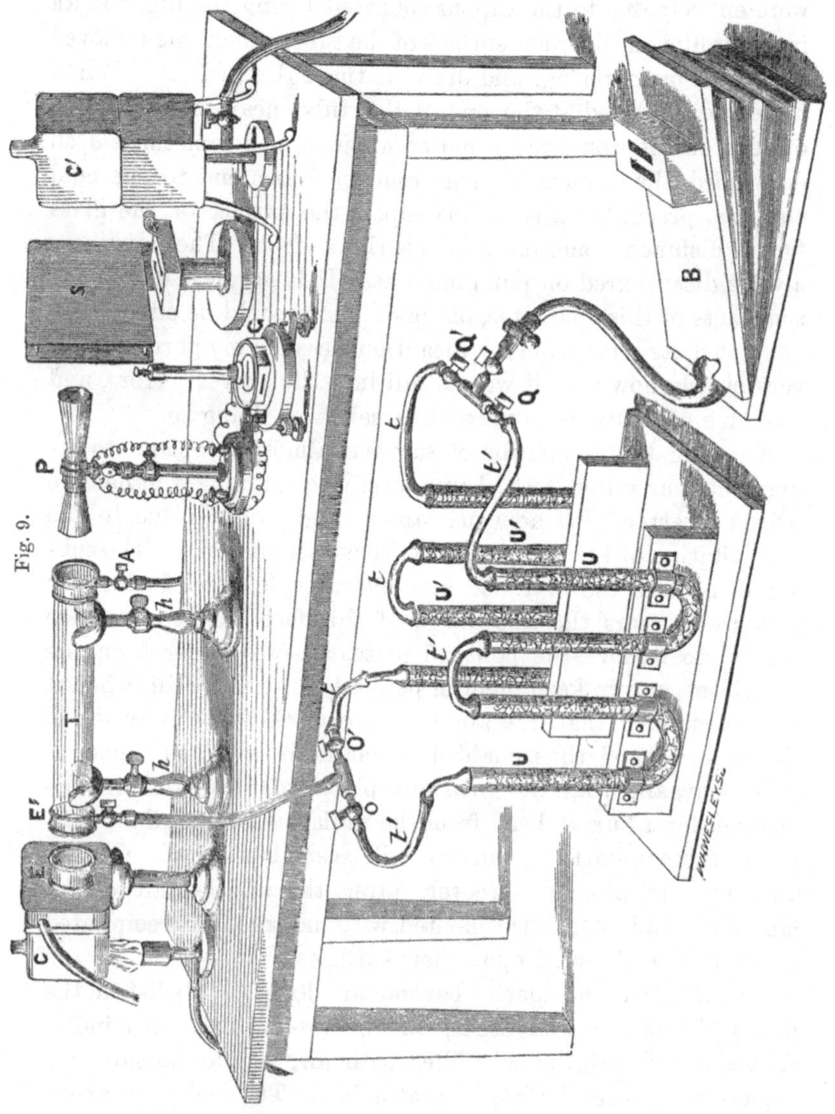

Fig. 9.

of the experimental tube was placed in connexion with an air-pump. The pile P, the screen S, and the compensating cube C′ were used as in the other experiments. E is the end of the front chamber, and C the source of heat. In some experiments I had the end E closed by a plate of rock-salt, in others it was allowed to remain open, a distance of about 12 inches intervening between the radiating surface and the open end E′ of the experimental tube.

Closing the cocks Q and O, and opening Q′ and O′, gentle pressure being applied to the bag B, a current of moist air was slowly discharged at the end E′ of the experimental tube. The pump in connexion with A was then worked, and thus by degrees the air was sucked into the tube T. The deflection of the galvanometer was 30°, when the moist air filled the tube as completely as the arrangement permitted,*—this deflection being due to the predominance of the compensating cube over the radiating source C.

The cocks Q′ and O′ were now closed, and Q and O opened; proceeding as before, a current of *dry* air was discharged at E′, and this air was drawn into the tube T in the manner just described. The moist air was thus displaced by dry; and, while the displacement was going on, the galvanometer was observed through the distant telescope. The needle soon began to sink, and slowly went down to zero, proving that a greater quantity of heat passed through the dry than through the moist air. The wet air was substituted for the dry, and the dry for the wet twenty times in succession, with the same constant result: the entrance of the humid air caused the needle to move from 0° to 30°, while the entrance of dry air caused it to fall from 30° to 0°. The air-pump was resorted to, because I found in attempting to displace the air by the direct force of the current from B, the temperature of the pile, or of the source of heat, was so affected by the fresh air as to confuse the result. I may remark that not only have I operated thus for days with aqueous vapour, *but every result obtained with vapours generally has been thus confirmed,* so that all doubt as to the applicability of the rock-salt plates to researches of this nature may, I think, be abandoned.†

* Still, of course, only partially.

† This proved to be source of error in Professor Magnus's subsequent experiments. See Memoir IV. It is sheer want of time that prevents me from describing more particularly the numerous experiments executed with open tubes.

§ 22.

Proposed Solution of Discrepancies.

Whence, then, arise those differences between Professor Magnus and myself? I am quite convinced that his experiments have been made with the utmost care which it is possible to bestow upon scientific work, and the differences between us are, in my opinion, to be referred to a radical defect in his apparatus. His desire to do away with plates of all kinds between his source of heat and his pile, caused him to bring his gas *into direct contact with his source of heat.* I was on the point of falling into the same error; but a series of experiments executed with reference to this point, so early as July 26, 1859, proved the accuracy of the results to be entirely compromised by bringing the gas to be examined into contact with the source of heat. In one experiment where this occurred I obtained an action forty times what I knew it ought to be, being thereby confirmed in my opinion as to the necessity of interposing a vacuous chamber in front of the experimental tube. Let me here record a few experiments made on the 4th of last November in connexion with this subject.

Having first made sure that the drying apparatus was in perfect condition—the air of the laboratory producing, when sent through it, an absorption of 1—this same dry air was sent into the front chamber, that is, into direct contact with the source. The galvanometer needle moved as it does in the case of absorbent gases, and at the end of two minutes it declared a loss of heat equivalent to an absorption of 50. The front chamber is 8 inches in length; the experimental tube 33 inches; hence a column of 8 inches, in contact with the radiating surface, produced at least fifty times the effect of a column more than four times as long when the air was separated from the radiating surface.

The foregoing experiment was made three times in succession, and after two minutes * the needle was found pointing to precisely the same degree; the lowering of the source of heat was perfectly constant and regular, and in all cases showed a loss equivalent to an absorption of 50.

It will be remembered that Professor Magnus obtained a

* The time of exposure of Professor Magnus's pile.

greater absorption with hydrogen than with either oxygen or air. This result is perfectly explained by reference to the quicker convection of this gas. I operated with hydrogen as with air, first satisfying myself that a column of the gas 33 inches long exercised an absorption less than unity: in fact it could not be measured. The same hydrogen introduced into the front chamber, and allowed to remain there for two minutes, caused a withdrawal of heat equivalent to an absorption of 65. Now the action of air in Professor Magnus's experiments is to that of hydrogen as

11·12 : 14·21,

or as

50 : 64,

while my results of convection are as

50 : 65.

The coincidence is so perfect that one is disposed to regard it as in part accidental.*

Substantially the same remarks apply to the experiments with the glass tube stopped with plates of glass 4 millimetres thick. According to Melloni, 61 per cent. of the rays of a Locatelli lamp are absorbed by a plate of glass only 2·6 millimetres thick. True, Professor Magnus surrounded his flame by a glass cylinder; and this, it may be urged, partially sifted the heat of the lamp before it reached the end of the tube. But in so doing the glass cylinder itself must become intensely heated; and to the heat of the cylinder the glass ends of the tube would be *opaque*; they would absorb it all. Cold air admitted into such a tube is exactly similar to cold air let into my front chamber; it chills the secondary sources of heat, and maintains that chill by convection. The heat applied may, in fact, be thus analysed:—1. We have a portion of the heat from the lamp passing without losing the radiant form through the tube direct to the pile; 2, a portion of that flux *arrested* by the first glass plate; 3, a smaller portion *arrested* by the second glass plate; 4, the heat *radiated* by the first glass plate towards the second, and wholly absorbed by the latter; 5, the heat radiated by this latter against the pile. This analysis enables us to clearly understand how Professor Magnus obtained an

* I do not think it accidental. This result is entirely in accordance with those of Count Rumford. [1872].

absorption of only 2½ per cent. with the tube blackened within, and as much as 14·75 per cent. with the unblackened one. With the unblackened tube, both the source of heat and the plate of glass nearest to it sent a copious flux down the tube to the plate at the opposite end; for here the oblique rays are in great part reflected by the interior surface. With the blackened tube this oblique radiation is cut off, the rays incident on the interior surface being absorbed. Hence the plate of glass adjacent to the pile must be much more intensely heated with the unblackened tube than with the blackened one. The difference in the amount of heat impinging on the pile-end plate in the respective cases is rendered very manifest by the experiments of Professor Magnus himself; who finds the heat transmitted by the uncoated tube to be twenty-six times that transmitted by the coated one. What, therefore, Professor Magnus ascribes to a change of quality by reflexion, is perfectly explained by reference to the greater heating, *and consequent greater chilling by the cold air*, of the plate of glass close to the pile.

The difference between Professor Magnus and myself as regards the action of aqueous vapour admits also of easy explanation. His effect being one of convection, and not of absorption, the quantity of vapour present in his experiments—probably not more than 1 per cent. of the volume of the gas, certainly not 2 per cent.—vanished as a convecting agent, in comparison with the air.

It is hardly necessary to repeat these reflexions with reference to the experiments of Dr. Franz. The mistaking of the chilling of his plates for absorption caused him to find no difference of effect when he doubled the length of his tube. With a tube 450 millimetres long, he found precisely the same absorption as with one of 900. He also found the action of carbonic acid to be the same as that of air, although at atmospheric pressures the action of the former is 90 times that of the latter.* He found the vapour of bromine more destructive to radiant heat

* The sensible equality of all the transparent gases and air was regarded as self-evident by Dr. Franz. 'It might be seen,' he writes, 'from the outset that no decided difference would be observed between them' (p. 342). Similarly, Professor Magnus, speaking of aqueous vapour, writes, 'Although it might be foreseen with certainty that the small amount of aqueous vapour in the air could have no influence on the radiation,' &c. (p. 43).

than nitrous acid gas, whereas the latter is beyond comparison the most destructive. The heat rendered latent by the evaporation of the bromine of course augmented the chill of his plates, and thus magnified the effect which in reality he was measuring.

§ 23.

Action of Atmospheric Envelope.—Possible Experimental Determination of the Temperature of Space.

As a dam built across a river causes a local deepening of the stream, so our atmosphere, thrown as a barrier across the terrestrial rays, produces a local heightening of the temperature at the earth's surface. This, of course, does not imply indefinite accumulation, any more than the river dam does, the quantity lost by terrestrial radiation being, finally, equal to the quantity received from the sun. The chief intercepting substance is the aqueous vapour of the atmosphere,* the oxygen and nitrogen of which the great mass of the atmosphere is composed being sensibly transparent to the calorific rays. Were the atmosphere cleansed of its vapour, the temperature of space would be directly open to us; and could we under present circumstances reach an elevation where the amount of that vapour is insensible, we might determine the temperature of space by direct experiment. Colonel, now General, Strachey has written an admirable paper on the aqueous vapour of the atmosphere,† in which he shows that the amount of vapour diminishes much more rapidly with the elevation than might be inferred from Dalton's law.

It might therefore be possible to reach a height where, by preserving one face of a thermo-electric pile at the temperature of the locality, the other, protected from all terrestrial radiation, and turned to the zenith, would assume the temperature of space,‡ while the consequent galvanometric deflection would

* The mildness of an island climate must be in part due to this cause. The direct tendency of the vapour is to check sudden fluctuations of temperature. Where it is absent, as at the surface of the moon, such fluctuations must be enormous. The face turned towards the sun drinks in the solar rays without let or hindrance, while the radiation of the face turned from the sun pours unchecked into space.

† *Phil. Mag.* S. 4. vol. xxiii. p. 152.

‡ A *well* of cold air would be formed within the conical reflector, the lowest stratum of the well sharing the temperature of the face of the pile.

give us the means of determining the difference in temperature between the two faces of the pile. Knowing, therefore, the temperature of the locality, we could infer from it the temperature of stellar space. Many eminent writers, it is true, have supposed the upper atmospheric regions to be colder than space, the temperature being lowered by the radiation of the aërial particles, just as the temperature of a grass-blade is lowered by radiation on a clear night. This notion must, I think, be abandoned; for experiment leads us to conclude that air, and particularly air in the higher atmospheric regions, behaves as a vacuum both as regards radiation and absorption.

§ 24.

Remarks on the Experimental Evidence of Gaseous Conduction.—Influence of Density on Convection.—Internal Friction of Air.

In his paper on the conduction of heat by gases, Professor Magnus has adduced some striking experiments to show that the cooling of an incandescent wire in hydrogen is not due to the convection of the gas. He finds that when the wire is enclosed in a narrow tube, with only a thin film of the gas surrounding it, and where therefore currents, in the ordinary sense, can hardly exist, the gas still exercises its cooling power. It had often occurred to me to make this experiment; and when intelligence of its successful performance by Professor Magnus first reached me I adopted his conclusion, that the cooling is due to conduction.

Reflexion, however, caused me to change this opinion. Suppose the wire to be stretched along the axis of a wide cylinder containing hydrogen, we should have convection, in the ordinary sense, on heating the wire. Where does the heat thus dispersed ultimately go? It is given up to the sides of the cylinder, and *if we narrow the cylinder we simply hasten the transfer.* The process of narrowing may continue till a tube like that used by Professor Magnus is the result; the convection between centre and sides will still continue, and produce the same cooling effect as before. Whether we assume conduction or convection on the part of the gas, the tube surrounding the wire must possess sufficient conducting power to carry the heat off, otherwise it would become incandescent itself by the accumulation of the heat.

The further reasoning of Professor Magnus in connexion with this subject is of extreme ingenuity. He contends that there is no reason why stronger currents should establish themselves in hydrogen than in other gases. Currents, he urges, are due to differences of density produced by the expansion of a portion of the gas by heat. But hydrogen actually expands *less* than other gases, and hence the differential action on which the currents depend is less in this gas than in the others. Professor Magnus alludes to the friction of the particles against each other, but considers this ineffective.

This reasoning leads us to the threshold of a question which might form the subject of a long and profitable investigation. The question is:—For a given difference of density, is not the mobility of hydrogen greater than that of the other gases? The experiments recorded in § 22, where different gases were brought into direct contact with the source of heat, seem to answer this question in the affirmative. I have had no time to pursue the question regarding hydrogen; but a few experiments have been made which show in a very striking manner the influence of density on the mobility of a gas.

Having first so purified atmospheric air as to render it sensibly neutral to radiant heat, I allowed 15 inches of it to enter the front chamber F (see *Frontispiece*), and there to come into contact with the source of heat. Convection, of course, immediately set in, and its amount was accurately measured by the quantity of heat withdrawn from the radiating surface; this, expressed in the units adopted throughout this memoir, was 62.

The quantity of gas in the front chamber was then doubled—in other words, increased to a whole atmosphere; the withdrawal of heat was expressed by the number 68.

In the last experiment we had double the number of atoms loading themselves with heat and carrying it away; if their motion had been as quick as that of the atoms when half an atmosphere was used, they would have withdrawn sensibly double the amount of heat; but the fact is that half an atmosphere carried off 62, while a whole atmosphere carried off 68; hence the absolute swiftness of the atoms in the case of the denser air must be very much less than in the case of the rarer. In fact, the amount of heat withdrawn will be proportional on the one hand to the number of carrying particles, and on the other to the velocity with which they move; hence if v and v' be these velocities, we have

$$\frac{62}{68}=\frac{v}{2v'}, \text{ or } \frac{v}{v'}=\frac{62}{34}.$$

Thus, while the atoms of the rarer gas travel 62 units in a second, those of the denser gas travel only 34.

This retardation can, I think, arise from nothing else than the resistance offered by the particles of the air to the motion of their fellows. It must be borne in mind that the smallness of the increment observed on doubling the amount of gas was not due to the partial exhaustion of the source of heat by the first half atmosphere of gas. The heat of the source was such that the withdrawal of 64 of our units could not sensibly affect the subsequent convection.

Here, then, we see what a powerful effect density, or the internal resistance which accompanies density, has on the mobility of a gas; and there is every reason to suppose that the mobility of hydrogen is due to the comparative absence, in its case, of internal friction. However this may be, the foregoing experiment enables us to draw some important inferences.

Local storms at great heights must be greatly facilitated by the mobility of the particles of the air. Storms are cases of convection on a large scale, and in our front chamber we had one in miniature.

In the summer of 1859 I was fortunate enough to induce Professor Frankland to accompany me to the summit of Mont Blanc, and to determine the comparative rates of combustion there and in the valley of Chamouni. Six composite candles were burnt for an hour at Chamouni, and the loss of weight determined. The same candles were lighted for the same time on the summit of the mountain, and the consumption again determined. Within the limits of error, the consumption above was equal to that below. The *light* below was immensely greater than that above, still the amount of stearine consumed in the two cases was sensibly the same. Professor Frankland surmised that this might be due to the greater mobility of the rarefied air, which allowed a freer interpenetration of the flame by the oxygen;* and the foregoing experiments show that the augmentation of mobility is just such as would account for the observed effect.

* The influence of interpenetration is well seen in the exposed gas-jets of London, particularly in the butchers' shops on a Saturday night. A gust of wind, which carries oxygen to the centre of a flame, suddenly deprives it of light. A simple and beautiful experiment consists of passing a lighted candle swiftly to and fro through the air; the white light reduces itself to a pale-blue band. Bunsen's burner is an illustration in the same line.

III.

ON THE RELATION OF RADIANT HEAT TO AQUEOUS VAPOUR.

ANALYSIS OF MEMOIR III.

In the analysis of Memoir II. the differences which had arisen between Professor Magnus and myself regarding the action of dry air on the one hand, and of aqueous vapour on the other, are briefly adverted to, and in the concluding sections of this memoir the subject is discussed and a solution of the discrepancies is offered. Professor Magnus had previously shown the danger arising from the hygroscopic character of rock-salt; and I, in the memoir referred to, replied by definite experiments to this objection.

In the summer of 1862 he came over to the International Exhibition and I had the great pleasure of spending a good deal of time in his genial company. This was a favourable opportunity for settling our differences, which were the subject of frequent conversation between us. He did me the pleasure to come to the Royal Institution to witness my experiments, and it was also his wish to show me his arrangements and to test them in my presence. This wish, however, his incessant occupations prevented him from carrying out.

It was first proved to his satisfaction that the method of compensation, regarding which I once observed him shake his head in doubt, was capable of the last degree of precision. In an experimental tube closed with rock-salt, I showed him the neutrality of dry air, and the activity of humid air. While the tube remained filled with the latter I removed the rock-salt plates and placed them in his hands for inspection. He looked at them, passed his finger and his dry handkerchief over them, and in the frankest manner exclaimed 'there is no moisture there.' I then repeated the experiments with an open tube, and over and over again, displacing moist air by dry, and dry air by moist, showed him by precise and concurrent measurements the constant difference subsisting between them.

I thought it due to him to pay strict attention to every objection he raised, whether in his published papers, his letters, or his conversation. He once mentioned to me his having found that a layer of air 12 inches deep sufficed to absorb all rays that air was capable of absorbing;* and he contended that the distance between the end of my experimental tube and my pile, owing to the length of its conical reflector, was sufficient to remove most of the heat taken up by air. This being the case, the neutrality of dry air followed in my experiments as a natural consequence. For, he rightly and ingeniously contended, assuming air to possess the alleged power, the fact that my pile stood *beyond* the experimental tube made no difference; because the introduction into the experimental tube of its charge of dry air merely transferred the absorption to a different part of the path traversed by the heat-rays, but did not alter the

* I understood him to say that he had prepared a paper for Poggendorff, in which this and other remarkable results were established. But I have never been able to find the paper.

amount of the absorption. He also one day drew my attention to the sunbeams slanting through the dusty air of London, and remarked good-humouredly, 'There is the source of your absorption.' My first care in 1862 was to meet these objections.

In the first experiment of the following memoir the reflector of the pile is placed *within* the experimental tube, the face of the pile being only $\frac{1}{20}$th of an inch distant from the plate of rock-salt. The distance, it was conceded, could produce no sensible absorption. The arrangement is also to be recommended because of the security it ensures against moisture; the heat being concentrated upon a small portion of the central area of the plate of salt. The results obtained with this arrangement were precisely the same as the former ones.

To meet the objection regarding London air, I sent special messengers to Hyde Park, Primrose Hill, Hampstead Heath, and Epsom Downs, and had air from these places. I made an expedition myself to the Isle of Wight, and carried home specimens of air from various parts of the island. The experiments made with all these samples of air entirely corroborated the previous ones.

London air, moreover, was purified and dried, until its action on a powerful beam of heat was insensible. It was then sent over fragments of glass moistened with distilled water. No smoke or dust could here mix with it; still its deportment was in perfect accord with the other experiments.

Dry smoke, thicker than it is ever seen in London streets, was then purposely sent into the experimental tube. Its action was found to be only a fraction of that of aqueous vapour.

I then sought to do away with the experimental tube itself, and to discharge dry air and moist alternately between the source of heat and the pile. The observed effect was small, but distinct. This experiment, however, which was of the most extreme delicacy, I should like to confirm by repetition.

The question whether condensation occurs on the interior of the experimental tube so as to diminish its reflective power is considered. Humid air is admitted in varying quantities, and it is found that the absorption of radiant heat is accurately proportional to the quantity of vapour present. Such proportionality, it is urged, could hardly arise from the supposed condensation.

The bearing of this property of aqueous vapour upon various problems and phenomena of meteorology is then pointed out. The great daily range of the thermometer in dry climates; the production of frost at night even in Sahara; the cold of the table-land of Asia; the contrast between day and night on mountains; the artificial production of ice in India; Leslie's significant remarks on his *æthrioscope*, where he shows that days of equal atmospheric clearness differ widely from each other as regards the power of the air to stop terrestrial radiation. All these observations are in harmony with, and are indeed explained, by this newly discovered property of aqueous vapour.

III.

ON THE RELATION OF RADIANT HEAT TO AQUEOUS VAPOUR.*

§ 1.

Objections to Rock-salt Plates considered.—New Experimental Arrangement.

I HAVE already given an account of experiments which brought to light the remarkable fact that the body of our atmosphere—that is to say, the mixture of oxygen and nitrogen of which it is composed—is a comparative vacuum to the calorific rays, its main absorbent constituent being the aqueous vapour which it contains. It is very important that the minds of meteorologists should be set at rest on this subject—that they should be able to apply, without misgiving, this newly revealed physical property of aqueous vapour; for it is certain to have numerous and important applications. I therefore thought it right to commence my investigations this year with a fresh and special series of experiments upon atmospheric vapour, which I have now the honour to submit to my readers.

Rock-salt is a hygroscopic substance. If we breathe on a polished surface of rock-salt, the affinity of the substance for the moisture of the breath causes the latter to spread over it in a film which exhibits brilliantly the colours of thin plates. The zones of colour shrink and finally disappear as the moisture evaporates. Visitors to the International Exhibition of this year may have witnessed how moist were the pieces of rock-salt exhibited in the Austrian and Hungarian Courts. This property of the substance has been referred to by Professor Magnus as a

* Received by the Royal Society November 20, and read before the Society December 18, 1862. *Philosophical Transactions,* part i. for 1863, and *Philosophical Magazine* for July 1863.

possible cause of error in my researches on aqueous vapour; a film of brine deposited on the surface of the salt would, he urges, produce the effect ascribed to the aqueous vapour. I will, in the first place, describe a method of experiment by which even an inexperienced operator may avoid all inconvenience of this kind.

In the plate which accompanies my former paper, the thermo-electric pile is figured with two conical reflectors, both outside the experimental tube; in my present experiments the reflector which faced the source of heat is placed *within* the experimental tube, its narrow aperture, which usually embraces the pile, abutting against the plate of rock-salt which stops the tube. Fig. 10 is a sketch of this end of the experimental tube.

Fig. 10.

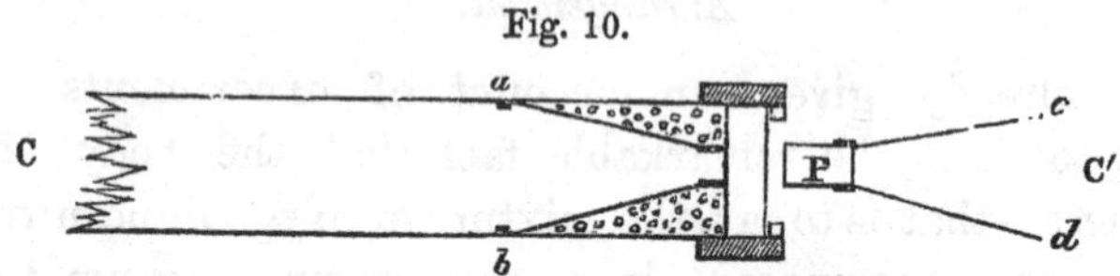

The edge of the inner reflector fits tightly against the interior surface of the tube at *a b*; *c d* is the diameter of the wide end of the outer reflector supposed to be turned towards the 'compensating cube' situated towards C′.* The naked face of the pile P is turned towards the plate of salt, being separated from the latter by an interval of about $\frac{1}{20}$th of an inch. The space between the outer surface of the interior reflector and the inner surface of the experimental tube is filled with fragments of freshly-fused chloride of calcium, intended to keep the circumferential portions of the plate of salt perfectly dry. The flux of heat coming from the source C, being converged upon the central portion of the salt, completely chases every trace of humidity from the surface on which it falls.

§ 2.

Objection to Employment of London Air considered.—Radiation through Air from Various Localities.

With this arrangement I repeated all my former experiments on humid and dry air. The result was the same as before. *On*

* I here assume an acquaintance with my last two memoirs, in which the method of compensation is described.

a day of average humidity the quantity of vapour diffused in London air produced upwards of 60 *times the absorption of the air itself.*

It had been suggested to me that the air of our laboratory might be impure; the suspended carbon particles in a London atmosphere had also been mentioned to me as a possible cause of the absorption ascribed to aqueous vapour. With regard to the first objection, I may say that the same results were obtained when the apparatus was removed to a large room at a distance from the laboratory; and with regard to the second cause of doubt, I met it by procuring air from the following places:—

1. Hyde Park.
2. Primrose Hill.
3. Hampstead Heath.
4. Epsom Race-Course.
5. A field near Newport, Isle of Wight.
6. St. Catharine's Down, Isle of Wight.
7. The sea-beach near Black Gang Chine.

The aqueous vapour of the air from these localities exerted absorptions from 60 *to* 70 *times that of the air in which the vapour was diffused.*

I then purposely experimented with smoke, by carrying air through a receiver in which ignited brown paper had been permitted to smoulder for a time, and drying it afterwards. It was easy, of course, in this way to intercept the calorific rays; but, adhering to the lengths of air actually experimented on, it was proved that, *even when the east wind blows, and pours the carbon of the city upon the West End of London, the heat intercepted by the suspended carbon particles is but a minute fraction of that absorbed by the aqueous vapour.*

Further, the air of the laboratory was so well purified that its absorption was less than unity; the purified air was then conducted through two U-tubes filled with fragments of clean glass moistened with distilled water. Its neutrality when dry proved that all prejudicial substances had been removed from the air; and in passing through the U-tubes it could have contracted nothing save the pure vapour of water. *The vapour thus carried into the experimental tube exerted an absorption* 90 *times as great as that of the air which carried it.*

I have had the pleasure of showing the experiments on atmospheric aqueous vapour to several distinguished men, and among others to Professor Magnus. After operating with common undried air, which showed its usual absorption, and while the undried air remained in the experimental tube, I removed the plates of rock-salt from the tube and submitted them to his inspection. They were as dry as polished rock-crystal or polished glass; their polish was undimmed by humidity; and a dry handkerchief placed over the finger and drawn across the plates left no trace behind it.*

Remark.—I would make one additional remark on the above experiments. A reference to the plate which accompanies the last two memoirs will show the thermo-electric pile standing, with its two conical reflectors, at some little distance from the end of the experimental tube. Hence, to reach the pile after it had quitted the tube, the heat had to pass through a length of air somewhat greater than the depth of the reflector. It has been suggested to me that the calorific rays may be entirely sifted in this interval—that all rays capable of being absorbed by air may be absorbed in the space intervening between the experimental tube and the adjacent face of the pile. If this were the case, then the filling of the experimental tube itself with dry air would produce no sensible absorption. Thus, it was imagined, the neutrality of dry air which my experiments revealed might be accounted for, and the difference between myself and Professor Magnus, who obtained an absorption of 12 per cent. for dry air, explained. But I think the hypothesis is disposed of by the foregoing experiments; for here the reflector which separated the pile from the tube no longer intervenes, and it cannot be supposed that in an interval of $\frac{1}{20}$th of an inch of air an absorption of 12 per cent. has taken

* The present number of the *Monatsbericht* of the Academy of Berlin contains an account of some experiments executed with plates of rock-salt by Professor Magnus. The plates which stopped the ends of a tube were so far wetted by humid air tha the moisture trickled from them in drops. As might be expected, the plates thus wetted cut off a large amount of heat. The experiments are quite correct, but they have no bearing on my results. In the earlier portions of my journal many similar cases are described. In fact, it is by making myself, in the first place, acquainted with the anomalies adduced that my results have been rendered secure. I may add that the communication just referred to was made to the Academy of Berlin before Professor Magnus had an opportunity of examining my rock-salt plates. I do not think he would now urge this objection against my mode of experiment.

place. If, however, a doubt on this point should exist, I can state that I have purposely sent radiant heat through an interval of 24 inches of dry air previous to permitting it to enter the experimental tube, and found all effects to be the same as when the beam had traversed 24 inches of a vacuum.

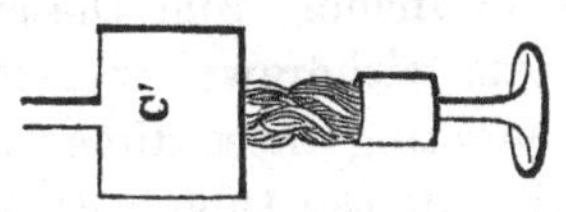

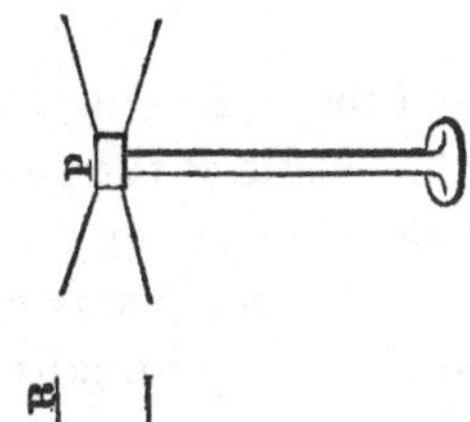

§ 3.

Radiation through Open Tubes.

In confirmation of the results obtained when the experimental tube was stopped by plates of rock-salt, the following experiments have been recently made with a tube in which no plates were used. S (fig. 11) is the source of heat, and S T the front chamber which in ordinary experiments is kept exhausted. This chamber is now left open. A B is the experimental tube, with both its ends also open. P is the thermo-electric pile, the anterior face of which receives rays from the source S, while its posterior surface is warmed by the rays from the compensating cube C′. At *c* and *d* are two stopcocks—that at *c* being connected with an india-rubber bag containing air, while that at *d* is connected with an air-pump.

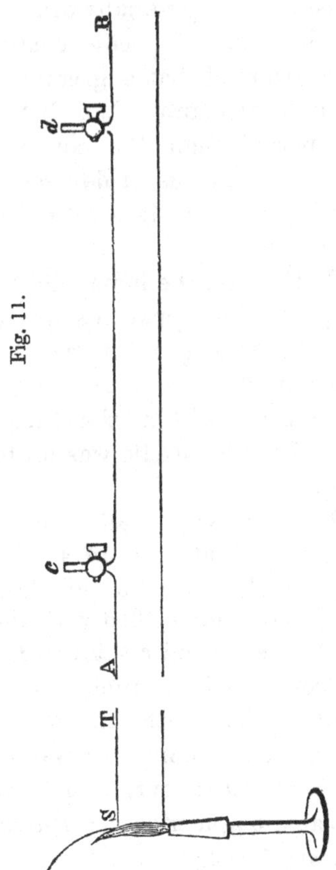

Fig. 11.

My aim in this arrangement was to introduce at pleasure, into the portion of the tube between *c* and *d*, dry air, the common laboratory air, or air artificially moistened. The point *c*, at which the air entered, was 18 inches from the source S; the point *d*, at which the air was withdrawn, was 12 inches from the face of the pile. By adopting these dimensions, and thus isolating the central portion of the tube, one kind of air may with ease and certainty be displaced by another without producing any agitation either at the source on the one hand, or at the pile on the other.

The tube A B being filled by the common air of the laboratory, and the needle of the galvanometer pointing steadily to zero, dry air was forced gently from the india-rubber bag through the cock *c*; the pump was gently worked at the same time, the dry air being thus gradually drawn towards *d*. On the entrance of the dry air, the needle commenced to move in a direction which showed that a greater quantity of heat was now passing through the tube than before. The dry air proved more transparent than the common air, and the final deflection thus obtained was 41 degrees. Here the needle stopped, and beyond this point it could not be moved by the further entrance of dry air.

Shutting off the india-rubber bag and stopping the action of the pump, the apparatus was abandoned to itself; the needle returned with great slowness to zero, thus indicating a correspondingly slow diffusion of the aqueous moisture through the dry air within the tube. By working the pump the descent of the needle was hastened, and it finally came to rest at zero.

Dry air was again admitted; the needle moved as before, and reached a final limit of 41 degrees; common air was again substituted, and the needle descended to zero.

The tube being filled with the common air of the laboratory, which was not quite saturated, and the needle pointing to zero, air from the india-rubber bag was now forced through two U-tubes filled with fragments of glass wetted with distilled water. The common air was thus displaced by air more fully charged with vapour. The needle moved in a direction which indicated augmented absorption; the deflection obtained in this way was 15 degrees.

These experiments have been repeated hundreds of times, and

on days widely distant from each other. I have also subjected them to the criticism of various eminent men, and altered the conditions in accordance with their suggestions. The result has been invariable. The entrance of each kind of air is always accompanied by its characteristic action. The needle is under the most complete control, its motions are steady and uniform. In short, *no experiments hitherto made with solids and liquids are more free from caprice, or more certain in their execution, than are the foregoing experiments with dry and humid air.*

The quantity of heat absorbed in the above experiments, expressed in hundredths of the total radiation, was found by screening off one of the sources of heat, and determining the full deflection produced by the other and equal source of heat.

By a careful calibration, repeatedly verified, this deflection was proved to correspond to 1,200 units of heat—the unit being, as before, the quantity of heat necessary to move the needle of the galvanometer from 0° to 1°. According to the same standard, a deflection of 41° corresponds to an absorption of 50 units. From these data we immediately calculate the number of rays per hundred absorbed by the aqueous vapour.

$$1200 : 100 = 50 : 4{\cdot}2.$$

An absorption of 4·2 per cent. was therefore effected by the atmospheric vapour which occupied the tube between the points *c* and *d*. Air *perfectly saturated* on the day in question gave an absorption of 5½ per cent.

§ 4.

Radiation through Closed Tubes.—The Quantity of Heat absorbed proportional to the Quantity of Humid Air.

These results were obtained in the month of September, and on the 27th of October I determined the absorption of aqueous vapour with the same tube when stopped with plates of rock-salt. Three successive experiments gave the deflections produced by the aqueous vapour as 46·6°, 46·4°, 46·8°. Of this concurrent character are all the experiments on the aqueous vapour of the air. The absorption corresponding to the mean deflection here is 66. The total radiation through the exhausted tube was on this day 1085; hence we have

$$1085 : 100 = 66 : 6{\cdot}1;$$

that is to say, the absorption of the aqueous vapour of the air contained in a tube 4 feet long, was on this day 6 per cent. of the total radiation.

The tube employed in these experiments was of brass, polished within; and it was suggested to me that the vapour of the moist air might have precipitated itself on the interior surface of the tube, thus diminishing its reflective power, and producing an effect apparently the same as absorption. In reply to this objection, I would remark that the air on many of the days on which the experiments were made was at least 25 per cent. under its point of saturation. It can hardly be supposed that air in this condition would deposit its vapour upon a polished metallic surface, against which, moreover, the rays from the source of heat were impinging. More than this, the absorption was exerted even when only a small fraction of an atmosphere was made use of, and found to be proportional to the quantity of atmospheric vapour present in the tube. The following table shows the absorptions of humid air at pressures varying from 5 to 30 inches:—

Pressures in inches	Absorption per 100	
	Observed	Calculated
5	16	16
10	32	32
15	49	48
20	64	64
25	82	80
30	98	96

The third column here is calculated on the assumption that the absorption, within the limits of the experiment, is sensibly proportional to the quantity of vapour in the tube. The agreement with observation is almost perfect. *It cannot be supposed that results so regular as these, agreeing so completely with those obtained with small quantities of other vapours, and even with small quantities of the permanent gases, can be due to the condensation of vapour on the surface of the tube.* When 5 inches were in the tube it had less than one-sixth of the quantity of vapour necessary to saturate the space. Condensation under these circumstances is not to be assumed, and more especially a condensation which should produce such regular effects as those above recorded.

§ 5.

Radiation through the Open Air.

The subject, however, is so important that I thought it worth while to make the following additional experiments:—

C (fig. 12) is a cube of boiling water, intended for our source

Fig. 12.

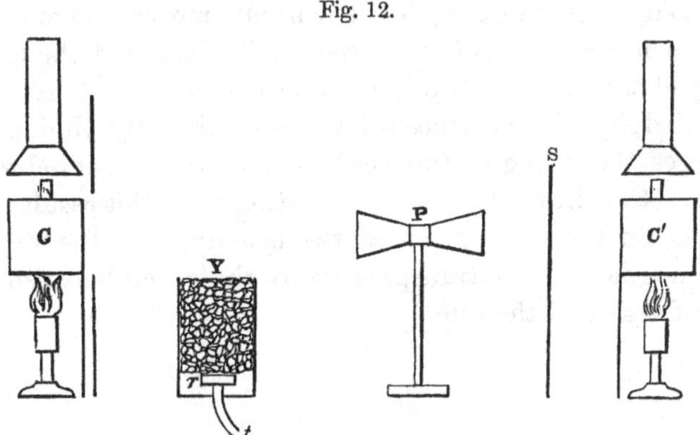

of heat; Y is a hollow brass cylinder, 3·5 inches in diameter and 7·5 inches in depth; P is the thermo-electric pile, and C′ the compensating cube; S is an adjusting screen, used to regulate the amount of heat falling on the posterior surface of the pile. The apparatus was entirely surrounded by boards, the space within being divided by tin screens into compartments which were loosely stuffed with paper or horsehair. The formation of air-currents near the cube or the pile was thus prevented, and irregular motions of the external air were intercepted. A roof, moreover, was bent over the pile, and this was flanked by sheets of tin. The action here sought I knew must be small, and hence the necessity of excluding every disturbing influence.

The cylinder Y was first filled with fragments of quartz moistened with distilled water. A rose burner *r* was placed at the bottom of the cylinder, and from it the tube *t* led to a bag containing air. The bag being subjected to gentle pressure, the air passed upwards amid the fragments of quartz, imbibing moisture from them, and finally discharged itself in the open space between the cube C and the pile. The needle moved and assumed a permanent deflection of 5 degrees, indicating that the opacity of the intervening space to the rays of heat was augmented by the discharge of the saturated air.

The moist quartz fragments were now removed, and the vessel Y was filled with fragments of the chloride of calcium. The rose burner being, as before, connected with the india-rubber bag, air was gently forced up among the calcium fragments and discharged in front of the pile. The needle moved and assumed a permanent deflection of 10 degrees, indicating that the transparency of the space between the pile and source of heat was augmented by the presence of the dry air. By timing the discharges the swing of the needle could be augmented to 20 degrees. Repetition showed no deviation from this result: the saturated air always augmented the opacity, and the dry air always augmented the transparency of the space between the source of heat and the pile.

§ 6.

Application of Results to Meteorology—Tropical Rains—Cumuli—Condensation by Mountains—Temperatures at Great Elevations—Thermometric Range in Australia, Tibet, and Sahara—Leslie's Observations—Melloni on Serein.

The power of aqueous vapour being thus established, meteorologists may, I think, apply the result without fear. That 10 per cent. of the entire terrestrial radiation is absorbed by the aqueous vapour which exists within ten feet of the earth's surface on a day of average humidity, is a moderate estimate. In warm weather and air approaching to saturation, the absorption would probably be considerably greater. This single fact at once suggests the importance of the established action as regards meteorology. I am persuaded that by means of it many difficulties will be solved, and many familiar effects, which we pass over without sufficient scrutiny because they are familiar, will have a novel interest attached to them by their connexion with the action of aqueous vapour on radiant heat. While leaving these applications to be made in all their fulness by meteorologists, I would refer, by way of illustration, to one or two points on which the experiments seem to bear.

And first it is to be remarked that the vapour which absorbs heat thus greedily radiates it very copiously. This fact must, I think, come powerfully into play in the tropical region of calms, where enormous quantities of vapour are raised by the sun, and discharged in deluges upon the earth. These have been assigned to the chilling consequent on the rarefaction of the ascending air. But if we consider the amount of heat liberated in the formation of those falling torrents, the chilling due to rarefaction will hardly account for the entire precipitation. The substance quits the earth as vapour, it returns to it as water; how has the latent heat of the vapour been disposed of? It has in great part, I think, been radiated into space. But the radiation which disposes of such enormous quantities of heat subsequent to condensation, is competent, in some measure at least, to dispose of the heat possessed prior to condensation, and must therefore hasten the act of condensation itself.

Aqueous vapour is a powerful radiant, but it is an equally

powerful absorbent, and its absorbent power is a maximum when the body which radiates into it is vapour like itself. Hence, when the vapour first quits the equatorial ocean and ascends, it finds, for a time, above it a screen of its own substance, into which it pours its heat, and by which that heat is intercepted and in part returned. Condensation in the lower regions of the atmosphere is thereby prevented. But as the mass ascends it passes through successive vapour-strata, which Strachey has shown to diminish far more speedily in density than the associated strata of air, until finally our ascending body of vapour finds itself lifted above the screen which for a time protected it. It now radiates freely into space, and condensation is the necessary consequence. The heat liberated by condensation is, in its turn, spent in space, and the mass thus deprived of its potential energy returns to the earth as water. To what precise extent this power of aqueous vapour as a radiant comes into play as a promoter of condensation, I will not now inquire; but it must be influential in producing the torrents which are so characteristic of the tropics.

The same remarks apply to the formation of cumuli in our own latitudes. They are the heads of columns of vapour which rise from the earth's surface and are condensed to cloud at a certain elevation. Thus the visible cloud forms the capital of an invisible pillar of saturated air. Certainly the top of the column, piercing the sea of vapour which hugs the earth, and offering itself to space, must lose heat by the radiation from its vapour, and in this act alone we should have the necessity for condensation. The 'vapour plane' must also depend, to a greater or less extent, on the chilling effects of radiation.

The action of mountains as condensers must, I think, be connected with these considerations. When a moist wind encounters a mountain-range it is tilted upwards, and condensation is no doubt to some extent due to the work performed by the expanding air; but the other cause cannot be neglected; for the air not only performs work, but it is lifted to a region where its vapour can freely lose its heat by radiation into space. During the absence of wet winds the mountains themselves also lose their heat by radiation, and are thus prepared for actual surface condensation. We must indeed take into account the fact that this radiant quality of water is persistent throughout its three states of exaggeration. As vapour it loses its heat and promotes condensation; as water it loses its heat and promotes congelation; as snow it loses its heat and renders the surfaces on which it rests more powerful refrigerators than they would otherwise be. The formation of a cloud before the air which contains it *touches* a cold mountain, and indeed the formation of a cloud anywhere over a cold tract of land, where the cloud is caused by the cold of the tract, is due to the radiation from the aqueous vapour. The uniformly diffused fog which sometimes fills the atmosphere in still weather may be due to cold generated by uniform radiation throughout the mass, and not to the mixture of currents of different temperatures. The cloud by which the track of the Nile and Ganges (and sometimes the rivers of our own country) may be followed on a clear morning is, I believe, due to the chilling of the saturated air above the river by radiation from its vapour.

Observation proves the radiation to augment as we ascend a mountain. Martins and Bravais, for example, found the lowering of a radiation-thermometer 5·7° C. at Chamouni; while on the Grand Plateau, under the same conditions, it was 13·4° C. The following remarkable passage from Hooker's 'Himalayan Journals,' 1st edit. vol. ii. p. 407, bears directly upon this point: 'From a multitude of desultory observations I conclude that, at 7,400 feet, 125·7° or 67° above the temperature of the air, is the average maximum effect of the sun's rays on a black-bulb thermometer. . . . . These results, though greatly above those obtained at Calcutta, are not much, if at all, above what may be observed on the plains of India [because of the dryness of the air.—J. T.]. The effect is much increased with the elevation. At 10,000 feet, in December, at 9 A.M. I saw the mercury mount to 132° [in the sun], with a difference [above the shaded air] of 94°, while the temperature of shaded snow hard by was 22°. At 13,100 feet, in January, at 9 A.M. it has stood at 98°, with a difference of 68·2°, and at 10 A.M. at 114°, with a difference of 81·4°, *whilst the radiating thermometer on the snow had fallen at sunrise to* 0·7°.' This enormous chilling is fully accounted for by the absence of aqueous vapour overhead. I never under any circumstances suffered so much from heat as in descending on a sunny day from the Corridor to the Grand Plateau of Mont Blanc. The air was perfectly still, and the sun literally blazed against my friend Mr. Hirst and myself. We were hip deep in snow; still the heat was unendurable. Immersion in the shadow of the Dôme du Goûté soon restored our powers, though the *air* of the shade was not sensibly colder than that through which the sunbeams passed.

Without quitting Europe we find places where, even when the day temperature is high, the hour before sunrise is intensely cold. I have often experienced this even in Germany; and the Hungarian peasants, if exposed at night, take care, in hot weather, to prepare for the nocturnal chill. The *range* of temperature augments with the dryness, and an 'excessive climate' is certainly in part caused by the absence of aqueous vapour.

Regarding Central Australia, Mr. Mitchell publishes extremely valuable tables of observations, from which we learn that, when the days are at the same time calm and clear, the daily thermometric range is exceedingly large. On the 2nd of March 1835 the temperature at noon was 68°, while that at sunrise next morning was 20°, showing a difference of 48°. The 7th and 8th were also clear and calm; the difference between noon and sunrise on the former day was 38°, while on the latter it was 41°. Indeed between April and September a range of 40° in clear weather was quite common—or more than double the amount observed in London at the corresponding season of the year.

A freedom of escape similar to that from bodies at great elevations would occur at any other level, were the vapour removed from the air above it. Hence the withdrawal of the sun from any region over which the atmosphere is dry must be followed by quick refrigeration. This is simply an *à priori* conclusion from the facts established by experiment; but, I believe, all the experience of meteorology confirms it. The winters in Tibet are almost unendurable from this cause. The isothermals dip deeply from the north into Central Asia during the winter, the earth's heat being wasted without impediment in space, and no sun existing sufficiently powerful to make good the loss. I believe the fact is well established that the desert of Sahara, which during the day is burning hot, is often extremely cold at night. This effect has been hitherto referred in a general way to the 'purity of the air:' but purity, as judged by the eye, is a very imperfect test of radiation, for the existence of large quantities of vapour is consistent with a transparent atmosphere. The purity really consists in the absence of aqueous vapour from those so-called rainless districts, which, when the sun is withdrawn, enables the hot surface of the earth to run speedily down to a freezing temperature.

On the most serene days the atmosphere may be charged with vapour; in the Alps, for example, it often happens that skies of extraordinary clearness are the harbingers of rain. On such days, no matter how pure the air may seem to the eye, terrestrial radiation is arrested. And here we have the simple explanation of an interesting fact noticed by Sir John Leslie, which has remained without explanation up to the present time. This eminent experimenter devised a modification of his differential thermometer, which he called an *Æthrioscope.* The instrument consisted of two bulbs united by a vertical tube, of a bore small enough to retain a little liquid index by its own adhesion. The lower bulb was protected by a metallic coating; the upper or sentient bulb was blackened, and was placed in a polished metal cup, which protected it completely from terrestrial radiation.

'This instrument,' says its inventor, 'will at all times during the day and night indicate an impression of cold shot downwards from the higher regions. . . . . But the cause of its variations does not always appear so obvious. Under a fine blue sky the *Æthrioscope* will sometimes indicate a cold of 50 millesimal degrees; yet on other days, *when the air is equally bright,* the effect is hardly 30°.' It is, I think, certain that these anomalies were due to differences in the amount of aqueous vapour in the air, which escaped the sense of vision. Indeed, Leslie himself connects the effect with aqueous vapour by the following remark: 'The pressure [? *presence*] of hygrometric moisture in the air probably affects the indications of the instrument.' In fact, the absence or presence of vapour opened or closed an invisible door for radiation from the 'sentient bulb' into space.

The following observation in reference to radiation-experiments with Pouillet's pyrheliometer, now also receives its explanation. 'In making such experiments,' says M. Schlagintweit, 'deviations in the transparency are often recognised which are totally inappreciable to the telescope or the naked eye, but which afterwards announce themselves in the presence of thin clouds,' &c.

In his beautiful essay on dew, Wells gives the true explanation of the formation of ice in India, by ascribing the effect to radiation. I think, however, his theory needs supplementing. Given the same day-temperature in England as at Benares, could we, even in clear weather, obtain a sufficient fall of temperature to produce ice? I think not. The interception of the calorific rays by our humid air would too much retard the chill. It is apparent, from the descriptions of the process, that a dry still air is the most favourable for the formation of the ice. The nights when it is formed in greatest abundance are those during which the dew is not copious. The flat pans used in the process are placed on dry straw, and if the straw become wet it must be removed. Wells accounts for this by saying that the wet straw is more dense than the dry, and hence more competent to transfer heat from the earth to the basins. This is hardly a satisfactory explanation; a better one seems to be that the evaporation from the moist straw, by throwing over the pans an atmosphere of aqueous vapour, checks the radiation and thus diminishes the cold.

Melloni, in his excellent paper 'On the Nocturnal Radiation of Bodies,' gives a theory of *serein*, an excessively fine rain which sometimes falls in a clear sky a few moments after sunset. Several authors, he says, attribute this effect to the cold resulting from radiation of the air, during the fine season, immediately on the departure of the sun. 'But,' writes Melloni, 'as no fact is yet known which distinctly proves the emissive power of pure transparent elastic fluids, it appears to me more conformable to the principles of natural philosophy to attribute this species of rain to the radiation and subsequent condensation of a thin veil of vesicular vapour distributed through the higher strata of the atmosphere.'* Now, however, that the power of aqueous vapour as a radiant is known, the difficulty experienced by Melloni disappears. The former hypothesis, however, though probably correct in ascribing the effect to radiation, was incorrect in ascribing it to the radiation of '*the air*.'

Dr. Hooker encourages me to hope that this newly discovered action may throw some light on the formation of hail. The

* Taylor's *Scientific Memoirs*, vol. v. p. 551.

wildest and vaguest theories are afloat upon this subject. But the same action which produces *serein* must, if augmented, freeze the minute rain, and the aggregation of the small particles thus frozen would form hail. Many kinds of hail that I have had an opportunity of examining could not be due to the freezing of *drops* of water, each hailstone being merely the ice of the drop. The 'stones' are granular aggregates, the components of which may be produced by the chill of radiation. I will not, however, dwell further on this subject, but will now commit the entire question to those who are more specially qualified for its pursuit.

IV.

ON THE PASSAGE OF RADIANT HEAT THROUGH DRY AND HUMID AIR.

ANALYSIS OF MEMOIR IV.

WITH Professor Magnus's visit to London, I thought the points of difference between us settled. The experiment with the open tube had evidently impressed him, and, with a view to repeating it, on his return to Berlin, he mounted an apparatus similar to mine. 'The result of this experiment,' he says, 'was so surprising, and so little in accordance with what I had found by other processes, that when I reached home I determined to repeat it.' He describes his apparatus and mode of using it, and thus states his results:—'With this arrangement I got, on allowing dry or moist air to flow through the tube, deflections of the galvanometer which corresponded to those described by Professor Tyndall. But I did not always get them; and what particularly surprised me was that the deflection of the needle did not correspond to an absorption of heat by its passage through moist air, but that, on the contrary, when the moist air was passed through the tube, the face of the pile turned towards the tube was found to be most heated. In order to clear up the already mentioned uncertainty of the experiment, I have repeated the blowing in of dry and moist air many hundred times; but in no single case was the deflection such as to indicate a greater absorption by moist air.'

'It would be out of place,' he continues, 'to relate the numerous experiments which were undertaken, partly in order to make myself master of the phenomena, and partly in order to explain the surprising contradiction between my results and the conclusions which Professor Tyndall has drawn from his experiments. I found, in the first place, that the deflection took place only when the air was driven in with a certain amount of force. It was found, further, that when the air was pressed in continuously, the deflection of the galvanometer was not maintained constant, but that the instrument gradually returned to its position of equilibrium. Hence it resulted that the air did not cause the deflections by absorption. I suspected that possibly moisture might be condensed on the internal surface of the tube, and that the heating effect might be thus produced; but this supposition was likewise found to be erroneous. It appears, on the other hand, that the phenomenon is caused by the absorption which takes place at the surface of the pile itself.'

Professor Magnus then proves in the most satisfactory manner that the deflections observed *in his experiments* were wholly due to the condensation of the aqueous vapour of the moist air on the surface of his pile, and its subsequent evaporation by dry air. 'We see,' he continues, 'by these results, how little fitted air is, while in motion, for experiments as to its power of absorption.'

Had Professor Magnus allowed himself the time requisite for testing it, he would not, I am persuaded, have offered this solution of the differences existing between us. 'Air in motion,' employed as long experience has taught me to employ it, yielded at the time here referred to perfectly certain and concurrent results. Nor were these results, as supposed by Professor Magnus, dependent on any accident as to the dimensions of the apparatus.

Numerous passages in these memoirs show how fatal to accuracy I regarded the bringing of air or any other gas into contact either with the source of heat or with the surface of the pile; it was not therefore likely that in the present case I should shut my eyes to so obvious a source of error. In Memoir III. (page 113) I actually referred to it. Special experiments, moreover, had assured me that no air from the experimental tube came near the pile; and in a thousand trials I never had the slightest difficulty in obtaining the results controverted by the foregoing experiments. They were perfectly precise, closely concurrent, and always attainable. Nevertheless, on reading Professor Magnus's paper I went once more over the ground, and verified all my former statements.

As, therefore, I could not add to my own certainty, and as the evidence of an independent observer seemed desirable, I handed over the whole apparatus to Dr. Frankland, who minutely tested every point involved in, or arising out of, the objection of Professor Magnus. He showed that the source of error signalised above had nothing to do with my experiments, making it evident that its introduction could only arise from instrumental defects. Professor Frankland expresses his conclusions in these words:—'After a careful scrutiny, I have been unable to detect any source of fallacy in these experiments; they therefore appear to me to prove conclusively that obscure radiant heat passes much more readily through dry than through moist air. In conclusion, I cannot but express my surprise and admiration at the precision and sharpness of the indications of your apparatus. Without having actually worked with it I should not have thought it possible to obtain these qualities in so high a degree in determinations of such extreme delicacy, and which are so well known to be exposed to numerous sources of derangement.' *

In the concluding portion of this memoir the effect of bringing the gases into contact with the source of heat is briefly considered, and the effect of contact with the pile is more fully developed. It is shown that deflections the same in kind as those which Professor Magnus ascribes to absorption, *must be produced by the contact of the air whether its temperature be higher or lower than that of the pile.*

The disturbance arising from the dynamic heating of the air, and of all surfaces against which it impinges, is also dwelt upon and illustrated.

In the communication of Professor Magnus above cited, a remark occurs which merits a word of explanation. 'Besides the defect,' he writes, 'arising from the hygroscopic character of the rock-salt plates, Professor Tyndall's method labours

* In a brief communication inserted in Poggendorff's *Annalen*, and translated in the *Philosophical Magazine*, vol. xxvii. p. 249, Professor Magnus thus speaks of Frankland's confirmation:—'Dr. Tyndall has had his experiments repeated by Dr. Frankland, in order, as he says, to prove that he had not mistaken cold for hot and hot for cold. Such a confirmation was, in my opinion, unnecessary. I have not implied an error of that kind, but have only said that, in repeating Dr. Tyndall's experiments, it has not even once happened to me to obtain the same result as he did.' Surely that an investigator of such acknowledged skill and caution was not only unable to obtain my results, but, in 'many hundred' experiments, had obtained others diametrically opposed to mine, was a sufficient justification of the desire to see my experiments verified.

under another difficulty. . . . The value obtained for dry air formed the unit for the determination of the other gases, all of which were compared in the same way with the vacuum. Therefore the smaller the difference between the dry atmospheric air and the vacuum, the greater the apparent absorptive power of the other gases. Hence, if this difference were to be equal to nothing, the absorption of the other gases would come out infinitely great.'

I would ask it to be remembered that my object in these inquiries was not to follow the track of my eminent predecessors, who made radiant heat the primary object of their thoughts, but rather to employ radiant heat as an explorer of molecular condition. What I aimed at specially in these first memoirs was to bring clearly into view the astonishing change in the relations of the luminiferous æther accompanying the act of chemical combination. I wanted to show the *physical* significance of an atomic theory which had been founded on purely *chemical* considerations. By making the substance of feeblest action my standard, and referring all the others to it, this purpose was carried out in the most direct and simple way. My tables, in fact, resemble those of the atomic weights, in which hydrogen, the lightest atom, is employed as a standard.

IV.

ON THE PASSAGE OF RADIANT HEAT THROUGH DRY AND HUMID AIR.*

It is known to the reader that Professor Magnus and myself have arrived at different conclusions regarding the action of dry air, and of the aqueous vapour diffused throughout our atmosphere, on radiant heat. Last autumn I had the pleasure of meeting my eminent friend in London; and soon after his arrival it was agreed upon between us to subject the points on which we differed to a searching examination. We accordingly met on several occasions in the laboratory of the Royal Institution, where every result that I had previously announced was reproduced in the presence of Professor Magnus. Facts were placed before him which he professed his inability to explain; but, like a cautious philosopher, he reserved his opinion. It was, however, proved that the results observed by us in common could not be ascribed to any defect of method or error of observation which it was then possible to point out. I wished very much to subject the most recent experiments of Professor Magnus to a similar examination, and he evinced an equal desire to show them to me. He began his arrangements, but it was not my good fortune to see them accomplished. In fact, coming to London as a visitor to the International Exhibition, the numerous other claims upon his time and attention were amply sufficient to prevent him from carrying out his own wishes and gratifying

In the latest number of Poggendorff's 'Annalen,' Professor Magnus has published a paper 'On the Diathermancy of Dry and Moist Air,' a translation of which is printed in the

* *Philosophical Magazine*, July 1863.

'Philosophical Magazine' for July 1863. From it I learn that the experiments on atmospheric vapour which struck him most in London were those performed with a tube open at both ends. The results thus obtained were so opposed to those obtained in another way by himself, that he returned to Berlin resolved to repeat my experiments. The paper just referred to contains an account of his researches, and a proposed explanation of my results.

Operating with an open tube, he displaced with a pair of bellows dry air by moist and moist air by dry, and obtained, though not always, deflections corresponding to mine. But he was particularly surprised to find that the direction in which the needle moved when moist air was blown into the tube, indicated, not a withdrawal of heat from the thermo-electric pile, but an augmentation of heat. When dry air was forced into the tube, the deflection observed, instead of indicating that a greater amount of heat fell upon the pile, showed, on the contrary, that the pile was chilled. These effects, he urges, are due to the absorption of aqueous vapour by the lampblack coating of the thermo-pile. This absorption, when moist air was blown against the instrument, rendered heat free; when dry air, on the other hand, was forced against it, the evaporation of the condensed vapour chilled the pile, and the deflection due to cold was observed. He wishes it, in short, to be inferred that in my experiments cold has been mistaken for heat, and heat for cold, and that effects have been ascribed to absorption which are really due to the condensation and evaporation of aqueous moisture at the surface of my thermo-electric pile.

To commit such an error, and to persist in it so long, would, I fear, leave me little claim to confidence as an experimenter. But the truth is that some years have elapsed since I became acquainted with the facts now urged against me by Professor Magnus. Experimenting years ago on dry and moist air with tubes which had been coated inside by lampblack or lined with blackened paper, I found, when moist air was introduced, the radiation from the interior surface so energetic as to compel me to abandon the coating. The promptness and energy with which these effects of condensation and evaporation are produced are remarkable. Dry air urged against the face of my pile on a day of average humidity drives the needle of my galvanometer through an arc of 196 degrees, and keeps it for a time pointing to nearly 90°, from which, while the air-current continues, it gradually sinks to zero. On simply stopping the current of dry air, the needle swings quickly to the other side of zero, passing through an arc of 120 degrees, this large deflection being produced by the sudden re-absorption of the atmospheric vapour when the dry air is intercepted. Air artificially moistened produces still larger deflections.

Such effects were well known and duly guarded against. Indeed, it is to me interesting to notice my own experience in this inquiry reproduced years subsequently in the experience of Professor Magnus. I never had the least doubt of the correctness of his results; but, for the most part, they have absolutely nothing to do with mine. We are equally successful in our efforts. His object, for example, is to bring the hygroscopic character of rock-salt into strong relief, and he succeeds in wetting the plates; my object is to avoid this source of disturbance, and I am equally successful. He, by blowing vigorously into his tube, urges the air against the face of his pile, and obtains the effects due to condensation and evaporation; I, by operating cautiously and permitting the air to enter the tube so slowly and at such distances from the source and from the pile that neither of them is affected by it, obtain the effects due to absorption. One great feature of the case, however, is, that while the results of Professor Magnus have been known to me for years, and while I can produce them on a large scale at any moment, he has not yet succeeded in reproducing mine. 'Never,' he writes, 'in a single instance has the deflection indicated a greater absorption by the humid air.'

After reading the last paper of Professor Magnus, I felt that it would be useless on my part to reiterate what had been already so often affirmed, and I therefore wished to subject my experiments to the scrutiny of an independent observer. Mr. Faraday had already seen those experiments, and it was purely my reluctance to give him trouble that prevented me from asking him to witness them again. Next to him I could hardly find a man whose testimony on such a subject will have greater weight than that of my colleague, Dr. Frankland; and he, at my request, kindly undertook to satisfy himself upon the points at issue. I mounted the apparatus, and left it entirely in his hands; and he has favoured me with the following account of his observations:—

'MY DEAR TYNDALL,—At your request I have made a number of experiments on the comparative transcalency of common air, and of air deprived of its moisture by contact with monohydrated sulphuric acid. The apparatus which I used was that described by you in the 'Philosophical Transactions' for December 1862. It was exclusively under my own control; and I arranged the details of manipulation in such a manner as appeared to me best calculated to eliminate all sources of error. My mode of operating was as follows:—The brass tube open at both ends, formed the conduit for a portion of the thermal radiation from the source of heat. These heat-rays, after passing through the tube, traversed several inches of intervening air-space before entering the cone of the thermo-electric pile, where they produced their effect, in opposition to that arising from another constant source of heat affecting the opposite face of the pile (the compensating cube). The differential action was indicated as usual by a delicate galvanometer. These arrangements being once for all made, I was able by means of an air-pump to introduce at pleasure into the tube either the ordinary air of the laboratory, dry air, or air rendered moist by passage over extensive surfaces of wet glass. At the commencement of the experiments the tube was of course full of the common air of the laboratory; the needle of the galvanometer marked 42°, and remained steady for a quarter of an hour within a degree of that point. I now interposed in the path of the rays entering the brass tube a sheet of tin-plate; the needle at once bounded from 42° up to 90°. It was thus evident that any obstruction to the passage of the rays of heat through the tube, or, in other words, any *cooling* of that face of the pile which was turned towards the tube, would be indicated by an increased deflection of the needle on the same side of zero, which I will call the − side, whilst a *heating* of the same face of the pile would be attended by a diminished deflection, or even by a passage of the needle to the opposite or + side of zero. The following are the results which I observed:—

Tube filled with	Permanent deflection of needle
	°
Common air	−42
Air dried by contact with monohydrated sulphuric acid and introduced gently into the open tube	+13
Common air which had spontaneously displaced the foregoing dry air	−43
Common air gently drawn in by air-pump	−43·3
Common air gently blown in from caoutchouc bag	−45
Same air gently blown in from caoutchouc bag, but dried by passage over sulphuric acid	+6
Air from same bag, but not dried	−46·5
Air of laboratory	−42
Air dried and introduced as before	+14
Air dried as before with sulphuric acid, but afterwards passed over fragments of glass moistened with water	−46
Common air gently drawn in by air-pump	−42·5

'At the conclusion of the experiments I found that the deflection due to the total radiation was 86°·2.

'I also saw the following experiments made by yourself when the ends of the brass tube were closed by plates of rock-salt:—

	Permanent deflection of needle
	°
Tube filled with dry air	+ 7
After exhaustion of tube	− 3
After admission of laboratory air	−42

'Rain was falling at the time these last determinations were made, and the air was very moist. On removing the plates of rock-salt from the tube they appeared to be quite dry; and after being breathed upon, the film of moisture soon disappeared and they recovered their previous lustre. I ought perhaps to mention that these experiments are not selected, they are the only ones I have made upon the subject, and they were performed in the sequence given above: after a very careful scrutiny I have been unable to detect any source of fallacy in them, and they therefore appear to me to prove conclusively that obscure radiant heat passes much more readily through dry than through moist air.

'In conclusion, I cannot but express my surprise and admiration at the precision and sharpness of the indications of your apparatus. Without having actually worked with it I should not have thought it possible to obtain these qualities in so high a degree in determinations of such extreme delicacy, and which are so well known to be exposed to numerous sources of derangement.

'Believe me, yours very truly,

'E. FRANKLAND.

'Royal Institution,
June 19, 1863.'

'P.S.—Since writing the foregoing letter, I have repeated the experiments there recorded without any source of heat at either end of the pile, in order to ascertain whether the introduction and withdrawal of dry air at all affected the galvanometer. The tube was first full of the common air of the laboratory, and the needle remained steadily at +12·5 for a quarter of an hour. A current of moistened air was now drawn through the tube for ten minutes in precisely the same manner as when the two sources of heat were employed, the needle being closely watched during the whole time. It oscillated between +12 and +13, but never passed these limits on either side. The current was now interrupted and the needle closely watched for five minutes: it remained perfectly steady at 12·5. A current of dried air was now conveyed through the tube for ten minutes; the needle oscillated as before between 12° and 13°.

'Thus far I operated on the air exactly as in the experiments recorded in the foregoing letter. I then quadrupled the velocity of the current through the tube, introducing in the first place dry air: the needle in a first experiment moved 6 degrees in the direction of cold; but on repeating the experiment with both dry and moist air no effect whatever was produced. I now removed the tube and delivered a gentle current of dry air into the cone of the pile; immediately the needle moved 90° in the direction of *cold*. The current was continued uninterruptedly for ten minutes, during which time the needle gradually returned to nearly its original position. The current of dry air being now stopped, the needle moved 40° in the direction of *heat*, returning again gradually and slowly to its normal position. The same temporary deflection for heat was also produced in an exalted degree when the dry current was immediately succeeded by a moist one.

'These supplementary experiments lead me to the following conclusions:—

'1st. The gentle currents of air which were caused to flow through the tube in the experiments detailed in my letter did not in any way disturb the results of those experiments, neither would they have done so in any material degree even had their velocity been quadrupled.

'2ndly. The impact of air drier than that previously in contact with the pile cools that face of the instrument with which it comes in contact, whilst the like impact of moister air produces the opposite effect.

'3rdly. It is, however, impossible to confound the effects obtained in the above experiments on transcalency with those produced by the impact of dry and moist air upon the face of the pile, because in the first place the former are *permanent*, whilst the latter are essentially *transitory*; and in the second place the deflections due to the impact of dry or moist air against the face of the pile are always *in the opposite direction* to those obtained by the interposition of the same kind of air in the path of radiant heat. Thus, if the heat-rays falling upon one face of the pile be made to traverse dry air, the needle will move in the direction of heat, but if the apparatus be so arranged as to cause the dry air to impinge upon the face of the pile, the effect due to the greater transcalency of the dry air would be at first more or less neutralized, or even altogether overborne, by the cooling influence due to evaporation at the surface of the pile so brought into contact with dry air.—E. F.

'Royal Institution,
June 20, 1863.'

In my remarks on the experiments of Professor Magnus, I had pointed out two sources of error in the method which he employed. One of these was the bringing of the cold gas to be examined into direct contact with his source of heat: and the other was the bringing of the same gas into direct contact with the face of his thermo-electric pile. In his last paper he urges, in reference to the first point, that my objections do not apply to his apparatus, because in it the column of air is heated at the top. This argument would be strictly valid if the heat could be applied with perfect uniformity to a perfectly horizontal plane, but in practice such perfection is not attainable. The top of Professor Magnus's recipient is dome-shaped, even where it is in perfect contact with the source of heat, while beyond the limits of this contact—that is to say, down the sides of the recipient—it is propagated more or less by conduction. Indeed Professor Magnus himself states that a portion of the heat effective in his experiments is derived from the glass thus warmed. 'The heating of the thermometer,' he writes, 'although due only to conduction and radiation, involves a very complicated process. Besides the direct heating through conduction and radiation, reflexion also takes place at the inner surface of the vessel. Further, *the portions of the surface adjacent to the vessel of boiling water are heated by conduction, and also radiate heat against the thermometer.*'* I have italicised the most important part of the passage. Now air in contact with such a surface is substantially in the same condition as in my front chamber, and such air, as I have shown, must more or less diminish the temperature of the surface exposed to it. If Professor Magnus fails to detect this with his new apparatus, it can, I think, only be due to its lack of the requisite delicacy. Without the actual numbers no safe opinion can be formed upon this point; the probability is that his total heat is so small that the lowering of the temperature of his source of heat by the admission of air into contact with it becomes infinitesimal.

An important difference between Professor Magnus and me consists in the high absorptive power which he ascribes to dry air. His absorption, in fact, is more than 140 times mine. I would here bespeak the reader's attention to an examination of the conditions in which Professor Magnus places his instruments. From his last figure, and also from a passage of his paper, it is to be inferred that *in his recent experiments* the air has free access to the two faces of his pile, the axis of which is vertical. The upper face is furnished with a conical reflector, while the lower one is provided with one of the cylindrical tubes which usually accompany the instrument. It will repay us to reflect for a moment on the processes involved in this arrangement. Professor Magnus keeps the space which contains his pile at a constant temperature of 15° C. Let us suppose the two faces of his pile to be at the same temperature, the radiation from the source being suspended, and the space around the pile a vacuum. Suppose, in the first instance, the temperature of the air outside to be lower than that of the pile, that the pile, in other words, is a warm body in comparison with the air; what will be the effect

* Poggendorff's *Annalen*, vol. cxii. p. 544.

of admitting the air into the vessel?* Manifestly on the upper face of the pile will rest a column of air, heated at its bottom by the surface on which it rests; convection will immediately set in, and heat will be continually abstracted from the face of the pile. At the lower face, on the contrary, an equal abstraction does not take place; for the air once warmed remains in contact with the face of the pile, convection here being almost *nil.* Thus a less amount of heat is abstracted from the lower than from the upper face of the pile, and hence the instrument, which before the entrance of the cool air produced no current, must, in virtue of the different action of this air on its two opposite faces, generate a current similar to what would be produced by the direct heating of the lower face of the pile.

A moment's reflection suffices to prove that precisely the same deflection is obtained when the external air is *hotter* than the pile. Supposing, as before, the temperature of both faces to be the same at the commencement, the needle of the associated galvanometer being at zero. When the warm air enters it is chilled by the upper face of the pile, contracts, and remains in contact with that face, forming in fact *a pool of heavy air* at the bottom of the reflector. The air chilled by the opposite face of the pile falls by its weight; its place being supplied by fresh warm air. It is therefore evident that the lower face of the pile will in this case be more heated by the air than the upper one; and hence we infer that whether the external air be colder than the pile, or hotter than the pile, the same galvanometric effect follows its introduction into the vessel.

Instead of supposing the pile to be in the first place of uniform temperature, let us imagine it exposed to the radiation from the source of heat. This makes the upper face warmer than the under one, and produces a deflection commensurate with the difference of temperature of the two faces. Let air now enter: it is manifest from the foregoing analysis that, whether this air is colder than the pile or hotter, its effect will be to render the lower face *relatively* warmer, and thus to diminish the deflection. If, moreover, the air be of the exact temperature of the upper face, it will warm the under one; if of the

* The reader, if he wish, may see a drawing of Professor Magnus's apparatus in the *Philosophical Magazine* for 1863, vol. xxvi.

exact temperature of the under face it will chill the upper one. If its temperature be the mean temperature of the mass of the pile, it will chill the upper face and warm the lower one at the same time. *No matter, then, what the temperature of the air may be when it enters the vessel, the effect of its contact with the pile is to diminish the deflection due to the radiation from the source of heat, and thus produce the same galvanometric effect as a true absorption.* How Professor Magnus releases his pile from this apparently inevitable action he does not inform us; and how he can distinguish between this effect, in which absorption has absolutely nothing to do, and one of real absorption, I am at a loss to imagine.

His apparatus will enable him to make this experiment in a far more unexceptionable manner. Let him place a second plate of salt across his tube immediately above his pile, and thus isolate it from the air which he intends to examine. He will then obtain the almost pure effect of radiation. Professor Magnus has actually made this experiment, and the result, expressed in his own words, is 'a hardly perceptible difference between dry air and a vacuum.'

It is scarcely necessary to repeat what I have already stated regarding the dynamic heating of the pile when the air enters. Professor Magnus never once refers to this effect, though he does refer, for the first time, in his last paper to the chilling consequent on pumping out. Had his apparatus been sufficiently delicate, the effect to which I refer must have long ago attracted his attention. Some conception of its magnitude may be formed from the following quotation from a paper laid before the Royal Society on the 18th of this month:—

'A brass tube 3 feet long and very slightly tarnished within was used for dynamic radiation. Dry air on entering the tube produced a deflection of 12 degrees. The tube was then polished within and the experiment repeated: the deflection by dry air was instantly reduced to 7·5 degrees.

'The rock-salt plate at the end of this tube was removed, and a lining of black paper 2 feet long was introduced within it. The tube was again closed, and the experiment of allowing dry air to enter it was repeated. The deflections in three successive experiments rose from 7·5° to

80°, 81°, 80°,

and this result might be obtained as long as the lining was permitted to remain within the tube.

'The plate of rock-salt was again removed, and the length of the lining was reduced to a foot; the dynamic radiation on the entrance of dry air produced in three successive experiments the deflections

76°, 74°, 75°.

'The plate was again removed and the lining reduced to three inches in length; the deflections obtained in two successive experiments were

66°, 65°.

'Finally, the lining was reduced to a ring only 1½ inch in width; the dynamic radiation from this small surface gave, in three successive trials, the deflections

60°, 56°, 56·5°.

'The lining was then entirely removed; and the deflection instantly fell to

7·5°.

'In the foregoing experiments the lining was first heated by the collision of the air, and it then radiated its heat through a thick plate of rock-salt against the pile. The effect of the heat was enfeebled by distance, by reflexion from the surface of the salt, and by partial absorption. Still we see the radiation thus weakened competent to drive the needle almost through the quadrant of a circle. Suppose, instead of being thus separated from the lining, *the face of the pile itself* [as in the experiments of Professor Magnus] to form part of the inner surface of the tube, receiving there the direct impact of the particles of air; of course the deflections then obtained would be far greater than the highest of those above recorded. I do not doubt the possibility of causing the needle of my galvanometer, subjected to such an action, to swing through an arc of 1,000 degrees; and it is my reluctance to derange the magnetism of my needle that prevents me from making the experiment.' *

Professor Magnus refers to the agreement which subsists

* When the pile was placed entirely within the tube (as Professor Magnus places it), a single stroke of the pump in exhausting drove the needle through an arc of 115 degrees.

between his results and mine in the case of the more powerfully acting gases, in proof of the correctness of his mode of experiment. The agreement, however, is not such as to warrant the conclusion drawn from it. The case may be illustrated by reference to a delicate chemists' balance as compared with one of those used in common life. Weighing *pounds*, both balances would roughly agree, but in weighing *milligrammes* the coarser balance would infallibly fail. I think it vain to expect a correct determination in any case requiring great delicacy with the apparatus which Professor Magnus employs.*

He again refers to the hygroscopic character of rock-salt. This is admitted. His experiments on this substance are quite correct; but they have no bearing upon mine. During our joint experiments, and while the humid air, whose absorption produced a deflection of 43 degrees of my galvanometer, was still in the experimental tube, the rock-salt plates were detached and placed in his hands. He saw no moisture, and he expressed himself satisfied that there was none. Professor Magnus finds another difficulty in the fact that I make air my unit, and refer the action of all other gases to this unit. There is, I submit, no more 'difficulty' here than in the tables of atomic weights, where hydrogen is taken as the unit. My object was, and is, *to make radiant heat an explorer of molecular condition;* and my results seem to me more instructive and emphatic as now presented than if I had followed the common method pursued by Professor Magnus. The difficulty referred to does not touch the method of experiment at all, but merely my way of presenting the results.

I may add that, in a paper recently presented to the Royal Society, the action of all the vapours which I have examined is compared with that of the liquids from which these vapours are derived. The order of absorption of vapours and liquids is precisely the same.† At the bottom of the list stands water, as the most opaque liquid examined. It would form a most remarkable exception to what, so far as I can see, is a *general law*, if the vapour of this liquid proved so ineffectual as the experiments of Professor Magnus make it. One word with reference

* This remark, I fear, was displeasing to Professor Magnus. His retort was warm. See *Philosophical Magazine*, 1864, vol. xxvii. p. 250.

† See Section 9 of Memoir V. and Section 5 of Memoir VI. of this collection.

to the importance of this subject. In a certain sense Professor Magnus is quite right in rating it low. It derives its importance from the fact that aqueous vapour is everywhere present in our atmosphere, and that, for the future, the proved action of this vapour must form one of the chief foundation-stones of the science of meteorology.

Royal Institution,
June 19, 1863.

V.

ON THE ABSORPTION AND RADIATION OF HEAT BY GASEOUS AND LIQUID MATTER.

ANALYSIS OF MEMOIR V.

Melloni determined the transmission of radiant heat through liquids and solids of different thicknesses: in the following investigation the same is effected for gases and vapours.

A new experimental arrangement is described, which permits of varying the thickness of the gaseous layers between the limits of one-hundredth of an inch and forty-nine inches; or in the ratio of 1 : 4900. With the stronger gases even the thinnest of these layers is shown to yield a measurable action.

The influence of a diathermanous envelope upon the temperature of a planet has been more than once adverted to in these memoirs. It is here proved experimentally that a layer of olefiant gas only two inches in thickness wrapping the earth, and allowing comparatively free passage to the solar rays, would intercept at least 33 per cent. of the terrestrial radiation.

It is also shown that an envelope of sulphuric-ether vapour two inches thick would stop at least 35 per cent. of the terrestrial radiation. In connexion with these results, the possible action of aqueous vapour is adverted to.

Experiments are then described where the experimental tube is divided into two chambers by a rock-salt partition; the radiation from both chambers, taken singly and together, being determined for both gases and vapours. The dividing rock-salt is made to occupy different positions within the tube, so as to render the two chambers sometimes of equal, and sometimes of unequal length.

The *sifting* of the heat by one chamber, and its influence on the transmission of the heat through the other, are pointed out.

New experiments on Dynamic Radiation are described, the arrangement permitting the dynamically heated gas or vapour to radiate through a vacuum, through a column of its own substance, or through a column of any other gas or vapour. The influence of coincidence in vibrating period is further illustrated by these experiments.

The effect of tarnish on the interior surface of the tube, or of an interior lining, upon the dynamic radiation from the surface is experimentally shown.

The variation of the dynamic radiation with the length of the radiating column is rendered manifest. The result has an important bearing on all experiments in which a thin stratum of moist air is the radiating source.

A first comparison is instituted between the transmission of heat through vapours, and through the liquids from which they are derived. From this comparison the inference may be drawn that when a liquid is a powerful absorber, the vapour of that liquid is sure also to be a powerful absorber. The relation here revealed is more fully developed in Memoir VI.

V.

ON THE ABSORPTION AND RADIATION OF HEAT BY GASEOUS AND LIQUID MATTER.*

INTRODUCTION.

THE Royal Society has already done me the honour of publishing in the 'Philosophical Transactions' various memoirs on the relations of radiant heat to the gaseous form of matter. In the first of these † it was shown that for heat emanating from the blackened surface of a cube filled with boiling water, a class of bodies which had been previously regarded as equally, and indeed, as far as laboratory experiments went, *perfectly* diathermic, exhibited vast differences both as regards radiation and absorption. At the common pressure of one atmosphere the absorptive energy of olefiant gas, for example, was found to be 290 times that of air, while when lower pressures were employed the ratio was still greater. The reciprocity of absorption and radiation on the part of gases was also experimentally established in this first investigation.

In the second inquiry ‡ I employed a different and more powerful source of heat, my desire being to bring out with still greater decision the differences which revealed themselves in the first investigation. By carefully purifying the transparent elementary gases, and thus reducing their action upon radiant heat, the difference between them and the more strongly acting compound gases was greatly augmented. In this second inquiry, for example, olefiant gas, at a pressure of one atmosphere, was

* Received and read at the Royal Society June 18, 1863. *Philosophical Transactions*, 1864, p. 201; *Philosophical Magazine*, August 1864.

† *Philosophical Transactions*, February 1861; and *Philosophical Magazine*, September 1861. Memoir I. of this series.

‡ *Philosophical Transactions*, January 1862; and *Philosophical Magazine*, October 1862. Memoir II. of this series.

shown to possess 970 times the absorptive energy of atmospheric air, while it was shown to be probable that, when pressures of $\frac{1}{30}$th of an atmosphere were compared, the absorption of olefiant gas is nearly 6,000 times that of air. A column of ammoniacal gas, moreover, 3 feet long, was found sensibly impervious to the heat employed in the inquiry, while the vapours of many of the volatile liquids were proved to be still more opaque to radiant heat than even the most powerfully acting permanent gases. In this second investigation, the discovery of dynamic radiation and absorption is also announced and illustrated, and the action of odours and of ozone on radiant heat is made the subject of experiment.

The third paper * of the series referred to was devoted to the examination of one particular vapour, which, on account of its universal diffusion, possesses an interest of its own—I mean, of course, the vapour of water. All the objections which had been urged against my results up to the time when the paper was written were here considered. I replied to each of them by definite experiments, removing them one by one; and finally placing, as I believe, beyond the pale of reasonable doubt the action of the aqueous vapour of our atmosphere. In this third paper, moreover, the facts established by experiment were applied to the explanation of various atmospheric phenomena.

§ 1.

Further Experiments on the Power of Gaseous Matter over Radiant Heat—New Apparatus—Absorption by gaseous Strata of different thicknesses.

In the present memoir an attempt is made to bring further into view both the power and the differences of power of gaseous bodies over radiant heat. Hitherto the gases and vapours operated on were introduced in succession into the same experimental tube, the heat being thus permitted to pass through the same thickness of different gases. The earlier part of the present inquiry is devoted to the examination of the transmission of radiant heat through different thicknesses of the same gaseous body. The brass tube with which my former experiments were

* *Philosophical Transactions*, December 1862; and *Philosophical Magazine*, July 1863. Memoir III. of this series.

conducted is composed of several pieces, which are screwed together when the tube is to be used as a whole; but they may be dismounted and used separately in lengths, varying from 2·8 inches to 49·4 inches. I wished, however, to operate upon gaseous strata much thinner than the thinnest of

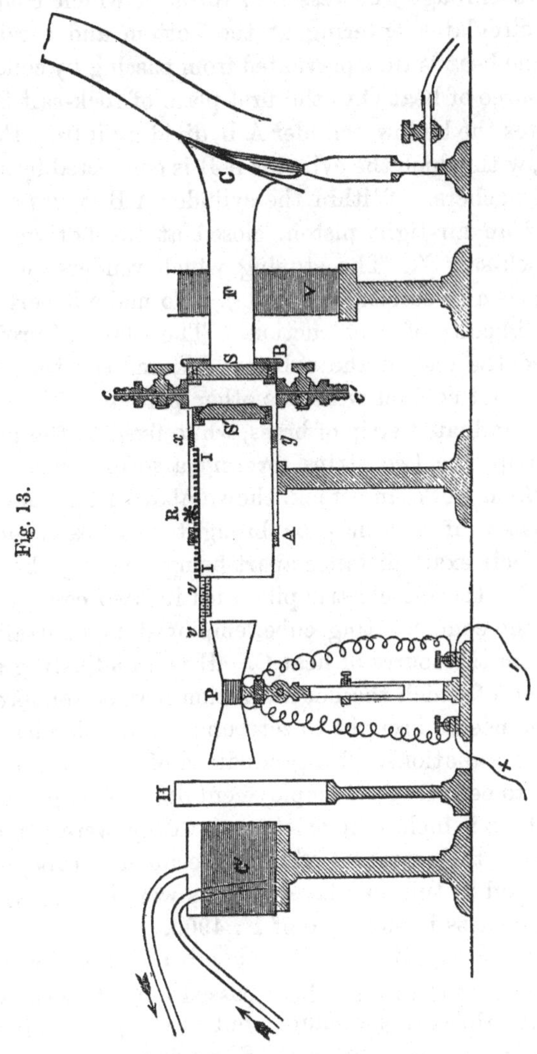

Fig. 13.

these; and for this purpose a special apparatus was devised, and, with much time and trouble, rendered at length practically effective.

The apparatus is sketched in fig. 13. C is the source of heat, consisting of a plate of copper against the back of which a steady sheet of flame is caused to play. The copper plate forms one end of the chamber F (the 'front chamber' of the former memoirs). This chamber, as in previous investigations, passes through the vessel V, through which cold water constantly circulates, entering at the bottom and escaping at the top. The heat is thus prevented from passing by conduction from the source of heat C to the first plate of rock-salt S. The plate S′ closes the hollow cylinder A B, dividing it from the front chamber F, with which the cylinder A B is connected by suitable screws and washers. Within the cylinder A B moves a second one, I I, as an air-tight piston, closed at the bottom by the plate of rock-salt S′. The stuffing which renders the piston air-tight is seen in section at *x* and *y*. To make it perfect was the main difficulty of construction. The plate S′ projects a little beyond the end of the cylinder I I, and can therefore be brought into flat contact with the other plate S. Fixed firmly to A B is a graduated strip of brass, while fixed to the piston is a second strip, the two strips forming a vernier, *v v*. By the pinion R, which works in a rack, shown above I I in the figure, the two plates of salt may be brought near each other or separated, their exact distance apart being given by the vernier *v v*. P is the thermo-electric pile with its two conical reflectors; C′ is the compensating cube, employed to neutralize the radiation from the source of heat C. H is an adjusting screen, by the motion of which the neutralization may be rendered perfect, and the needle brought to zero under the influence of the two opposing radiations. The graduation of the vernier was so arranged as to permit of the employment of layers of gas varying from 0·01 to 2·8 inches in thickness. They were afterwards continued with the segments of the experimental tube, already referred to, and in this way layers of gas were examined which varied in thickness in the ratio of 1 : 4900.

In my former experiments the chamber F was always kept exhausted, so that the rays of heat passed immediately from the source of heat through a vacuum; but in the present instance, fearing the strain upon the plate S, fearing also the possible intrusion of a small quantity of the gas under examination into the front chamber F, if the latter were kept exhausted, and having proved that a length of 8 inches of dry air exerts no sensible action on the rays of heat, I had no scruple in filling the chamber F with dry air. Its absorption was *nil*, and it merely had the effect of lowering in an infinitesimal degree, by convection, the temperature of the source of heat. The two stopcocks c and c' stand exactly opposite to the junction of the two plates of salt, S S′, when they are in contact, and when they are drawn apart these cocks are in communication with the space between the plates.

After many trials to secure the best mode of experiment, the following one was adopted:—The holder containing the gas to be examined was connected by an india-rubber tube with the cock c', the other cock c being at the same time left open. The piston was then moved by the screw R until the requisite distance between the plates was obtained. This space being filled with dry air, the radiations on the two faces of the pile were equalized, and the needle brought to zero. The cock of the gas-holder was now opened, and by gentle water pressure the gas was forced first through a drying apparatus, and then into the space between the plates of salt. The air was quickly displaced, and a layer of the gas substituted for it. When the layer of gas possessed any sensible absorbing power, the equilibrium of the two sources of heat was destroyed; the source C′ triumphed, and from the deflection due to its preponderance the exact proportion of heat intercepted by the gas could be calculated.

When oxygen, hydrogen, or nitrogen was substituted for atmospheric air, no change in the position of the galvanometer-needle occurred; but when any one of the compound gases was allowed to occupy the space between the plates, a measurable deflection ensued. The plates of rock-salt were not so smooth, nor was their parallelism so perfect as entirely to exclude the gas when they were in contact. Hence a stratum of gas sufficient, though but of filmy thickness, to effect a sensible absorption, could find its way between the plates even when they touched each other. On this account the first distance in the following tables was always really a little more than 0·01 of an inch.

TABLE I.—*Carbonic Oxide.*

Thickness of gas	Absorption in hundredths of the total radiation	Thickness of gas	Absorption in hundredths of the total radiation
0·01 of an inch	0·2	0·4 of an inch	3·5
0·02 „	0·5	0·5 „	3·8
0·03 „	0·7	0·6 „	4·0
0·04 „	0·9	1 „	5·1
0·06 „	1·4	1·5 „	6·1
0·1 „	1·6	2 „	6·8
0·3 „	3		

TABLE II.—*Carbonic Acid.*

Thickness of gas	Absorption	Thickness of gas	Absorption
0·01 of an inch	0·86	0·4 of an inch	5·3
0·02 „	1·2	0·5 „	5·7
0·03 „	1·5	0·6 „	5·9
0·04 „	1·9	0·7 „	6
0·05 „	2·1	0·8 „	6·1
0·06 „	2·3	0·9 „	6·2
0·1 „	3·3	1 „	6·3
0·2 „	4·1	1·5 „	7
0·3 „	4·8	2 „	7·6

TABLE III.—*Nitrous Oxide.*

Thickness of gas	Absorption	Thickness of gas	Absorption
0·01 of an inch	1·48	0·4 of an inch	10·20
0·02 „	2·33	0·5 „	11
0·03 „	3·80	0·6 „	11·70
0·04 „	4	0·8 „	12·17
0·05 „	4·20	1 „	12·80
0·1 „	6	1·5 „	14·20
0·2 „	7·77	2 „	15·7

TABLE IV.—*Olefiant Gas.*

Thickness of gas	Absorption	Thickness of gas	Absorption
0·01 of an inch	1·80	0·5 of an inch	23·30
0·02 „	3·08	1 „	26·33
0·05 „	5·37	2 „	32·80
0·1 „	9·14		

§ 2.

Effect of an Atmospheric Shell of Gas or Vapour two inches thick upon the Temperature of a Planet.

We here find that a layer of olefiant gas only two inches in thickness intercepts nearly 33 per cent. of the radiation from our source of heat. Were our globe encircled by a shell of olefiant gas of this thickness, the shell would offer a scarcely sensible obstacle to the passage of the solar rays earthward, but it would intercept, and in great part return, 33 per cent. of the terrestrial radiation. Under such a canopy, trifling as it may appear, the surface of the earth would be kept at a stifling temperature. The possible influence of an atmospheric envelope on the temperature of a planet is here most forcibly illustrated.

The only *vapour* examined with this piston apparatus is that of sulphuric ether. Glass fragments were placed in a U-tube and wetted with the ether. Through the tube dry air was gently forced, whence it passed, vapour-laden, into the space between the rock-salt plates, S S′. The following table contains the results :—

TABLE V.—*Air saturated with the Vapour of Sulphuric Ether.*

Thickness of vapour	Absorption in hundredths of the total radiation	Thickness of vapour	Absorption in hundredths of the total radiation
0·05 of an inch	2·07	0·8 of an inch	21·0
0·1 „	4·6	1·5 „	34·6
0·2 „	8·7	2 „	35·1
0·4 „	14·3		

Comparing these results with those obtained with olefiant gas, we find for thicknesses of 0·05 of an inch and 2 inches respectively the following absorptions :—

Olefiant gas		Sulphuric ether	
Thickness of 0·05 . .	5·37	Thickness of 0·05 . .	2·07
Thickness of 2 inches .	32·80	Thickness of 2 inches .	35·1

Sulphuric-ether vapour, therefore, commences with an absorption much lower than that of olefiant gas, and ends with a higher absorption. This is quite in accordance with the result established in my second memoir,* where it has been shown that while in a short tube the absorption effected by the sparsely scattered molecules of a vapour is far less than that of a gas at the pressure of an atmosphere, in a long tube the gas is excelled by the vapour. Still more impressively than that of olefiant gas, the deportment of sulphuric ether shows what mighty changes of climate might be brought about by the introduction into the

* *Philosophical Transactions*, part i. 1862; and *Philosophical Magazine*, vol. xxiv. p. 343. Memoir II. of this series; § 14.

earth's atmosphere of an almost infinitesimal amount of a powerful vapour. And if *aqueous vapour* can be shown to be thus powerful, the effect of its withdrawal from our atmosphere may be inferred.

§ 3.

New Method of Experiment and its results—Division of Experimental Tube into two chambers—Transmission of Radiant Heat through Gases in one or both.

The inquiry was extended to greater thicknesses of gas, by means of the composite brass experimental tube already referred to. The arrangement adopted, however, was peculiar, being expressly intended to check the experiments, which, under my supervision, were for the most part made by my assistants. The source of heat and the front chamber remained as usual, a plate of rock-salt dividing, as in my previous investigations, the front chamber from the experimental tube. The distant end of the tube was also stopped by a plate of salt; but, instead of permitting it to remain continuous from beginning to end, the experimental tube was divided, by a third plate of rock-salt, into two air-tight compartments. Thus the rays of heat from the source had to pass through *three* distinct chambers, and through three plates of salt. The first chamber was always kept filled with perfectly dry air, while either or both of the other chambers could be filled at pleasure with the gas or vapour to be examined. For the sake of convenience, I will call the compartment of the experimental tube nearest to the front chamber the *first* chamber, and the compartment nearest to the pile the *second* chamber, the term 'front chamber' being, as before, restricted to that nearest to the source of heat. The arrangement is sketched in outline in fig. 14.

The entire length of the tube was 49·4 inches, and this length was maintained throughout the whole of the experiments. The only change consisted in the shifting of the plate of salt, S′, which formed the partition between the first and second chambers. Commencing with a first chamber of 2·8 inches long, and a second chamber 46·4 inches long, the former was gradually augmented, and the latter equally diminished. The actual course of experiment was this:—The first and second chambers being thoroughly cleansed and exhausted, the needle was brought to zero by the equalization of the radiations on

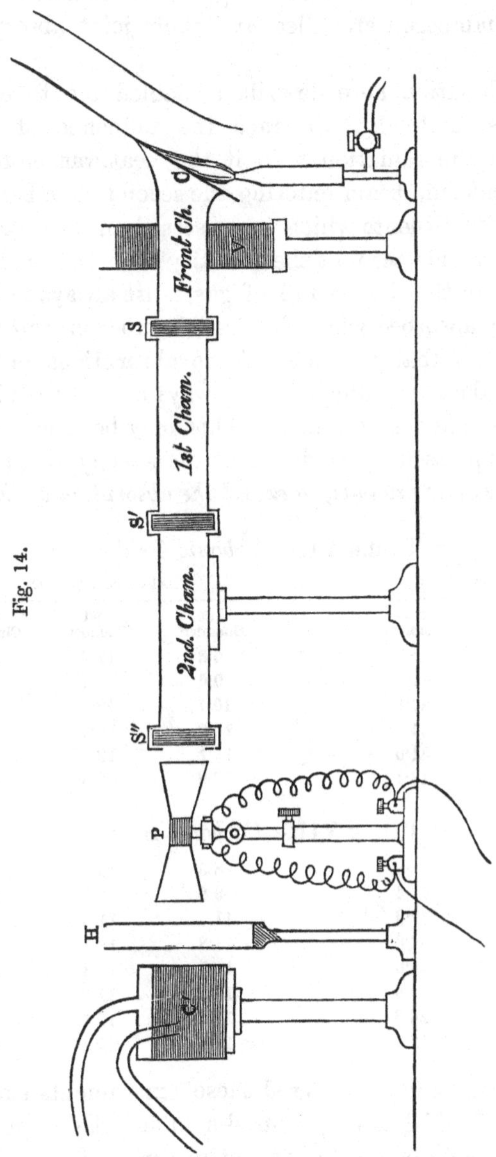

Fig. 14.

the opposite faces of the pile. Into the first chamber the gas or vapour to be examined was then introduced, and its absorption determined. This accomplished, the first chamber was cleansed, and the gas or vapour was introduced into the second chamber, its absorption there being also determined. Finally, both the chambers were filled and their joint absorption was determined.

The combination here described enabled me to check the experiments, and also to trace the influence of the first chamber on the radiation. In it the heat was more or less sifted, the calorific beam entering the second chamber deprived of certain constituents which it possessed on its entrance into the first. On this account the quantity absorbed in the second chamber when the first is full of gas must always fall short of the quantity absorbed when the first chamber is empty. From this it follows that the sum of the absorptions of the two chambers, taken separately, must always exceed the absorption of the tube taken as a whole. This may be briefly and conveniently expressed by saying that *the sum of the absorptions ought, on theoretic grounds, to exceed the absorption of the sum.*

TABLE VI.—*Carbonic Oxide.*

Length		Absorption per 100		
1st Chamber	2nd Chamber	1st Chamber	2nd Chamber	Both Chambers
2·8	46·6	6·8	12·9	12·9
8	41·4	9·6	12·2	12·9
12·2	37·2	10·7	12·2	12·9
15·4	34	10.9	12·2	13·4
17·8	31·6	11·1	12	13·3
36·3	13·1	12·6	10·3	13·4

TABLE VII.—*Carbonic Acid.*

2·8	46·6	8·6	13·8	13·3
8	41·4	9·9	12·7	13·0
12·2	37·2	11	11·4	13
15·4	34·0	11·8	12·1	13·9
23·8	25·6	11·7	11·4	13·1
23·8	25·6	11·2	11·2	12·6
23·8	25·6	10·4	10·5	12
36·3	13·1	11·6	10	12·3

Various causes have rendered these experiments exceedingly laborious. Could I have procured a sufficiently large quantity of gas in a single holder for an entire series of experiments, it would not have been difficult to obtain concurrent results, but the slight variations in quality of the same gas generated at different times tell upon the results and render perfect uniformity extremely difficult of attainment. The approximate constancy of the numbers in the third column is, however, a guarantee that the determinations are not very wide of the truth. Irregularities, however, are revealed. Some remarkable ones occur in the case of carbonic acid, with the chambers 23·8 and 25·6—the absorptions in the first chamber varying in this instance from 11·7 to 10·4, and in the second chamber from 11·4 to 10·5, and in both chambers from 13·1 to 12·0. The gas which gave the largest of these results was generated from marble and hydrochloric acid; the next was obtained from chalk and sulphuric acid, and the gas which gave the smallest result was obtained from bicarbonate of soda and sulphuric acid. The slight differences accompanying these different modes of generation made themselves felt in the manner recorded in the table.

TABLE VIII.—*Nitrous Oxide.*

Length		Absorption per 100		
1st Chamber	2nd Chamber	1st Chamber	2nd Chamber	Both Chambers
2·8	46·6	16·1	32·9	33·9
12·2	37·2	23·1	30	32
15·4	34	23·6	29·6	32
17·8	31·6	26·2	29·6	32·7

The differences arising from different modes of generation are most strikingly illustrated by the powerful gases. Dr. Frankland, for example, was kind enough to superintend for me the making of a large holder of olefiant gas by the so-called 'continuous process,' in which the *vapour* of alcohol is led through sulphuric acid diluted with its own volume of water. The following results were obtained:—

TABLE IX.—*Olefiant Gas.*

Length		Absorption per 100		
1st Chamber	2nd Chamber	1st Chamber	2nd Chamber	Both Chambers
2·8	46·6	34·6	66·1	67·7
8	41·4	44·2	65·3	67·5
15·4	34	53·6	62·3	67

The agreement of the absorption of both chambers, the sum of which was the constant quantity of 49·4 inches, must be regarded as satisfactory; and this is the general character of the results as long as we adhere to gas generated in the same way. On the other hand, olefiant gas produced by mixing the *liquid* alcohol with sulphuric acid and applying heat to the mixture, gave the results recorded in the following table:—

TABLE X.—*Olefiant Gas.*

Length		Absorption per 100		
1st Chamber	2nd Chamber	1st Chamber	2nd Chamber	Both Chambers
12·2	37·2	54·8	70	76·3
15·4	34	59·1	72·7	77·1
19·8	29·6	67·8	70·4	77
23·8	25·6	69·2	70·2	77·6
36·3	13·1	72·8	60·3	78·8

Here the joint absorption of the two chambers is about 10 per cent higher than that of the gas generated under Dr. Frankland's superintendence.

§ 4.

Influence of 'Sifting' by Gaseous Media.

A few remarks on these results may be introduced here. In the case of carbonic oxide (Table VI.), we see that while a length of 2·8 inches of gas is competent, when acting alone, to intercept 6·8 per cent. of the radiant heat, the cutting off of this length from a tube 49·4 inches long, or, what is the same, the addition of this length to a tube 46·6 inches long, makes no sensible change in its absorption. The second chamber absorbs as much as both. The same remark applies to carbonic acid, and it is also true within the limits of error for nitrous oxide and olefiant gas. Indeed it is only when 8 inches or more of the column have been cut away that the difference begins to make itself felt. Thus, in carbonic oxide, the absorption of a length of 41·4 being 12·2, that of a chamber 49·4, or 8 inches longer, is only 12·9, making a difference of only 0·7 per cent., *while the same thickness of* 8 *inches acting singly on the gas produces an absorption of* 9·6 *per cent.* So also with regard to carbonic acid; a tube 41·4 absorbing 12·7 per cent., a tube 49·4 absorbs only 13·0 per cent., making a difference of only 0·3 per cent. In the case of olefiant gas also (Table IX.), while a distance of 8 inches absorbs, acting singly, 44 per cent., the addition of 8 inches to a tube already 41·4 inches long raises the absorption only from 65·3 to 67·5, or 2·2 per cent. The reason is plain. In a length of 41·4 the rays capable of being absorbed by the gas are so much diminished, so few in fact remain to be attacked, that an additional 8 inches of gas produce a scarcely sensible effect. Similar considerations explain the fact that, while by augmenting the length of the first chamber from 2·8 inches to 15·4 inches we increase the absorption of olefiant gas nearly 20 per cent., the shortening of the second chamber by precisely the same amount effects a diminution of barely 4 per cent. of the absorption. *All these results conspire to prove the heterogeneous character of the radiation from a source heated to about 250° C.*

The 'sum of the absorptions' and the 'absorption of the sum,' placed side by side, exhibit the influence of sifting in an instructive manner. Tables VI., VII., VIII., IX., and X., thus treated, give the following comparative numbers:—

TABLE XI.—*Carbonic Oxide.*

Length of Chambers		Sum of Absorptions	Absorption of Sum
2·8	46·6	19·7	12·9
8	41·1	21·8	12·9
12·2	37·2	22·9	12·9
15·4	34	23·1	13·4
17·8	31·6	23·1	13·3
36·3	13·1	22·9	13·4
	Means . .	22·3	13·1

TABLE XII.—*Carbonic Acid.*

2·8	46·6	22·4	13·3
8	41·4	22·6	13
12·2	37·2	22·4	13
15·4	34	23·9	13·9
23·8	25·6	23·1	13·1
36·3	13·1	21·6	12·3
	Means . .	22·6	13·1

TABLE XIII.—*Nitrous Oxide.*

2·8	46·6	49	33·9
12·2	37·2	53·1	32
15·4	34	53·2	32
17·8	31·6	55·8	32·7
	Means . .	52·8	32·7

TABLE XIV.—*Olefiant Gas.*

Length of Chambers		Sum of Absorptions	Absorption of Sum
2·8	46·6	100·7	67·7
8	41·4	109·5	67·5
12·2	37·2	109·4	65
15·4	34	115·9	67
	Means .	. 108·9	66·8

TABLE XV.—*Olefiant Gas.*

12·2	37·2	124·8	76·3
15·4	34	131·8	77·1
19·8	29·6	138·2	77
23·8	25·6	139·4	77·6
36·3	13·1	133·1	78·8
	Means .	. 133·4	77·3

The conclusion that the sum of the absorptions is greater than the absorption of the sum is here amply verified. The tables also show the ratio of the sum of the absorptions to the absorption of the sum to be practically constant for all these gases. Dividing the first mean by the second in the respective cases, we have the following quotients :—

Carbonic oxide	1·70
Carbonic acid	1·72
Nitrous oxide	1·61
Olefiant gas (mean of both) . . .	1·68

The sum of the absorptions ought to be a maximum when the two chambers are of equal length. For, let them be unequal, one of them being in excess of half the length of the tube, and let us consider the action of this *excess.* Placed after the half-length, it receives the rays which have already traversed that half; placed after the shorter length, it receives the rays which have traversed the shorter length. In the former case, therefore, the 'excess' will absorb less than in the latter, because the rays in the former case have been more thoroughly sifted before the heat reaches the excess. From this it is clear that, more is gained in the way of absorption by attaching the excess to the short length of the tube than to the half-length; in other words, the sum of the absorptions, when the tube is divided into two equal parts, is a maximum. This reason is approximately verified by the experiments. As one length augments, and the other diminishes, we constantly approach the limit when the sum of the absorptions and the absorption of the sum are equal to each other. The effect of proximity to this limit is exhibited in the first experiment in each of the series; where the lengths of the compartments are very unequal, and the sum of the absorptions is, in general, a minimum.

§ 5.

Application of Method to Vapours.

After the absorption by the permanent gases had been in this way examined, I passed on to the examination of vapours. They were all used at a common pressure of 0·5 of an inch of mercury, or about $\frac{1}{60}$th of an atmosphere. The liquid which yielded the vapour was enclosed in the flasks described in my previous memoirs, and the pure vapour was allowed to enter the respective compartments of the experimental tube without the slightest ebullition. The following series of tables contains the results thus obtained:—

TABLE XVI.—*Bisulphide of Carbon. Pressure* 0·5 *of an inch.*

Length		Absorption per 100		
1st Chamber	2nd Chamber	1st Chamber	2nd Chamber	Both Chambers
2·8	46·6	3·6	7·6	7·6
8	41·4	4·4	7·3	7·6
15·4	34	5·7	6	7·5
17·8	31·6	5·8	6·4	7·5
23·8	25·6	6·7	6	7·8

TABLE XVII.—*Chloroform. Pressure* 0·5 *of an inch.*

1st Chamber	2nd Chamber	1st Chamber	2nd Chamber	Both Chambers
2·8	46·6	5·5	15·9	16·3
8	41·4	9·2	15·6	16·8
12·2	37·2	10·5	14·8	17·1
15·4	34	11·6	14·1	16·9
23·8	25·6	15	14	18·4
36·3	13·1	15·6	10·9	17·2

TABLE XVIII.—*Benzol. Pressure* 0·5 *of an inch.*

1st Chamber	2nd Chamber	1st Chamber	2nd Chamber	Both Chambers
2·8	46·6	4	20	20·6
8	41·4	8·4	17·3	20·4
12·2	37·2	9·8	16·5	19
17·8	31·6	11·9	15·7	20·1
23·8	25·6	14·3	15·1	21·0

TABLE XIX.—*Iodide of Ethyl. Pressure* 0·5 *of an inch.*

Length		Absorption per 100		
1st Chamber	2nd Chamber	1st Chamber	2nd Chamber	Both Chambers
2·8	46·6	7·1	23·5	25·4
8	41·4	9·1	21·1	23·3
12·2	37·2	12·8	20·5	25·2
15·4	34	14·6	20·8	25·2
17·8	31·6	15·8	20	25·5

TABLE XX.—*Alcohol. Pressure* 0·5 *of an inch.*

2·8	46·6	11·7	46·1	46·1
8	41·4	18·5	43·6	47
12·2	37·2	26	44·1	47·5
15·4	34	32·1	41·1	47
17·8	31·6	32·4	40	47·6

TABLE XXI.—*Alcohol. Pressure* 0·1 *of an inch.*

8	41·4	8	22·2	24·9
15·4	34	12·1	20	24·7
17·8	31·6	13·1	19·7	25·7
23·8	25·6	14·8	18·4	25·2
36·3	13·1	19·1	13·8	25·1

TABLE XXII.—*Sulphuric Ether. Pressure* 0·5 *of an inch.*

2·8	46·6	14·8	50	51·6
8	41·4	23·9	51	53·9
12·2	37·2	30·9	48·8	53·6
15·4	34	34	47·8	53·1

TABLE XXIII.—*Acetic Ether. Pressure* 0·5 *of an inch.*

2·8	46·6	17	60·2	62·9
8	41·4	30·7	58·1	64·6
12·2	37·2	41·6	55·1	64·2
15·4	34	44·4	55·5	62·4
23·8	25·6	50·9	52·7	64·7
36·3	13·1	58·1	42·6	64·8

TABLE XXIV.—*Formic Ether. Pressure* 0·5 *of an inch.*

2·8	46·6	17·4	63	64·4
8	41·4	33·3	59·1	63·4
17·8	31·6	40	48·4	60·3
23·8	25·6	45·6	47·2	60·2

In the following tables the sum of the absorptions is compared with the absorption of the sum in the case of vapours.

TABLE XXV.—*Bisulphide of Carbon.* *Pressure* 0·5 *of an inch.*

Length of Chambers		Sum of Absorptions	Absorption of Sum
2·8	46·6	11·2	7·6
8	41·4	11·7	7·6
15·4	34	11·7	7·5
17·8	31·6	12·2	7·5
23·8	25·6	12·7	7·8
	Means . .	11·9	7·6

TABLE XXVI.—*Chloroform.* *Pressure* 0·5 *of an inch.*

2·8	46·6	21·4	16·3
8	41·4	24·8	16·8
12·2	37·2	25·3	17·1
15·4	34	25·2	16·9
23·8	25·6	29	18·4
36·3	13·1	26·5	17·2
	Means . .	25·36	17·1

TABLE XXVII.—*Benzol.* *Pressure* 0·5 *of an inch.*

2·8	46·6	24	20·6
8	41·4	25·7	20·4
12·2	37·2	26·3	19
17·8	31·6	27·6	20·1
23·8	25·6	29·4	21
	Means . .	26·6	20·2

TABLE XXVIII.—*Iodide of Ethyl.* *Pressure* 0·5 *of an inch.*

2·8	46·6	30·6	25·4
8	41·4	30·2	23·3
12·2	37·2	33·3	25·2
15·4	34	35·4	25·2
17·8	31·6	35·8	25·2
	Means . .	33·1	24·9

TABLE XXIX.—*Alcohol.* *Pressure* 0·5 *of an inch.*

2·8	46·6	57·8	46·1
8	41·4	62·1	47
12·2	37·2	70·1	47·5
15·4	34	73·2	47
17·8	31·6	72·4	47·6
	Means . .	67·1	47

Table XXX.—*Alcohol. Pressure* 0·1 *of an inch.*

Length of Chambers		Sum of Absorptions	Absorption of Sum
8	41·4	30·2	24·9
15·4	34	32·1	24·7
17·8	31·6	32·8	25·7
23·8	25·6	33·2	25·2
36·3	13·1	32·9	25·1
	Means . .	32·2	25·1

Table XXXI.—*Sulphuric Ether. Pressure* 0·5 *of an inch.*

Length of Chambers		Sum of Absorptions	Absorption of Sum
2·8	46·6	64·8	51·6
8	41·4	74·9	53·9
12·2	37·2	79·7	53·6
15·4	34	81·8	53·1
	Means . .	75·3	53·05

Table XXXII.—*Formic Ether. Pressure* 0·5 *of an inch.*

Length of Chambers		Sum of Absorptions	Absorption of Sum
2·8	46·6	80·4	64·4
8	41·4	82·4	63·4
17·8	31·6	88·4	60·3
23·8	25·6	92·8	60·2
	Means . .	86	62·07

Table XXXIII.—*Acetic Ether. Pressure* 0·5 *of an inch.*

Length of Chambers		Sum of Absorptions	Absorption of Sum
2·8	46·6	77·2	62·9
8	41·4	88·8	64·6
12·2	37·2	96·7	64·2
15·4	34	99·9	62·4
23·8	25·6	103·6	64·7
36·3	13·1	100·7	64·8
	Means . .	94·5	63·9

In the case of vapours, the difference between the sum of the absorptions and the absorption of the sum is, in general, less than in the case of gases. This resolves itself into the proposition that for equal lengths, within the limits of these experiments, the sifting power of the gas is greater than that of the vapour. The reason of this is that the vapours are examined at a pressure of an atmosphere and the vapours at a pressure of $\frac{1}{60}$th of an atmosphere. Thus, as before proved,* no matter

* Section 14, Memoir II.

how powerful the individual molecules may be, their distance asunder renders a thin layer of them a comparatively open sieve.

§ 6.

New Experiments on Dynamic Radiation.—Radiation of Dynamically heated Gas through the same Gas, or through other Gases.

The entrance of a gas into an exhausted vessel is known to be accompanied by the generation of heat; and the gas thus warmed, if a radiator, will emit the heat generated. Conversely, on exhausting a vessel containing any gas, the gas is chilled, and thus an external body, which prior to the act of exhaustion possesses the same temperature as the gas, becomes, on the first stroke of the pump, a warm body with reference to the gas remaining in the vessel. If the body be separated from the cooled gas by a diathermic partition, it will radiate into the gas and be more or less chilled. It was shown in my second memoir that this dynamic warming and chilling of a gas or vapour furnished a practical means of determining, without any source of heat external to the gaseous body itself, both its radiative and absorptive energy, the terms dynamic radiation and dynamic absorption being then for the first time introduced to express this newly-discovered action.

During the last half-year a considerable number of experiments have been made in illustration of the manner in which dynamic radiation may be applied in researches on radiant heat. A few of these I will here describe. The source of heat was abolished; one end, S, (fig. 15) of the experimental tube was stopped by a plate of polished metal, the other, S″, by a transparent plate of rock-salt, while the space between the ends was divided into two compartments by a second plate of salt, S′. The thermo-electric pile, P, occupied its usual position at the end of the tube, the compensating cube, however, being abandoned. For the sake of convenient reference, I will call the compartment of the tube most distant from the pile the *first* chamber, and that adjacent to the pile the *second* chamber.

Both chambers being exhausted and the needle at zero, the gas was allowed to enter the first through a gauge-cock which made its time of entry 40 seconds; the second chamber at the same time being preserved a vacuum. The gas on entering was dynamically heated, and radiated its heat to the pile through the vacuous second chamber. The needle moved, and the limit of its excursion was noted. The first chamber was then exhausted and carefully cleansed with dry air. The second chamber was filled with the same gas, not yet with a view to its dynamic radiation, but to examine its effect upon the heat radiated from the first chamber. The needle being at zero, the gas was again permitted to enter the first chamber

Fig. 15.

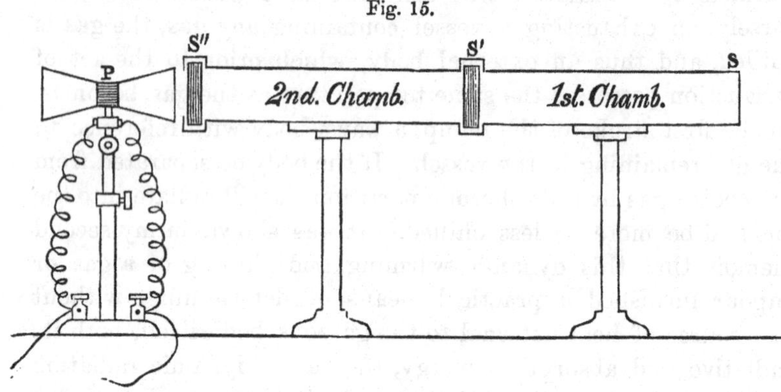

exactly as in the other experiment, the only difference being, that in the first case the heat passed through a vacuum to the pile, while in the second it had to pass through a column of the same gas as that from which it emanated. The reciprocity of radiation and absorption could thus be illustrated in a novel and interesting manner. In this way, in fact, the absorption exerted by any gas not only on heat radiated from the same gas, but from any other gas, may be determined. Finally, the apparatus being cleansed and the needle at zero, the gas was permitted to enter the second chamber, and its dynamic radiation from this chamber determined. The intermediate plate of salt, S′, was shifted, as in the former experiments, so as to alter the lengths of the two chambers, the sum of both lengths remaining constant as before.

In the following tables the three columns bracketed under the head of 'Deflection,' contain the arcs through which the needle moved in the three cases: (1) when the radiation from the gas in the first chamber passed through the empty second chamber; (2) when the radiation from the first chamber passed through the occupied second chamber; and (3) when the radiation proceeded from the second chamber.

Dynamic Radiation of Gases.

TABLE XXXIV.—*Carbonic Oxide.*

Length		Deflection		
1st Chamber	2nd Chamber	By 1st Chamber, 2nd Chamber empty	By 1st Chamber, Gas in 2nd Chamber	By 2nd Chamber
		°	°	°
2·8	46·6	1	0	28
15·4	34	3 8	2·1	24·4
36·3	13·1	13·7	6·3	16·6

TABLE XXXV.—*Carbonic Acid.*

2·8	46·6	1	0	33·6
15·4	34	3·7	1·25	23·3
36·3	13·1	16·8	6·6	17·5

TABLE XXXVI.—*Nitrous Oxide.*

2·8	46·6	1	0·2	44·5
15·4	34	4 3	1·2	31·7
36·3	13·1	19·5	6·2	22

TABLE XXXVII.—*Olefiant Gas.*

15·4	34	11·9	1	68
23·8	25·6	22 8	3	
36·3	13·1	59	10·4	65

The gases, it will be observed, exhibit a gradually increasing power of dynamic radiation from carbonic oxide up to olefiant gas. This is most clearly illustrated by reference to the results obtained in the respective cases with the first length of the second chamber. They are as follows:—

	°
Carbonic oxide	28
Carbonic acid	33·6
Nitrous oxide	44·5
Olefiant gas	68

Its proximity to the pile, and the fact of its having to cross but one plate of salt, make the radiation from the second chamber much greater than that of the first.

All the tables show that as the length of the chamber increases, the dynamic radiation of the gas contained in it increases; and as the length diminishes, the radiation diminishes. They also show how powerfully the gas in the second chamber acts upon the radiation from the first. With carbonic oxide, the introduction of a column of gas 13·1 inches long reduces the deflection from 13·7° to 6·3°, or more than 50 per cent. In the other cases the reduction is still greater. With carbonic acid it is reduced from 16·8 to 6·6; with nitrous oxide it is reduced from 19·5 to 6·2. Nor is this residual deflection, 6·2°, entirely due to the transparency of the gas, to heat emitted *by the gas*. No matter how well polished the experimental tube may be, there is always a certain radiation from its interior surface when the gas enters it. With perfectly dry air this radiation amounts to 8 or 9 degrees. Thus the radiation is composite, in part emanating from the molecules in the first chamber, and in part emanating from the surface of the tube. To these latter, the gas in the second chamber would be much more permeable than to the former; and to these latter, I believe, the residual deflection of 6 degrees, or thereabouts, is mainly due. That this number turns up so often, although the radiations from the various gases differ so considerably, is in harmony with the supposition just made. In the case of carbonic oxide, for example, the deflection is reduced from 13·7° to 6·3°, while in the case of nitrous oxide it is reduced from 19·5° to 6·2°; in the case of olefiant gas it is reduced from 59° to 10·4°, while in other experiments (not here recorded) the deflection by olefiant gas was reduced from 44° to 6°.

§ 7.

Influence of Tarnish, or of a Lining on the Interior Surface of Experimental Tube.—Dynamic Radiation from the Surface.

As may be expected, this radiation from the interior surface augments with the tarnish of the surface, but the extent to which it may be increased is hardly sufficiently known. *Indeed the gravest errors are possible in experiments of this nature if the influence of the interior be overlooked or misunderstood.* An experiment or two will illustrate this more forcibly than any words of mine.

A brass tube 3 feet long, and very slightly tarnished within, was used for dynamic radiation. Dry air on entering the tube produced a deflection of 12 degrees. The tube was then polished within, and the experiment repeated; the action of dry air was instantly reduced to 7·5 degrees.

The rock-salt plate at the end of the tube was then removed, and a lining of black paper 2 feet long was introduced. The tube was again closed, and the experiment of allowing dry air to enter it repeated. The deflections observed in three successive experiments were

80°, 81°, 80°.

This result might be obtained as long as the lining continued within the tube.

The plate of rock-salt was again removed, and the length of the lining was reduced to a foot; the dynamic radiation on the entrance of dry air in three successive experiments gave the deflections

76°, 74°, 75°.

The plate was again removed and the lining reduced to 3 inches; the deflections obtained in two successive experiments were

66°, 65°.

Finally, the lining was reduced to a ring only 1½ inch in width; the dynamic radiation from this small surface gave in two successive trials the deflections

56°, 56·5°.

The lining was then entirely removed, and the deflection instantly fell to

7·5°.

A coating of lampblack within the tube produced the same effect as the paper lining; common writing-paper was almost equally effective; a coating of varnish also produced large deflections, and the mere oxidation of the interior surface of the tube is also very effective.

In the above experiments the lining was first heated, and it then radiated its heat through a thick plate of rock-salt against the pile. The effect was enfeebled by distance, by reflexion from the surfaces of the salt, and by partial absorption. Still we see that the radiation thus weakened was competent to drive the needle almost through the quadrant of a circle. If, instead of being thus detached, *the face of the pile itself* had formed part of the interior surface of the experimental tube, the deflections would, of course, be far greater than the highest of those here recorded. I do not entertain a doubt that, by the dynamic heating of the surface of the pile, the needle of my galvanometer could be caused to whirl through an arc of 1000 degrees. Assuredly an arrangement subject to disturbances, or masking disturbances, such as these cannot be suitable in experiments in which the most refined delicacy is absolutely necessary.

§ 8.

Radiation of dynamically heated Vapour through the same Vapour and through a Vacuum.—Influence of Length of Radiating Column.—Different Effects of Length on Gases and Vapours.

Experiments similar to those recorded in § 6 were also made with vapours. Both chambers of the experimental tube being exhausted, the vapour was permitted to enter the first, and dry air to follow it. The air thus dynamically heated warmed the vapour, and the vapour radiated its heat against the pile. As in the case of gases, the heat passed in the first experiment through a vacuous second chamber, in the second experiment through the same chamber when it contained 0·5 of an inch of the same vapour as that from which the rays issued, while a third experiment was made to determine the dynamic radiation from the second chamber. The following tables contain the results:—

Dynamic Radiation of Vapours.

TABLE XXXVIII.—*Bisulphide of Carbon. Pressure* 0·5 *of an inch.*

Length		Deflection		
1st Chamber	2nd Chamber	By 1st Chamber, 2nd Chamber empty	By 1st Chamber, Vapour in 2nd Chamber	By 2nd Chamber
15·4	34	2·4°	1·6°	14·2°
36·3	13·1	9·75	5 5	9

TABLE XXXIX.—*Benzol.* *Pressure* 0·5 *of an inch.*

Length		Deflection		
1st Chamber	2nd Chamber	By 1st Chamber, 2nd Chamber empty	By 1st Chamber, Vapour in 2nd Chamber	By 2nd Chamber
15·4	34	3°	1·1°	34°
36·3	13·1	21·6	11·9	15·1

TABLE XL.—*Iodide of Ethyl.* *Pressure* 0·5 *of an inch.*

15·4	34	3·4	2·7	38·8
36·3	13·1	25·4	13·8	19

TABLE XLI.—*Chloroform.* *Pressure* 0·5 *of an inch.*

15·4	34	4·5	2·1	41
36·3	13·1	22·3	10	19

TABLE XLII.—*Alcohol.* *Pressure* 0·5 *of an inch.*

15·4	34	4·9	2	53·8
36·3	13·1	33·8	16·9	34·9

TABLE XLIII.—*Alcohol.* *Pressure* 0·1 *of an inch.*

15·4	34	2	1·3	35·7
36·3	13·1	21·8	16·2	11·5

TABLE XLIV.—*Boracic Ether.* *Pressure* 0·1 *of an inch.*

15·4	34	6·3	2·1	61
36·3	13·1	29·1	15·7	31·6

TABLE XLV.—*Formic Ether.* *Pressure* 0·5 *of an inch.*

15·4	34	6·3	2·5	68
36·3	13·1	46	23·8	41

TABLE XLVI.—*Sulphuric Ether.* *Pressure* 0·5 *of an inch.*

15·4	34	5·6	2·5	68
36·3	13·1	45·3	22·4	36·5

TABLE XLVII.—*Acetic Ether.* *Pressure* 0·5 *of an inch.*

15·4	34	5·7	1	73·9
36·3	13·1	49·1	22	41

Collecting the radiations from the second chamber for the lengths 34 inches and 13·1 inches into a single table, we see at a glance how the radiation is affected by the length.

TABLE XLVIII.

	Dynamic Radiation of various Vapours at 0·5-inch pressure and a common thickness of	
	34 inches	13·1 inches
Bisulphide of carbon	14°·2	9°
Benzol	34	15·1
Iodide of ethyl	38·8	19
Chloroform	41	19
Alcohol	53·8	34·9
Sulphuric ether	68	36·5
Formic ether	68	41
Acetic ether	73·9	41
	At a pressure of 0·1′ of an inch	
Alcohol	35·7	11·5
Boracic ether	61	31·6

The extraordinary energy of boracic ether as a radiant may be inferred from the last experiment. Although attenuated to $\frac{1}{300}$th of an atmosphere, its thinly scattered molecules are able to urge the needle through an arc of 61 degrees, and this merely by the warmth generated on the entrance of dry air into a vacuum.

Arranging the gases in the same manner, we have the following results:—

TABLE XLIX.

	Dynamic Radiation of Gases at 1 atm. pressure, and a common thickness of	
	34 inches	13·1 inches
Carbonic oxide	24°·4	16°·6
Carbonic acid	23·3	17·5
Nitrous oxide	31·7	22
Olefiant gas	68	65

As remarked in Memoir II. § 14, a greater length is available for radiation in the case of the vapour than in the case of the gas, because the radiation from the hinder portion of the column of vapour is less interfered with by the molecules in front. By shortening the column therefore we diminish, and by lengthening it we promote, the radiation from the vapour more than that from the gas. Thus while a shortening of the column from 34 inches to 13·1 causes a fall in the case of carbonic oxide only from 23·3° to 17·5°, the same amount of shortening causes benzol vapour to fall from 34° to 15·1°, a much greater diminution. So also as regards olefiant gas, a shortening of the radiating column from 34 inches to 13·1 inches causes a fall in the deflection only from 68° to 65°; the same diminution produces with sulphuric ether a fall from 68° to 36·5°; and with acetic ether from 73·9° to 41°. In the 34-inch long column, moreover, acetic ether beats olefiant gas, but in the 13-inch column the gas beats the vapour.

A series of experiments of this nature executed last autumn, though not free from irregularities, is nevertheless worth recording here. A brass experimental tube, slightly tarnished within, 49·4 inches long, and divided into two chambers, each 24·7 inches long, was employed, and with the following results:—

TABLE L.—*Dynamic Radiation of Vapours.*

	Deflection		
	By 1st Chamber, 2nd Chamber empty	By 1st Chamber, Vapour in 2nd Chamber	By 2nd Chamber
	°	°	°
Bisulphide of carbon	8·2	5·8	21·2
Benzol	20	12·4	45·9
Chloroform	24·3	10·9	55·2
Iodide of ethyl	27·5	14·7	55·3
Alcohol	42·7	22·3	69
Sulphuric ether	46·3	21·7	80·5
Formic ether	47·5	19·8	79·5
Propionate of ethyl	49·8	25	82·3
Acetic ether	53·3	30	82·1

§ 9.

First Comparison of the Actions of Liquids and their Vapours upon Radiant Heat.

To ascertain whether the action of these vapours bears any significant relation to that of the liquids from which they are derived, the transmission of radiant heat through those liquids was examined. The naked flame of an oil lamp was used as a source, and the liquids were enclosed in rock-salt cells. Thus the total radiation from the lamp, with the exception of the minute fraction absorbed by the rock-salt, was brought to bear upon the liquid.

In the following table the liquids are arranged in the order of their powers of transmission :—

TABLE LI.

Name of Liquid	Transmission in Hundredths of the Radiation
Bisulphide of carbon	83
Bisulphide of carbon saturated with sulphur	82
Bisulphide of carbon saturated with iodine	81
Bromine	77
Chloroform	73
Iodide of methyl	69
Benzol	60
Iodide of ethyl	57
Amylene	50
Sulphuric ether	41
Acetic ether	34
Formic ether	33
Alcohol	30
Water saturated with rock-salt	26

These results are but approximate, and the source of heat has been changed, still it is impossible to regard them without feeling how purely the act of absorption is a *molecular* act, and *that when a liquid is a powerful absorber the vapour of that liquid is sure also to be a powerful absorber.*

To experiment with water, it was necessary to saturate it with the salt of which the cell was formed, but the absorptive energy is due solely to the water. We might infer from this alone, were no experiments made on the aqueous vapour of the atmosphere, that that vapour must exert a powerful action upon terrestrial radiation. In fact, in all the statements that I have hitherto made its action has been underrated.

The deportment of the elements sulphur and iodine, dissolved in bisulphide of carbon, is in striking harmony with the other results which these researches have made known regarding the action of elementary bodies. The saturation of the bisulphide by sulphur scarcely affects the transmission, while a quantity of iodine sufficient to convert the liquid from one of perfect transparency to one of almost perfect opacity to light, produces a diminution of only 2 per cent. of the radiation. This shows that the heat really used in these experiments consists almost wholly of the obscure rays of the lamp. The deportment of bromine is also very instructive. The liquid is very dense, and so opaque as to cut off the luminous rays of the lamp; still it transmits 77 per cent. of the total radiation. It stands in point of diathermancy above every compound liquid in the list, except bisulphide of carbon. This latter substance is the rock-salt of liquids.

It is worth remarking *that the obscure rays of a luminous source have a much greater power of penetration in the case of the liquids here examined than the rays from an obscure source, however close to incandescence.*

Before a strict comparison can be made between vapours and liquids, they must both be examined by heat *of the same quality*, and arrangements have been already made with which I hope to obtain more complete and accurate results than those above recorded.

VI.

CONTRIBUTIONS TO MOLECULAR PHYSICS.

ANALYSIS OF MEMOIR VI.

The comparison of vapours and liquids, roughly illustrated towards the conclusion of the last inquiry, prompted a more accurate examination of this important question. The present memoir begins with the description of the necessary apparatus. A source of heat of perfectly definite quality is obtained by sending through a platinum spiral a voltaic current of a constant strength. Rock-salt cells are also devised wherein the liquids are enclosed, and the action upon radiant heat of eleven different liquids in five different thicknesses is determined.

The action of the vapours of those liquids upon radiant heat of the same quality is then determined.

In each series of experiments the different liquids are first compared at a common thickness, and the vapours at a common pressure. A striking general agreement was established, the order of absorption in the two cases being *almost* the same.

Fresh experiments are then executed in which the liquids and their vapours are rendered proportional to each other in quantity. Thus compared, the order of absorption in the two cases is proved to be identical.

It is hence inferred that the position of any vapour as an absorber of radiant heat is determined by the position of its liquid. *No experiment has ever been made to shake the validity of this inference.* From the deportment of water, therefore, the deportment of its vapour may be inferred, and the influence of this agent in the phenomena of meteorology anticipated and understood.

The special bearing of chemical constitution on absorption is made the subject of a brief section.

In the spring of 1862 coloured elementary liquids, embracing bromine and dissolved iodine, were examined and found exceedingly pervious to the obscure heat-rays. In the present investigation, special experiments are made with a view of accurately determining the diathermancy of dissolved iodine. It is shown to be *practically perfect* for obscure rays.

The influence of the temperature of the source upon the penetrative power of heat-rays has been the subject of frequent discussion among philosophers. The arrangement adopted in this inquiry, which permits of our varying the temperature at will, *while retaining throughout the same vibrating atoms*, enables us to put this question in a clearer light than usual. The platinum spiral is employed at a barely visible heat, at a bright red heat, at a white heat, and also at a heat close to fusion; the variations of transmissive power consequent upon these changes of temperature being recorded.

Experiments are described which show some very singular shiftings of the diathermic position of vapours consequent on varying the source of heat. Starting with a low heat, formic ether proves a more powerful absorbent than sulphuric ether; but as the temperature augments they become equal, and at a white heat their positions are reversed. When the source is a blackened cube of boiling water, formic ether also decidedly excels sulphuric in absorbent power.

Chloroform, at all temperatures of the platinum source, shows itself a feebler absorber than iodide of methyl; but when the source is a cube coated with lampblack it becomes the more powerful absorbent. The differences in the quality of the emission from different solid bodies of the same temperature are thus strikingly illustrated, various peculiarities of molecular vibration being at the same time brought into view.

The source is next changed to flames of different kinds; luminous gas-jets, the pale blue flame of Bunsen's burner, hydrogen flames, and carbonic-oxide flames being successively invoked. The radiation from the incandescent carbon particles of the gas-flame is contrasted with that from white-hot platinum. Other inversions of diathermic position are made manifest. When a luminous gas-jet is the source, bisulphide of carbon is decidedly more diathermic than chloroform; when a Bunsen's flame is the source, or even when the gas-flame is much reduced in size and brilliancy, chloroform is decidedly more diathermic than bisulphide of carbon. The removal of the white-hot carbon particles from the gas-flame more than doubles the relative diathermancy of the chloroform. When, moreover, a carbonic-oxide flame is the source, sulphuric ether excels formic ether in absorption; but when the source is a hydrogen flame the formic ether excels the sulphuric.

In every case here mentioned the deportment of the vapour was compared with that of its liquid, and it was found that every change in the position of the one was accompanied by a corresponding change in the position of the other.

The opacity of a gas or vapour to radiation from its own molecules has been frequently adverted to in these memoirs. To determine whether the radiation from a hydrogen flame, where we have aqueous vapour in a state of incandescence, is intercepted with peculiar energy by the cold aqueous vapour of the atmosphere, special experiments were instituted. The result justified the trial; for on a day when less than 6 per cent. of the heat from a red-hot platinum wire was absorbed by moist air, as nearly as possible three times this quantity was absorbed of the radiation from a hydrogen flame. *The mere plunging a platinum spiral into the hydrogen flame doubled the transmission through humid air.*

The radiation from a carbonic-oxide flame through carbonic acid is a case of still more striking interest. Among the compound gases, carbonic acid is one of the feeblest absorbers. But when it receives the rays from its own substance heated to incandescence, as in the carbonic-oxide flame, its opacity is astounding. It far transcends olefiant gas, which for all ordinary radiations far transcends it. So energetic is the action that *the carbonic acid expired from the lungs and freed of its moisture intercepts fully 50 per cent. of the radiation from the carbonic-oxide flame.* The method of experiment may be turned to account as a powerful and delicate test of the amount of the exhaled carbonic acid.

It is then remarked that the deportment of aqueous vapour towards the hydrogen flame and of carbonic acid towards the carbonic-oxide flame, prove the molecules of the two flames, the one having a temperature of 3259° C., and the other a temperature of 3042° C., to vibrate in synchronism with the molecules of cold aqueous vapour and cold carbonic acid.

Throughout the memoir this conception of discord and accord, between the periods of incident waves and of the molecules on which they are incident, is kept constantly in view; and by its aid a number of phenomena which have hitherto withstood explanation are reduced to order and clearness. Some very interesting results, published by Melloni and Knoblauch, are thus explained.

The diathermancy of bodies opaque to light is still further illustrated. Forty-one per cent. of the radiation from a hydrogen flame is shown to be transmitted by opaque soot; and 99 per cent. by opaque iodine.

The radiation of flames through liquids is made the subject of varied experiment; the change, and meaning of the change, caused by the introduction of solid bodies into flames, being demonstrated.

The following reasoning on the facts established in this memoir leads up to a conclusion of theoretic importance. From the singular opacity of water to the radiation of a hydrogen flame, the synchronism of the molecules of the liquid with those of the flame is inferred.

But from its opacity to the ultra-red undulations the synchronism of water with the longer waves of the spectrum may also be inferred; hence the emission from the hydrogen flame, which synchronises with water, must be mainly ultra-red.

It therefore follows that when a platinum wire is plunged into hydrogen and caused to glow with a white heat; or when the oxyhydrogen jet raises lime to dazzling incandescence, the light is produced by converting the long unvisual waves of the hydrogen and oxyhydrogen flames into shorter and visual ones.

Whether we have a vibrating atom of ponderable matter or a vibrating particle of the luminiferous æther, the principle, it is contended, is the same. And it is inferred that obscure *radiant heat*, if it could be rendered sufficiently intense, would, like the hydrogen flame, be competent to raise bodies to incandescence; in other words, to change its vibrating period in the direction of *augmented refrangibility*.*

The memoir concludes with some remarks on the connexion of radiation and conduction, and experiments are adduced which indicate that they are reciprocal phenomena.

* The late sagacious Dr. William Allen Miller drew this inference, and published it in 1855. Until informed of it by himself after the publication of the present memoir, I was ignorant of his having done so. The conclusion here arrived at with regard to radiant heat is abundantly proved in Memoir VIII.

VI.

CONTRIBUTIONS TO MOLECULAR PHYSICS.

The Bakerian Lecture delivered before the Royal Society, March 17, 1864.*

§ 1.

Preliminary Considerations.—Description of Apparatus.

THE natural philosophy of the future must, I imagine, mainly consist in the investigation of the relations which subsist between the ordinary matter of the universe and the æther in which this matter is immersed. Regarding the motions of the æther itself, as illustrated by the phenomena of reflexion, refraction, interference, and diffraction, the optical investigations of the last half-century have left nothing to be desired; but as regards the atoms and molecules which take up, and from which issue, the undulations of light and heat, and the relations of those atoms and molecules to the medium which they move, and by which they are set in motion, these investigations teach us little. To come closer to the origin of the æthereal waves —to get, if possible, some experimental hold of the oscillating atoms themselves—has been the main object of the researches in which I have been engaged for the last five years. In these researches radiant heat has been used as an instrument for exploring molecular condition, and this also is the object kept constantly in view throughout the investigation, the results of which I have now the honour to submit to the scientific public.

* *Philosophical Transactions* for 1864, p. 327. *Philosophical Magazine,* December 1864.

The first part of these researches is devoted to the more complete examination of a subject which was briefly touched upon at the conclusion of the last memoir—namely, the action of liquids, as compared with that of their vapours, upon radiant heat. The differences which exist between different gaseous molecules, as regards their power of emitting and absorbing radiant heat, have been already amply illustrated. When a gas is condensed to a liquid, the molecules approach and grapple with each other by forces which are insensible as long as the gaseous state is maintained. But though thus condensed and enthralled, the æther still surrounds the molecules. If, then, the powers of radiation and absorption depend mainly upon them individually, we may expect that the deportment towards radiant heat which experiment establishes in the case of molecules in a state of gaseous freedom, will maintain itself after the molecule has relinquished its freedom and formed part of a liquid. If, on the other hand, the state of aggregation be of paramount importance, we may expect to find on the part of liquids a deportment altogether different from that of the vapours from which they are derived.

Melloni, it is well known, examined the diathermancy of various liquids, but he employed for this purpose the flame of an oil-lamp, covered by a glass chimney. His liquids, moreover, were contained in glass cells; hence the radiation from the source was profoundly modified before it entered the liquid at all, for the glass was impervious to a considerable part of the radiation. It was not only my wish to interfere as little as possible with the primitive emission, but it was also my aim to compare the action of liquids with that of their vapours, examined in a tube stopped with plates of rock-salt. I therefore devised an apparatus in which layers of liquid of variable thickness could be enclosed between two polished plates of rock-salt. It was skilfully constructed for me by Mr. Becker, and the same two plates have already done service in more than six hundred experiments.

The apparatus consists of the following parts:—A B C (fig. 16) is a plate of brass, 3·4 inches long, 2·1 inches wide, and 0·3 of an inch thick. Into it, at its corners, are rigidly fixed four upright pillars, furnished at the top with threads, for the reception of the nuts *q r s t*. D E F is a second plate of brass of the same size as the former, and pierced with holes at its four corners, so as to enable it to slip over the four columns of the plate A B C. Both these plates are perforated by circular apertures,

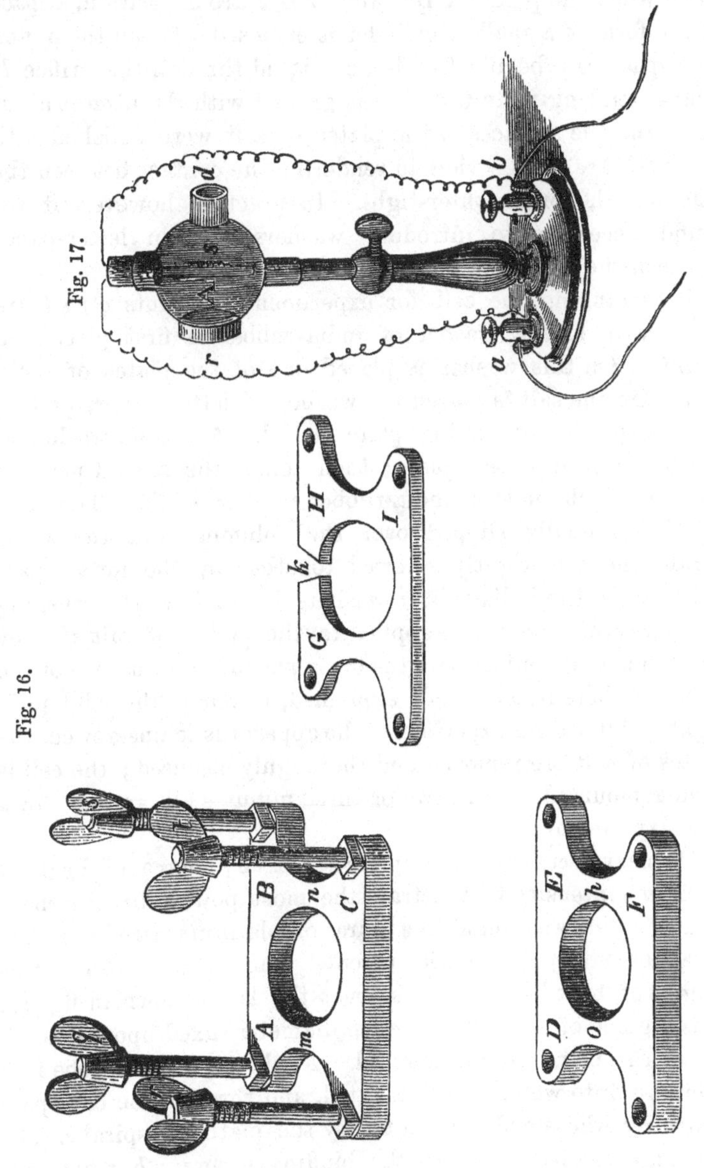

Fig. 17.

Fig. 16.

m n and *o p*, 1·35 inch in diameter. G H I is a third plate of brass of the same area as D E F, and, like it, having its centre and its corners perforated. It is intended to come between the two plates of rock-salt, which are to form the walls of the cell, and its thickness determines that of the liquid layer. Thus when the plates A B C and D E F are in position, a space of the form of a shallow cylinder is enclosed between them, and this space can be filled with any liquid through the orifice *k*. The separating plate G H I was ground with the utmost accuracy, and the surfaces of the plates of salt were polished with extreme care, with a view to rendering the contact between the salt and the brass water-tight. In practice, however, it was found necessary to introduce washers of thin letter-paper between the plates of salt and the separating plate.

In arranging the cell for experiment, the nuts *q r s t* are unscrewed, and a washer of india-rubber is first placed on A B C. On this washer is placed one of the plates of rock-salt. On the salt is placed the washer of letter-paper, and on this again the separating plate G H I. A second washer of paper is placed on this plate; then comes the second plate of salt, on which another india-rubber washer is laid. The plate D E F is finally slipped over the columns, and the whole arrangement is tightly screwed together by the nuts *q r s t*. The use of the india-rubber washers is to relieve the crushing pressure which would be applied to the plates of salt if they were in actual contact with the brass plates; and the use of the paper washers is, as already explained, to render the cell liquid-tight. After each experiment, the apparatus is unscrewed, the plates of salt are removed and thoroughly cleansed; the cell is then remounted, and in two or three minutes all is ready for a new experiment.

My next necessity was a perfectly steady source of heat, of sufficient intensity to penetrate the most powerfully absorbent liquid. This was found in a spiral of platinum wire, rendered incandescent by an electric current. The frequent use of this source of heat led me to construct the lamp shown in fig. 17. A is a globe of glass 3 inches in diameter, fixed upon a stand, which can be raised and lowered. At the top of the globe is a tubulure, into which a cork is fitted, and through the cork pass two wires whose ends are united by the platinum spiral *s*. The wires are carried down to the binding-screws *a b*, which are fixed in the foot of the stand, so that when the instrument is attached to the battery no strain is ever exerted on the wires which carry the spiral. The ends of the thick wire to which the spiral is attached are also of stout platinum; for when it was attached to copper wires, unsteadiness was ultimately introduced through oxidation. The heat from the incandescent spiral issues by the opening *d*, which is an inch and a half in diameter. Behind the spiral, finally, is a metallic reflector, *r*, which augments the flux of heat without sensibly changing its quality. In the open air the red-hot spiral is a capricious source of heat; but surrounded by its glass globe its steadiness is admirable.

The whole experimental arrangement will be immediately understood from the sketch given in fig. 18. A is the platinum lamp just described, heated by a current from a Grove's battery of five cells. It is necessary that this lamp should remain perfectly constant throughout the day; and to keep it so, a tangent galvanometer and a rheocord are introduced into the circuit.

In front of the spiral, and surrounding the tubulure of its globe, is the tube B with an interior reflecting surface, through which the heat passes to the rock-salt cell C. This cell is placed on a little stage soldered to the back of the perforated screen S S′, so that the heat, after having crossed the cell, passes through the hole in the screen, and afterwards impinges on the thermo-electric pile P. The pile is placed at some distance from the screen S S′, so as to render the temperature of the cell C itself of no account. C′ is the compensating cube, containing water kept boiling by steam from the pipe *p*. Between the cube C′ and the pile P is the screen Q, which regulates the amount of heat falling on the posterior face of the pile. The whole arrangement is here exposed, but in practice the pile P and the cube C′ are carefully protected from the capricious action of the surrounding air.

The experiments are thus performed. The empty rock-salt cell C being placed on its stage, a double silvered screen (not shown in the figure) is first introduced between the end of the tube B and the cell C—the radiation from the spiral being thus totally cut off, and the pile subjected to the action of the cube C′ alone. By means of the screen Q, the total heat to be adopted throughout the series of experiments is obtained: say that it is sufficient to produce a galvanometric deflection of 50 degrees. The double screen used to intercept the radiation from

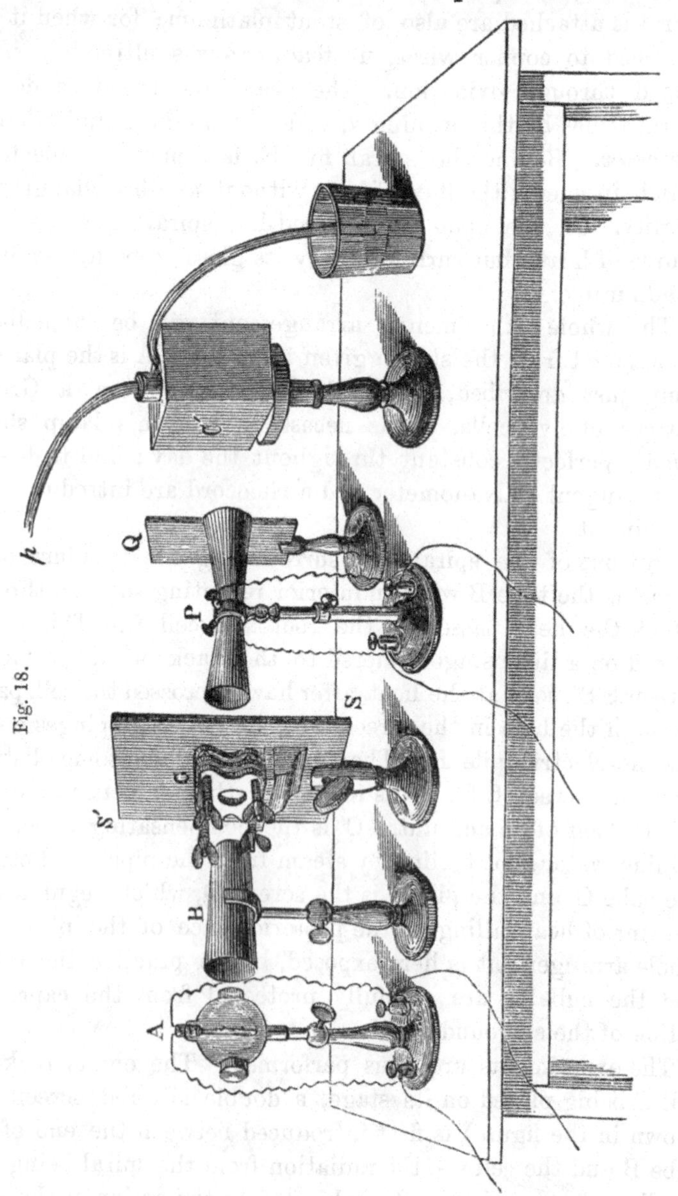

Fig. 18.

the spiral is then gradually withdrawn until this radiation completely neutralizes that from the cube, and the needle of the galvanometer points steadily to zero. The position of the double screen, once fixed, remains subsequently unchanged, the slight and slow alteration of the source of heat being neutralized by the rheocord. Thus the rays in the first instance pass from the spiral through the empty rock-salt cell. A small funnel, supported by a suitable stand, dips into the aperture leading into the cell, and through this the liquid is poured. The introduction of the liquid destroys the previous equilibrium, the galvanometer needle moves, and finally assumes a steady deflection; and from this deflection we can immediately calculate the quantity of heat absorbed by the liquid, and express it in hundredths of the entire radiation.

For example, the empty cell being placed upon its stand, and the needle being at 0°, the introduction of iodide of methyl produced a deflection of 30·8°. The total radiation on this occasion was 44·2°. Taking the force necessary to move the needle from 0° to 1° as our unit, the deflection 30·8° corresponds to 32 such units, while the deflection 44·2° corresponds to 58·3 such units. Hence the statement

$$58{\cdot}3 : 100 = 32 : 54{\cdot}9,$$

which gives an absorption of 54·9 per cent. for a layer of liquid iodide of methyl 0·07 of an inch in thickness.

§ 2.

Absorption of Radiant Heat of a certain Quality by eleven different Liquids at five different Thicknesses.

The following table contains the results obtained in this manner with the respective liquids there mentioned. It embraces both the deflection produced by the introduction of the liquid, and the quantity per cent. intercepted of the entire radiation.

It has been intimated to me that the publication of such details as would enable a reader to judge of the precision attainable by my apparatus would be desirable. In this paper, I, to some extent, endeavour to meet this desire, without, however, quitting the ordinary course of experiment.

TABLE I.—*Radiation of Heat through Liquids.*

Source of heat: *red-hot platinum spiral.*

Thickness of liquid layer, 0·07 of an inch.

Name of Liquid	Deflection	Absorption per 100
Iodide of methyl	33°·5	53·7
Iodide of ethyl	35·5	58·7
Benzol	37·5	64·4
Amylene	39·5	70·7
Sulphuric ether	41	75·4
Acetic ether	41·5	76·9
Formic ether	42·4	80
Alcohol	43·5	84·2
Water	44·7	90·5
Total heat	46·7	100

In these experiments a far less delicate galvanometer was employed than that used in my former researches. The experiments were made on September 29, 1863, and on the following day they were repeated with the same result.

On October 28 my most delicate galvanometer was at liberty, and with it I executed the experiments performed with the coarser one. The following are the results:—

TABLE II.—*Radiation of Heat through Liquids.*

Source of heat: *red-hot platinum spiral.*

Thickness of liquid layer, 0·07 of an inch.

Name of Liquid	Deflection	Absorption per 100	
Bisulphide of carbon	9°	12·5	
Chloroform	25·2	35	
Iodide of methyl	36	53·2	53·7
Ditto, strongly coloured with iodine	36	53·2	
Iodide of ethyl	38·2	59	58·7
Benzol	39·2	62·5	64·4
Amylene	42	73·6	70·7
Sulphuric ether	42·6	76·1	75·4
Acetic ether	43·4	78	76·9
Formic ether	43·3	79	80
Alcohol	44·4	83·6	84·2
Water	45·6	88·8	90·5
Total heat	48	100	

Beside the results obtained with the delicate galvanometer are placed those obtained with the coarser one. It is not my object to push these measurements to the last degree of nicety; otherwise the satisfactory agreement here exhibited might be made still more exact.

The following series of tables contain the results obtained with liquid layers of various thicknesses, employing throughout my most delicate galvanometer:—

Table III.—*Radiation of Heat through Liquids.*

Source of heat: *red-hot platinum spiral.*

Thickness of liquid layer, 0·02 of an inch.

Name of Liquid	Deflection	Absorption per 100
Bisulphide of carbon	4°	5·5
Chloroform	12	16·6
Iodide of methyl	26	36·1
Iodide of ethyl	27·5	38·2
Benzol	31·3	43·4
Amylene	38	58·3
Boracic ether	39	61·8
Sulphuric ether	39·5	63·3
Formic ether	40	65·2
Alcohol	40·5	67·3
Water	43·7	80·7
Total heat	48	100

Table IV.—*Radiation of Heat through Liquids.*

Source of heat: *red-hot platinum spiral.*

Thickness of liquid layer, 0·04 of an inch.

Name of Liquid	Deflection	Absorption per 100
Bisulphide of carbon	6°·1	8·4
Chloroform	18	25
Iodide of methyl	33	46·5
Iodide of ethyl	35	50·7
Benzol	37	55·7
Amylene	40	65·2
Boracic ether	41	69·4
Sulphuric ether	42	73·5
Acetic ether	42·1	74
Formic ether	42·5	76·3
Alcohol	43·2	78·6
Water	45	86·1
Total heat	48	100

TABLE V.—*Radiation of Heat through Liquids.*

Source of heat: *red-hot platinum spiral.*

Thickness of liquid layer, 0·14 of an inch.

Name of Liquid	Deflection	Absorption per 100
Bisulphide of carbon	11°	15·2
Chloroform	28·6	40
Iodide of methyl	40	65·2
Iodide of ethyl	40·9	69
Benzol	41·5	71·5
Amylene	43	77·7
Sulphuric ether	43·2	78·6
Acetic ether	44	82
Formic ether	44·5	84
Alcohol	44·8	85·3
Water	46·3	91·8
Total heat	48	100

TABLE VI.—*Radiation of Heat through Liquids.*

Source of heat: *red-hot platinum spiral.*

Thickness of liquid layer, 0·27 of an inch.

Name of Liquid	Deflection	Absorption per 100
Bisulphide of carbon	12°·5	17·3
Chloroform	32·3	44·8
Iodide of methyl	40·8	68·6
Iodide of ethyl	41·5	71·5
Benzol	42	73·6
Amylene	44·1	82·3
Sulphuric ether	44·8	85·2
Acetic ether	45	86·1
Formic ether	45·2	87
Alcohol	45·7	89·1
Water	46·1	91
Total heat	48	100

The foregoing results are collected together in the following table:—

TABLE VII.—*Absorption of Heat by Liquids.*

Source of heat: *Platinum Spiral heated to a bright redness by a voltaic current.*

Liquid	Thickness of liquid in parts of an inch				
	0·02	0·04	0·07	0·14	0·27
Bisulphide of carbon . .	5·5	8·4	12·5	15·2	17·3
Chloroform	16·6	25·0	35·0	40·0	44·8
Iodide of methyl . . .	36·1	46·5	53·2	65·2	68·6
Iodide of ethyl . . .	38·2	50·7	59·0	69·0	71·5
Benzol	43·4	55·7	62·5	71·5	73·6
Amylene	58·3	65·2	73·6	77·7	82·3
Sulphuric ether . . .	63·3	73·5	76·1	78·6	85·2
Acetic ether	......	74·0	78·0	82·0	86·1
Formic ether	65·2	76·3	79·0	84·0	87·0
Alcohol	67·3	78·6	83·6	85·3	89·1
Water	80·7	86·1	88·8	91·8	91·0

Had it been desirable to push these measurements to the utmost limit of accuracy, I should have repeated each experiment, and taken the mean of the determinations. But considering the way in which the different thicknesses check each other, an inspection of the table must produce the conviction that the results express, within small limits of error, the action of the bodies mentioned. The *order* of absorption is certainly that here shown.

§ 3.

Absorption of Radiant Heat of the same quality by the Vapours of these Liquids at a common Pressure.

As liquids, then, those bodies are shown to possess very different capacities of intercepting the heat emitted by our radiating source; and we have next to inquire whether these differences continue after the molecules have been released from the bond of cohesion. We must, of course, test vapours and liquids by waves of the same period; and this our mode of experiment renders easy of accomplishment. The heat generated in a wire by a current of a given strength being invariable, it was only necessary, by means of the tangent compass and rheocord, to keep the current constant from day to day in order to obtain, both as regards quantity and quality, an invariable source of heat.

The liquids from which the vapours were derived were placed in small long flasks, a separate flask being devoted to each. The air above the liquid and within it being first carefully removed by an air-pump, the flask was attached to the experimental tube. This was of brass, 49·6 inches long, and 2·4 inches in diameter, its two ends being stopped by plates of rock-salt. Its interior surface was polished. At the commencement of each experiment, the tube having been thoroughly cleansed and exhausted, the needle stood at zero.* The cock of the flask containing the volatile liquid was then carefully turned on, and the vapour allowed slowly to enter the experimental tube. The barometer attached to the tube was finely graduated, and the descent of the mercurial column was observed through a magnifying lens. When a pressure of 0·5 of an inch was obtained, the vapour was cut off and the permanent deflection of the needle noted. Knowing the total heat, the absorption in 100ths of the entire radiation could be at once deduced from the deflection. The following table contains the results of a series of experiments made with the platinum spiral as source:—

TABLE VIII.—*Radiation of Heat through Vapours.*

Source of heat: *red-hot Platinum Spiral.*

Pressure, 0·5 of an inch.

Name of Vapour	Deflection	Absorption per 100
Bisulphide of carbon	16·5°	4·7
Chloroform	22·8	6·5
Iodide of methyl	33	9·6
Iodide of ethyl	45	17·7
Benzol	48	20·6
Amylene	55·3	27·5
Alcohol	55·7	28·1
Formic ether	58·2	31·4
Sulphuric ether	58·5	31·9
Acetic ether	59·9	34·6
Total heat	78	100

§ 4.

Order of Absorption of Liquids at a common Thickness, and Vapours at a common Pressure.

We are now in a condition to compare the action of a series of volatile liquids at a common thickness with that of the

* It is hardly necessary to remark that the principle of compensation described in my former memoirs was employed here also.

vapours of those liquids at a common pressure upon radiant heat.

Commencing with the substance of the lowest absorptive energy, and proceeding to the highest, we have the following order of absorption :—

Liquids	Vapours
Bisulphide of carbon.	Bisulphide of carbon.
Chloroform.	Chloroform.
Iodide of methyl.	Iodide of methyl.
Iodide of ethyl.	Iodide of ethyl.
Benzol.	Benzol.
Amylene.	Amylene.
Sulphuric ether.	Alcohol.
Acetic ether.	Formic ether.
Formic ether.	Sulphuric ether.
Alcohol.	Acetic ether.
Water.	

Here, as far as amylene, the order of absorption is the same for both liquids and vapours. But from amylene downwards, though strong liquid absorption is in a general way paralleled by strong vapour absorption, the order of both is not the same. There is not the slightest doubt that next to water alcohol is the most powerful absorber in the list of liquids; but there is just as little doubt that the position which it occupies in the list of vapours is the correct one. This has been established by reiterated experiments. Acetic ether, on the other hand, though certainly the most energetic absorber in the state of vapour, falls behind both formic ether and alcohol in the liquid state. Still, on the whole, I think it is impossible to contemplate these results without arriving at the conclusion, that the act of absorption is in the main molecular, and that the molecule maintains its power as an absorber and radiator when it changes its state of aggregation. Should, however, any doubt linger as to the correctness of this conclusion, it will speedily disappear.

§ 5.

Order of Absorption of Liquids and Vapours in proportional Quantities.

A moment's reflexion will show that the comparison instituted in the last section is not a strict one. We have taken the liquids at a common thickness, and the vapours at a common volume and pressure. But if the layers of liquid employed were turned bodily into vapour, the volumes obtained would *not* be the same. Hence the quantities of matter traversed by the radiant heat are neither equal nor proportional to each other in the two cases; and to render the comparison strict they ought to be proportional. It is easy, of course, to make them so; for the liquids being examined at a constant volume, their specific gravities give us the relative quantities of matter traversed by the radiant heat, and from these and the vapour-densities we can immediately deduce the corresponding volumes of the vapour. Calling the quantity of matter q, the vapour-density d, and the volume V, we have

$$Vd = q,$$

or

$$V = \frac{q}{d}.$$

Dividing, therefore, the specific gravities of our liquids by the densities of their vapours, we obtain a series of volumes proportional to the masses of liquid employed. The densities of both liquids and vapours are given in the following table:—

IX.—*Table of Densities.*

	Vapour	Liquid
Bisulphide of carbon	2·63	1·27
Chloroform	4·13	1·48
Iodide of methyl	4·90	2·24
Iodide of ethyl	5·39	1·95
Benzol	2·69	0·85
Amylene	2·42	0·64
Alcohol	1·59	0·79
Sulphuric ether	2·56	0·71
Formic ether	2·56	0·91
Acetic ether	3·04	0·89
Water	0·63	1
Air	1	

Substituting for q the numbers of the second column, and for d those of the first, we obtain the following series of vapour volumes, whose weights are proportional to the masses of liquid employed:—

X.—*Table of Proportional Volumes.*

Bisulphide of carbon	0·48
Chloroform	0·36
Iodide of methyl	0·46
Iodide of ethyl	0·36
Benzol	0·32
Amylene	0·26
Alcohol	0·50
Sulphuric ether	0·28
Formic ether	0·36
Acetic ether	0·29
Water	1·60

Employing the vapours in the volumes here indicated, the following results were obtained:—

TABLE XI.—*Radiation of Heat through Vapours.*

Mass of Vapour proportional to Mass of Liquid.

Name of Vapour	Pressure in parts of an inch	Deflection	Absorption per 100
		°	
Bisulphide of carbon	0·48	{8·4, 8·5}	4·3
Chloroform	0·36	{13, 13}	6·6
Iodide of methyl	0·46	{20, 20·4}	10·2
Iodide of ethyl	0·36	{30·6, 30·6}	15
Benzol	0·32	{33·4, 33·1}	16·8
Amylene	0·26	37·7	19
Sulphuric ether	0·28	{42·5, 42·6}	21·5
Acetic ether	0·29	{44, 44}	22·2
Formic ether	0·36	{44·5, 44·7}	22·5
Alcohol	0·50	{45, 44·9}	22·7

Here the discrepancies revealed by our former series of experiments entirely disappear, and it is proved that for heat of the same quality the order of absorption for liquids and their vapours is the same. We may therefore safely infer that the position of a vapour as an absorber or radiator is determined by that of the liquid from which it is derived. Granting the validity of this inference, the position of *water* fixes that of *aqueous vapour*. From the first seven tables of this memoir, or from the *résumé* of results in Table VII., it will be seen that for all thicknesses water exceeds the other liquids in the energy of its absorption. Hence, if no single experiment on the vapour of water existed, we should be compelled to conclude, from the deportment of its liquid, that, weight for weight, aqueous vapour transcends all others in absorptive power. Add to this the direct and multiplied experiments by which the action of this substance on radiant heat has been established, and we have before us a body of evidence sufficient, I trust, to set this question for ever at rest, and to induce the meteorologist to apply without misgiving the radiant and absorbent property of aqueous vapour to the phenomena of his science.

§ 6.

Remarks on the Chemical Constitution of Bodies with reference to their Powers of Absorption.

The order and relative powers of absorption of our vapours, when equal volumes are compared, are given in Table VIII.: the chemical formulæ of the substances, and the number of atoms which the molecules embrace, are as follows:—

	Formula	Number of Atoms in Molecules
Bisulphide of Carbon	$C\,S^2$	3
Chloroform	$C\,H\,Cl^3$	5
Iodide of methyl	$C\,H^3\,I$	5
Iodide of ethyl	$C^2\,H^5\,I$	8
Benzol	$C^6\,H^6$	12
Amylene	$C^5\,H^{10}$	15
Alcohol	$C^2\,H^6\,O$	9
Formic ether	$C^3\,H^6\,O^2$	11
Sulphuric ether	$C^4\,H^{10}\,O$	15
Acetic ether	$C^4\,H^8\,O^2$	14
Boracic ether	$B\,C^6\,H^{15}\,O^3$	25

Here for the first six vapours, the radiant and absorbent powers augment with the number of atoms contained in the molecules. Alcohol and amylene vapours, however, are nearly alike in absorptive power, the molecule of amylene containing 15 atoms while that of alcohol embraces only 9. But in alcohol we have a third element introduced, which is absent in the amylene; the oxygen of the alcohol gives its molecule a character which enables it to transcend the amylene molecule, though the latter contains the greater number of atoms. Here the idea of *quality* superadds itself to that of number. Acetic ether also has a smaller number of atoms in its molecule than sulphuric ether; the latter, however, has but one atom of oxygen, while the former has two. Formic ether and sulphuric ether are almost identical in their absorptive powers for the heat here employed; still formic ether has but 11 atoms in its molecule, while sulphuric has 15. But formic ether possesses two atoms of oxygen, while sulphuric possesses only one. Two things seem influential on the absorbent and radiant power, which may be expressed by the terms *multitude* and *complexity*. As a molecule of multitude, amylene, for example, exceeds alcohol; as a molecule of complexity, alcohol exceeds amylene; and in this case, as regards radiant and absorbent power, the complexity is more than a match for the multitude. The same remarks apply to sulphuric and formic ether: the former excels in multitude, the latter in complexity, the excess in the one case almost exactly balancing that in the other. Adding two atoms of hydrogen and one of carbon to formic ether, we obtain acetic ether, and by this addition the balance is turned; for though acetic ether falls short of sulphuric ether in multitude, it transcends it in absorbent and radiant power. Outstanding from all others, when equal volumes are compared, and signalizing itself by the magnitude of its absorption, we have boracic ether, each molecule of which embraces no less than 25 atoms. The time now at my disposal enables me to do little more than glance at these singular facts; but, in passing, I must direct the attention of chemists to the water-molecule: its power as a radiant and an absorbent is perfectly unprecedented and anomalous, if the usually recognized formula be correct.

§ 7.

Transmission of Radiant Heat through Bodies opaque to Light.—Remarks on the Physical Cause of Transparency and Opacity.

In Table II. a fact is revealed which is worth a little further attention. The measurements there recorded show that the absorption of a layer of iodide of methyl, strongly coloured with iodine (which had been liberated by the action of light) was precisely the same as that of a perfectly transparent layer of the liquid. The iodine, which produced so marked an effect on light, did not sensibly affect the radiant heat emitted by the platinum spiral. Here are the numbers:—

	Absorption per 100
Iodide of methyl (transparent)	53·2
Iodide of methyl (strongly coloured with iodine) . .	53·2

In this case, the incandescent platinum spiral, or a bright flame, was visible when looked at through the liquid; I therefore intentionally deepened the colour (a rich brown), by adding iodine, until the layer was able to cut off wholly the light of a brilliant jet of gas. The transparency of the liquid to the radiant heat was not sensibly affected by the addition of the iodine. The luminous heat was of course cut off; but this, as compared with the whole radiation, was so small as to be insensible in the experiments.

It is known that iodine dissolves freely in the bisulphide of carbon, the colour of the solution in thin layers being a splendid purple; but in layers of moderate thickness it may be rendered perfectly opaque to light. I dissolved in the liquid a quantity of the iodine sufficient, when introduced into a cell 0·07 of an inch in width, to cut off wholly the light of the most brilliant gas-flame. Comparing the opaque solution with the transparent bisulphide, the following results were obtained:—

	Deflection	Absorption per 100
Bisulphide of carbon (opaque) . . .	9°	12·5
Bisulphide of carbon (transparent). . .	9°	12·5

Here the presence of a quantity of iodine, perfectly opaque to a brilliant light, was without measurable effect upon the heat emanating from our platinum spiral. The liquid was sensibly thickened by the quantity of iodine dissolved in it.

The same liquid was placed in a cell 0·27 of an inch in width; that is to say, a solution which was perfectly opaque to light at a thickness of 0·07, was employed in a layer of nearly four times this thickness. Here are the results:—

	Deflection	Absorption per 100
Bisulphide of carbon (transparent). . .	13·6°	18·8
Bisulphide of carbon (opaque) . . .	13·7°	19

The difference between the two measurements lies within the limits of possible error.

Bisulphide of carbon is commonly used to fill hollow prisms, when considerable dispersion is desired in the decomposition of white light. Such prisms, filled with the opaque solution, intercept entirely the luminous part of the spectrum, but allow the ultra-red rays free passage. A heat-spectrum of the sun, or of the electric light may be thus obtained entirely separated from the luminous one. By means of a prism of the transparent bisulphide, I determined the position of the spectrum of the electric light upon a screen, and behind the screen placed a thermo-electric pile, so that when the screen was removed the ultra-red rays fell upon the pile. I then substituted an opaque prism for the transparent one: there was no visible spectrum on the screen; but its removal at once demonstrated the existence of an invisible spectrum by the thermo-electric current which it generated, and which was powerful enough to dash violently aside the needles of a large lecture-room galvanometer.

To what, then, are we to ascribe the deportment of iodine towards luminous and obscure heat? The difference between both qualities of heat is simply one of period. In the one case the waves which convey the energy are short and of rapid recurrence; in the other case they are long and of slow recurrence; the former are intercepted by the iodine, and the latter are allowed to pass. Why? There can, I think, be only one answer to this question—that the intercepted waves are those whose periods coincide with those of the atoms of the dissolved iodine. Supposing waves of any period to impinge upon an assemblage of molecules of any other period, it is, I think, physically certain that a *tremor* of greater or less intensity will be set up among the molecules; but for the motion to *accumulate* so as to produce sensible absorption, coincidence of period is necessary. Briefly defined, therefore, transparency is synonymous with *discord*, while opacity is synonymous with *accord* between the periods of the waves of æther and those of the molecules of the body on which they impinge. The transparency, then, of our solution of iodine to the ultra-red undulations demonstrates the incompetency of its atoms to vibrate in unison with the longer waves.

This simple conception will, I think, be found sufficient to conduct us with intellectual clearness through a multitude of otherwise perplexing phenomena. It may of course be applied immediately to that numerous class of bodies which are transparent to light, but opaque in a greater or less degree to radiant heat. Water, for example, is an eminent example of this class of bodies: while it allows the luminous rays to pass with freedom, it is highly opaque to all radiations emanating from obscure sources of heat. A layer of this substance one-twentieth of an inch thick is competent, as Melloni has shown, to intercept all rays issuing from bodies heated under incandescence. Hence we may infer that, throughout the range of the visible spectrum, the periods of the water-molecules are in discord with those of the æthereal waves, while beyond the red we have coincidence between both.

What is true of water is, of course, true in a less degree of glass, alum, calcareous spar, and of the various liquids named in the first section of this paper. They are all in discord with the visible spectrum; they are all more or less in accord with the ultra-red undulations of the spectrum.

Thus also as regards lampblack: the blackness of the substance is due to the accord which reigns between the oscillating periods of its atoms and those of the waves embraced within the limits of the visible spectrum. The substance which is thus impervious to the luminous rays is moreover the very one from which the whitest light of our lamps is derived. It can absorb all the rays of the visible spectrum; it can also emit them. But though in a far less degree than iodine, lampblack is also to some extent transparent to the longer undulations. Melloni was the first to prove this; and an experiment described in a former memoir proved that 30 per cent. of the radiation from an obscure source of heat found its way through a layer of lampblack which cut off totally the light of the most brilliant jet of gas. I shall have occasion to show that, for certain sources of heat of long period, between 40 and 50 per cent. of the entire radiation is transmitted by a layer of lampblack which is perfectly opaque to our most brilliant artificial lights. Hence, in the case of lampblack, while accord exists between the periods of its atoms and those of the light-exciting waves, discord, to a considerable extent, exists between the periods of the same atoms and those of the ultra-red undulations.

§ 8.

Influence of the Temperature of the Source of Heat on the Transmission of Radiant Heat.

To obtain sources of heat of different temperatures, Melloni resorted to lamps, to spirals heated to incandescence by the flame of alcohol, to copper laminæ heated by flame, and to the surfaces of vessels containing boiling water. No conclusions regarding temperature can, as will afterwards be shown, be drawn from such experiments; but by means of the platinum spiral we can go through all those changes of temperature, *retaining throughout the same vibrating atoms*, and we can therefore investigate how the alteration of the rate of vibration affects the rate of absorption. The following series of experiments were executed on the 9th of October, with a platinum spiral raised to barely visible redness, and vapours at a pressure of 0·5 of an inch:—

TABLE XII.—*Radiation of Heat through Vapours.*

Source of Heat: *Platinum Spiral barely visible in the dark.*

Name of Vapour	Deflection	Absorption per 100
Bisulphide of carbon	7·5°	6·5
Bisulphide of carbon	7·45	6·4
Chloroform	10·5	9·1
Chloroform	10·5	9·1
Iodide of methyl	14·5	12·5
Iodide of methyl	14·5	12·5
Iodide of ethyl	24·2	20·9
Iodide of ethyl	24·5	21·1
Benzol	31·0	26·7
Benzol	30	25·9
Amylene	37·6	35·6
Amylene	37·8	35·9
Sulphuric ether	41·1	43·4
Sulphuric ether	41	43·4
Formic ether	41·7	45
Formic ether	41·8	45·3
Acetic ether	43·6	49·8
Acetic ether	43·4	49·3

On the 10th of October the following results were obtained with the same platinum spiral, raised to a white heat:—

TABLE XIII.—*Radiation of Heat through Vapours.*

Source of heat: *White-hot Platinum Spiral.*

Name of Vapour	Deflection	Absorption per 100
Bisulphide of carbon	3·5°	2·9
Bisulphide of carbon	3·4	2·8
Chloroform	6·7	5·6
Chloroform	6·7	5·6
Iodide of methyl	9·2	7·7
Iodide of methyl	9·4	7·9
Iodide of ethyl	15·4	13
Iodide of ethyl	15	12·6
Benzol	19·3	16·6
Benzol	19	16·4
Total heat	59·2	100
Amylene	27·6	22·6
Amylene	27·7	22·7
Formic ether	30·5	25
Formic ether	30·7	25·2
Sulphuric ether	31·4	25·7
Sulphuric ether	31·7	26·0
Acetic ether	33	27
Acetic ether	33·2	27·3
Total heat	60	100

With the same spiral, brought still nearer to its point of fusion, and with four of the vapours, the following results were obtained:—

TABLE XIV.—*Radiation of Heat through Vapours.*

Source of heat: *Platinum Spiral at an intense white heat.*

Name of Vapour	Deflection	Absorption per 100
Bisulphide of carbon	14·5°	2·5
Bisulphide of carbon	14·5	2·5
Chloroform	23	3·9
Chloroform	23	3·9
Formic ether	60·4	21·3
Formic ether	60·5	21·3
Sulphuric ether	62·3	23·6
Sulphuric ether	62·5	23·8
Total heat	82·7	100

In the experiments recorded in the foregoing table, a total heat of 82·7°, or 588 units, was employed; and to test whether the absorption calculated from this high total agreed with the absorptions calculated from a low total, a portion of the current was diverted, the branch passing through the galvanometer producing a deflection of 49·4°. This corresponds to 77 units. The source of heat, it will be observed, is here quite unchanged; the rays are of the same quality, and pass through the tube in the same quantity as before; but in the one case the absorption is calculated from the deflection among the high degrees, and in the other case it is calculated from deflections among the low degrees of the galvanometer.

The experiments were limited to formic and sulphuric ether, with the following results:—

	Deflection	Absorption per 100	Absorption from Table XIV.
Formic ether	17·7°	23	21·3
Sulphuric ether	19·1	24·8	23·7

The agreement is such as to prove that no material error can have crept into the calibration.

Placing the results obtained with the respective sources side by side, the influence of temperature on the transmission comes out in a very decided manner.

TABLE XV.—*Absorption of Heat by Vapours.*

Pressure, 0·5 of an inch.

Name of Vapour	Source of heat: *Platinum Spiral*			
	Barely visible	Bright red	White-hot	Near fusion
Bisulphide of carbon . .	6·5	4·7	2·9	2·5
Chloroform	9·1	6·3	5·6	3·9
Iodide of methyl . . .	12·5	9·6	7·8	
Iodide of ethyl . . .	21	17·7	12·8	
Benzol	26·3	20·6	16·5	
Amylene	35·8	27·5	22·7	
Sulphuric ether . . .	43·4	31·4	25·9	23·7
Formic ether	45·2	31·9	25·1	21·3
Acetic ether	49·6	34·6	27·2	

The gradual augmentation of penetrative power as the temperature is augmented is here very manifest. By raising the spiral from a barely visible heat to an intense white heat, we reduce the absorption, in the cases of bisulphide of carbon and chloroform, to less than one-half. At barely visible redness, moreover, 56·6 and 54·8 per 100 pass through sulphuric and formic ether respectively; while, of the intensely white-hot spiral, 76·3 and 78·7 per 100 pass through the same vapours. By augmenting the temperature of solid platinum, we introduce into the radiation waves of shorter period, which, being in discord with the periods of the vapours, get more easily through them.

What becomes of the more slowly recurrent vibrations as the more rapid ones are introduced? Do the latter take the place of the former? This question is answered by experiments made with an opaque solution of iodine, and with lampblack. As the temperature of the platinum spiral increases from a dark heat to the most intense white heat, the absolute quantity transmitted through both these bodies steadily augments. But this heat is wholly obscure, for both the solution and the lampblack intercept all the luminous heat. Hence the conclusion that the augmentation of temperature which introduces the shorter waves augments at the same time the amplitude of the longer ones, and hence also the inference that *a body like the sun must of necessity include in its radiation waves of the same period as those emitted by obscure bodies.*

§ 9.

Changes of Diathermancy through Changes of Temperature.—Radiation from Lampblack at 100° *C. compared with that from white-hot Platinum.*

Running the eye along the numbers which express the absorptions of sulphuric and formic ether in Table XV., we find that, for the lowest heat, the absorption of the latter exceeds that of the former; for a bright red heat they are nearly equal, the formic still retaining a slight predominance; at a white heat, however, the sulphuric slips in advance, and at the heat near fusion its predominance is decided. I have tested this result in various ways, and by multiplied experiments, and placed it beyond doubt. We may at once infer from it that the capacity of the molecule of formic ether to enter into rapid vibration is less than that of sulphuric. By augmenting the temperature of the spiral we produce vibrations of quicker periods, and the more of these that are introduced, the more transparent, in comparison with sulphuric ether, does formic ether become. Thus its 'complexity' tells upon the vibrating periods of the formic ether, the atom of oxygen which it possesses in excess of the sulphuric ether rendering it a more sluggish vibrator. Experiments made with a source of 212° Fahr. establish more decidedly the preponderance of the formic ether for slow vibrations.

TABLE XVI.—*Radiation through Vapours.*

Source of heat: *Leslie's Cube, coated with Lampblack.*

Temperature, 212° Fahr.

Name of Vapour	Absorption per 100
Bisulphide of carbon	6·4
Iodide of methyl	18·4
Chloroform	19·5
Sulphuric ether	54·8
Formic ether	60·9

For heat issuing from this source, the absorption of formic ether is 6·1 per cent. in excess of that of sulphuric.

Deeming the result worthy of rigid confirmation, I once more determined the order of absorption:—

TABLE XVII.

Source of heat: *Leslie's Cube, coated with Lampblack.*

Temperature, 212° Fahr.

Name of Vapour	Deflections
Bisulphide of carbon	9°·3
Iodide of methyl	25
Chloroform	26·5
Sulphuric ether	47·3 }
Sulphuric ether	47·7 }
Formic ether	49·7 }
Formic ether	49·9 }

When the absorptions were calculated from these deflections, that of formic ether was found to be 6·3 per cent. in excess of that of sulphuric. In the last table the excess was 6·1.

But in both Tables XVI. and XVII. we notice another case of reversal. In all the experiments with the platinum spiral recorded in Table XV., chloroform showed itself less energetic as an absorber than iodide of methyl; but in Tables XVI. and XVII. chloroform proved to be decidedly the more powerful of the two. Cases of this kind have, in my estimation, a peculiar significance, and I therefore took care to verify them. Three different series of experiments with the vapours in question were therefore executed, with the following results:—

TABLE XVIII.—*Radiation through Vapours.*

Source of heat: *Blackened Cube of Boiling Water.*

Name of Vapour	Absorptions per 100 I.	II.	III.
Bisulphide of carbon	6·4	6·6	
Iodide of methyl	18·4	18·8	18·3
Chloroform	19·5	21·6	20·6
Sulphuric ether	54·8	54·1	53·2
Formic ether	60·9	60·4	60

Were it essential to my purpose, I should certainly be able to cause even the small differences which here show themselves to disappear. But the agreement is such as to place the correctness of the experiments beyond doubt. *It will be seen that, contrary to the results obtained with a white-hot spiral, in all three cases, where a blackened cube of boiling water was the source, chloroform exceeds iodide of methyl, and formic ether exceeds sulphuric in absorbent power.*

To clench the demonstration, I once more resorted to the white-hot spiral, and obtained the following results:—

TABLE XIX.—*Radiation through Vapours.*

Source of heat: *White-hot Platinum Spiral.*

Name of Vapour	Deflection	Absorption per 100
Chloroform	9°·8	4·5
Chloroform	9·5	4·5
Iodide of methyl	16	7·3
Iodide of methyl	15·8	7·3
Formic ether	42·1	24·2
Formic ether	42·3	24·5
Sulphuric ether	43·6	26·3
Sulphuric ether	43·5	26·2
Total heat	70·9	100

Here chloroform retreats once more behind iodide of methyl, and formic ether behind sulphuric.

§ 10.

Changes of Diathermancy through Change of Source of Heat.—Radiation from Platinum and from Lampblack at the same Temperature.

The positions of sulphuric and formic ether are reversed within the range of the experiments made with the platinum spiral, but this is not the case with the chloroform and the iodide of methyl. Even when the spiral was at a barely visible heat, the iodide was decidedly the most opaque of the two. The same result was obtained with a spiral heated below redness, as proved by the following figures:—

Name of Vapour	Deflection	Absorption per 100
	°	
Chloroform	8·5	12·14
Chloroform	8·5	12·14
Iodide of methyl	10	14·28
Iodide of methyl	10	14·28
Total heat	47·3	100

Here the iodide is still predominant. Is it, then, a question of *temperature* merely? or is there a special flux emitted by the lampblack, to which chloroform is particularly opaque? In other words, is there a special accord between the rates of vibration of lampblack and chloroform? To answer this question I operated thus:—The platinum spiral was heated by only two cells, and the strength of this current was lowered by the introduction of resistance. When decidedly below a red heat, the spiral was plunged into boiling water. Bubbles of steam issued from it, proving that its temperature was above 212° Fahr. By augmenting the resistance its heat was lowered, until it was no longer competent to produce the least ebullition. It was then withdrawn from the water, and employed as a source: the following are the results:—

TABLE XX.—*Radiation through Vapours.*

Source of heat: *Platinum Spiral at* 100° *C.*

Name of Vapour	Deflection	Absorption per 100
	°	
Bisulphide of carbon	5·7	7·03
Chloroform	14	16·8
Iodide of methyl	15·3	18

No reversal was here obtained. The temperature was then reduced so that the total heat fell from 81 units to 59 units; but not even in this case (when the temperature was considerably below that of boiling water) could the reversal be obtained. The absorptions approach each other, but the iodide has still the advantage of the chloroform. Here are the numbers:—

TABLE XXI.—*Radiation through Vapours.*

Source of heat: *Platinum Spiral, heated under* 100° *C.*

Name of Vapour	Deflection	Absorption per 100
Bisulphide of carbon	5°·2	9·2
Chloroform	10	17·3
Iodide of methyl	10·8	18·2

It is not, therefore, temperature alone which determines the inversion: the experiments prove that there is a greater synchronism between the vibrating periods of chloroform and lampblack than between those of chloroform and platinum raised to the temperature of the lampblack. It is seen, however, that as the temperature of the platinum falls, the opacity of the chloroform increases more quickly than that of the iodide: with an intensely white-hot spiral, as shown in Table XXI., the absorption of chloroform is to that of the iodide as 100 : 162, while with the spiral heated to a temperature of 212° Fahr., the ratio of the absorption is as 100 : 105.

§ 11.

Radiation from Flames through Vapours.—Further Changes of Diathermancy.

We have hitherto occupied ourselves with the radiation from heated solids: I will now pass on to the examination of the radiation from flames. The first experiments were made with a steady jet of gas issuing from a small circular burner, the flame being long and tapering. The top and bottom of the flame were excluded, and its most brilliant portion was chosen as the source of heat. The following results were obtained:—

TABLE XXII.—*Radiation of Heat through Vapours.*

Source of heat: *A highly luminous Jet of Gas.*

Name of Vapour	Deflection	Absorption per 100	White-hot Spiral
Bisulphide of carbon . . .	8°·9	9·8	2·9
Chloroform	10·9	12	5·6
Iodide of methyl	15·4	16·5	7·8
Iodide of ethyl	17·7	19·5	12·8
Benzol	20	22	16·5
Amylene	27·5	30·2	22·7
Formic ether	31·5	34·6	25·1
Sulphuric ether	32·5	35·7	25·9
Acetic ether	34·2	38·7	27·2
Total heat	53·8	100	

To facilitate the comparison of the white-hot carbon with the white-hot platinum, I have here placed beside the results in the last table those recorded in Table XIII. The emission from the flame is thus proved to be far more powerfully absorbed than the emission from the spiral. Doubtless, however, the carbon, in reaching incandescence, passes through lower stages of temperature, and in those stages emits heat more in accord with the vapours. It is also mixed with the vapour of water and carbonic acid, both contributing their quota to the total radiation. It is therefore probable that the greater accord between the periods of the flame and those of the vapours is due to the slower periods of the substances which are unavoidably mixed with the incandescent carbon.

The next source of heat employed was the flame of a Bunsen's burner, the temperature of which is known to be very high. The flame was of a pale-blue colour, and emitted a very feeble light. The following results were obtained :—

TABLE XXIII.—*Radiation of Heat through Vapours.*

Source of heat: *Pale-blue Flame of Bunsen's Burner.*

Name of Vapour	Deflection	Absorption per 100	From Table XXII. Luminous Jet of Gas
Chloroform	5°	6·2	12
Bisulphide of carbon	9	11·1	9·8
Iodide of ethyl	11·3	14	19·5
Benzol	14·5	17·9	22
Amylene	19·6	24·2	30·2
Sulphuric ether	25·8	31·9	35·7
Formic ether	27	33·3	34·6
Acetic ether	29·4	36·3	38·7
Total heat	50·6	100	100

Comparing Tables XXII. and XXIII., we see that the radiation from the Bunsen's burner is, on the whole, less powerfully absorbed than that from the luminous gas jet. In some cases, as in that of formic ether, they come very close to each other; in the case of amylene and a few other substances they differ more markedly. But an extremely interesting case of reversal here shows itself. Bisulphide of carbon, instead of being first, stands decidedly below chloroform. With the luminous jet, the absorption of bisulphide of carbon is to that of chloroform as 100 : 122, while with the flame of Bunsen's burner the ratio is 100 : 56; *the removal of the carbon from the flame more than doubles the relative transparency of the chloroform.* The case is of too much interest to be passed over without verification. Here is the result obtained with a different total heat:—

Source of heat: *Pale-blue Flame of Bunsen's Burner.*

	Deflection	Absorption per 100
Chloroform	16°·5	8·4
Chloroform	16	8·2
Bisulphide of carbon	19	9·7
Bisulphide of carbon	19·4	9·9
Total heat	68·4	100

And again, with an intermediate total heat:—

Source of heat: *Pale-blue Flame of Bunsen's Burner.*

	Deflection	Absorption per 100
Chloroform	10°·2	8·4
Chloroform	10	8·4
Bisulphide of carbon	12	9·8
Bisulphide of carbon	11·8	9·7
Total heat	60	100

There is therefore no doubt that, while in the case of a platinum spiral at all temperatures, of a luminous gas-flame, and, more especially, of lampblack heated to 212° Fahr. the absorption of chloroform exceeds that of bisulphide of carbon, for the flame of a Bunsen's burner the bisulphide is the more powerful absorber of the two. The absorptive energy of the chloroform, as shown in Table XVIII., is more than three times that of the bisulphide, while in Table XXIII. the action of the bisulphide is nearly half as much again as that of the chloroform. We have here, moreover, another instance of the reversal of formic and sulphuric ether. For the luminous jet the sulphuric ether is decidedly the more opaque; for the flame of Bunsen's burner it is excelled in opacity by the formic.

The total heat radiated from the flame of Bunsen's burner is greatly less than that radiated when the incandescent carbon is present in the flame. The moment the air is permitted to mix with the luminous flame, the radiation falls so considerably that the diminution is at once detected, even by the hand or face brought near the flame.

§ 12.

Radiation of Hydrogen Flame through Dry and Humid Air.—Influence of Vibrating Period on the Absorption.

The main radiating bodies in the flame of a Bunsen's burner are, no doubt, aqueous vapour and carbonic acid. Highly heated nitrogen is also present, which may produce a sensible effect: the unburnt gas, moreover, in proximity with the flame, and warmed by it, may contribute to the radiation, even before it unites with the atmospheric oxygen. But the main source of the radiation is, no doubt, the aqueous vapour and the carbonic acid. I wished to separate these two constituents, and to study them separately. The radiation of aqueous vapour could be obtained from a flame of pure hydrogen, while that of carbonic acid could be obtained from an ignited jet of carbonic oxide. To me the radiation from the hydrogen flame possessed a peculiar interest; for, notwithstanding the high temperature of such a flame, I thought it likely that the accord between its periods of vibration and those of the cool aqueous vapour of the atmosphere might be such as to cause the atmospheric vapour to exert a special absorbent power. The following experiments test this surmise:—

TABLE XXIV.—*Radiation through Atmospheric Air.*

Source of heat: *A Hydrogen Flame.*

	Deflection	Absorption per 100
Dry air	0°	0
Undried air	21·5	17·20
Total heat	60·4	100

Thus, in a polished tube 4 feet long, the aqueous vapour of our laboratory air absorbed 17 per cent. of the radiation from the hydrogen flame. Of the radiation of a platinum spiral, heated by electricity to a degree of incandescence not greater than that obtainable by plunging a wire into the hydrogen flame, the undried air of the laboratory absorbed

5·8 per cent.,

or one-third of the quantity absorbed when the flame of hydrogen was employed.

The plunging of a spiral of platinum wire into the flame reduces its temperature; but it at the same time introduces vibrations which are not in accord with those of aqueous vapour: the absorption by ordinary undried air of heat emitted by this composite source amounted to

8·6 per cent.

On humid days the absorption of the rays emitted by a hydrogen flame exceeds even the above large figure. Employing the same experimental tube and a new burner, the experiments were repeated some days subsequently, with the following result:—

TABLE XXV.—*Radiation through Air.*

Source of heat: *Hydrogen Flame.*

	Absorption per 100
Dry air	0
Undried air	20·3
Total heat	100

The undried air here made use of embraced the carbonic acid of the atmosphere; the air was afterwards conducted through a tube containing a solution of caustic potash, in which the carbonic acid was intercepted, while the air charged itself with a little additional moisture. The absorption then observed amounted to

20·3 per cent.

of the entire radiation. The exact agreement of this with the last result is, of course, an accident; the additional humidity of the air derived from the solution of potash happened to compensate for the action of the carbonic acid withdrawn.

§ 13.

Radiation of Carbonic-oxide Flame through Dry and Humid Air, and through Carbonic Acid Gas.—Further illustration of Influence of Vibrating Period.

The other component of the flame of Bunsen's burner is carbonic acid; and the radiation of this substance is immediately obtained from a flame of carbonic oxide. With the air of the laboratory the following results were obtained:—

TABLE XXVI.—*Radiation through Atmospheric Air.*

Source of heat: *Carbonic-oxide Flame (very small).*

	Deflection	Absorption per 100
Dry air	0°	0
Undried air	10	16·1
Total heat		100

Of the heat emitted by carbonic acid, 16 per cent. was absorbed by the common air of the laboratory. After the air had been passed through sulphuric acid, the aqueous vapour being thus removed while the carbonic acid remained, the absorption was

13·8 per cent.

An india-rubber bag was filled from the lungs; it contained therefore both the aqueous vapour and the carbonic acid of the breath. It was then conducted through a drying apparatus, the mixed air and carbonic acid being permitted to enter the experimental tube. The following results were obtained:—

TABLE XXVII.—*Air from the Lungs containing CO^2.*

Source of heat: *Carbonic-oxide Flame.*

Pressure in inches	Deflection	Absorption per 100
1	7°·2	12
3	15	25
5	20	33·3
30	30·8	50
Total heat		100

Thus the tube filled with dry air from the lungs intercepted 50 per cent. of the entire radiation from a carbonic-oxide flame. It is quite manifest that we have here a means of testing with surpassing delicacy the amount of carbonic acid emitted under various circumstances in the act of expiration.*

That pure carbonic acid is highly opaque to the radiation from the carbonic-oxide flame, is forcibly evidenced by the results recorded in the following table:—

* My late assistant, Mr. W. F. Barrett, subsequently carried out this notion. See article 'On a Physical Analysis of the Human Breath,' *Philosophical Magazine*, vol. xxviii. p. 108.

TABLE XXVIII.—*Radiation through dry Carbonic Acid.*

Source of heat: *Carbonic-oxide Flame.*

Pressure in inches	Deflection	Absorption per 100
1	33°·7	53
2	37	61·7
3	38·6	66·9
4	39·4	70
5	40	72·3
10	41·4	78·7

About four months subsequent to the performance of these experiments they were repeated, using as a source of heat a much smaller flame of carbonic oxide. The absorptions were found somewhat less, but still very high. They follow in the next table.

TABLE XXIX.—*Radiation through dry Carbonic Acid.*

Source of heat: *Small Carbonic-oxide Flame.*

Pressure in inches	Deflection	Absorption per 100
1	17°·3	48
2	20	55·5
3	21·7	60·3
4	22·8	65·1
5	24	68·6
10	26	74·3

For the rays emanating from the heated solids employed in all my former researches, carbonic acid proved to be one of the most feeble of gaseous absorbers; but here, when the waves sent into it emanate from molecules of its own substance, its absorbent energy is enormous. The thirtieth of an atmosphere of the gas cuts off half the entire radiation; while at a pressure of 4 inches, nearly 70 per cent. is intercepted.

§ 14.

Comparative Radiation of Carbonic-oxide Flame through Carbonic Acid Gas and Olefiant Gas.

The energy of olefiant gas, both as an absorbent and a radiant, is well known to the reader of these memoirs; for the solid sources of heat just referred to, its power is incomparably greater, while for the radiation from the carbonic-oxide flame, its power is far feebler than that of carbonic acid. This is proved by the following experiments :—

TABLE XXX.—*Radiation through dry Olefiant Gas.*

Source of heat : *Carbonic-oxide Flame.*

Pressure in inches	Deflection	Absorption per 100
1	17°	24·2
2	26	37·1
4	33	49·1
Total heat . . .	47·3	100

Four months subsequent to the performance of the above experiments, a second series were made with olefiant gas, and the following results obtained :—

TABLE XXXI.—*Radiation through dry Olefiant Gas.*

Source of heat : *Small Carbonic-oxide Flame.*

Pressure in inches	Deflection	Absorption per 100	From Table XXIX.
1	11°·4	23·2	48
2	17	34·7	55·5
3	21·6	44	60·3
4	24·8	50·6	65·1
5	27	55·1	68·6
10	32·1	65·5	74·3

Besides the absorption by olefiant gas, I have placed that by carbonic acid derived from Table XXIX. The superior power of the acid is most decided in the smaller pressures ; at a pressure of an inch it is twice that of the olefiant gas. The substances approach each other more closely as the quantity of gas augments. Here, in fact, both of them approach perfect opacity; and as they draw near to this common limit, their absorptions, as a matter of course, approximate.

§ 15.

Radiation of Hydrogen Flame through Carbonic Acid Gas and Olefiant Gas.

A comparison of these results with the radiation of a hydrogen flame through carbonic acid gas and olefiant gas respectively, brings out with great distinctness the differences of the *radiant qualities* of the two flames.

TABLE XXXII.—*Radiation through Carbonic Acid Gas.*

Source of heat: *Hydrogen Flame.*

Pressure in inches	Deflection	Absorption per 100
1	5°·5	7·4
2	9·5	12·8
4	11	14·9
30	19	25·7
Total heat . . .	48·5	100

TABLE XXXIII.—*Radiation through Olefiant Gas.*

Source of heat: *Hydrogen Flame.*

Pressure in inches	Deflection	Absorption per 100	From Table XXXII.
1	12°	16·2	7·4
2	18	24·3	12·8
4	24	32·4	14·9
30	38·5	58·8	25·7
Total heat . . .	48·5	100	100

A comparison of the last two columns, one of which is transferred from Table XXXII., proves the absorption of the rays from a hydrogen flame by olefiant gas to be about twice that of carbonic acid; while, when the source of heat was a carbonic-oxide flame, the absorption by carbonic acid at small pressures was more than twice that of olefiant gas.

The temperature of a hydrogen flame, as calculated by Bunsen, is 3259° C., while that of a carbonic-oxide flame is 3042° C. The foregoing experiments demonstrate that accord subsists between the oscillating periods of these sources of heat and the periods of aqueous vapour and carbonic acid at a temperature of 15° C. The heat of these flames *goes to augment the amplitude, and not to quicken the vibration.*

§ 16.

Radiation of Carbonic-oxide Flame through Carbonic Oxide, and of Bisulphide-of-Carbon Flame through Sulphurous Acid.

Sent through carbonic oxide, the radiation from the carbonic-oxide flame gave the following absorptions:—

TABLE XXXIV.—*Radiation through Carbonic Oxide.*

Source of heat: *Carbonic-oxide Flame.*

Pressure in inches	Deflection	Absorption per 100
1	18°	29
2	27	43·5
4	34	56·4
10	37·3	65·5

The absorptive energy is here high—higher, indeed, than that of olefiant gas; but it falls considerably short of that of carbonic acid. This result shows that the main radiant in the flame is its *product* of combustion, and not the carbonic oxide heated *prior* to combustion.

To examine the radiation through sulphurous acid of a flame whose product of combustion is sulphurous acid, I resorted to the flame of bisulphide of carbon. Here, however, we had carbonic acid mixed with the sulphurous acid of the flame. Of the heat radiated by this composite source of heat, the absorption by an atmosphere of sulphurous acid amounted to

60 per cent.

The gas was sent from its generating retort through drying-tubes of sulphuric acid into a glass experimental tube 2·8 feet long. The comparative shortness of the tube, and the mixed character of the radiation, rendered the absorption less than it would have been had a source of heat of pure sulphurous acid and a tube as long as that used in the other experiments been employed.

§ 17.

Radiation of the Flames of Carbonic Oxide and Hydrogen through Sulphuric and Formic Ether Vapours.—Reversal of Order of Absorption.

To test the comparative penetrative powers of the two sources of heat I subsequently caused the radiation from the carbonic-oxide flame to pass through the vapours of formic and sulphuric ether at a common pressure of 0·5 of an inch with the following results:—

Table XXXV.

Source of heat: *Carbonic-oxide Flame.*

	Deflection	Absorption per 100
Formic ether	14°·5	25·8
Sulphuric ether	18	32·1
Total heat	43	100

Table XXXVI.

Source of heat: *Hydrogen Flame.*

	Deflection	Absorption per 100
Sulphuric ether	32°	42·2
Formic ether	35	49·3
Total heat	48·5	100

We here find that, in the case of every one of the four vapours, the synchronism with hot aqueous vapour is greater than with hot carbonic acid. The temperature of the hydrogen flame is higher than that of the carbonic oxide; but the radiation from the more intense source of heat, instead of possessing the greatest penetrative power, is the most copiously absorbed. It has been already proved that, for waves of slow period, formic ether is more absorbent than sulphuric ether; while for waves of rapid period, the sulphuric ether is the more powerful absorber. For the radiation from hot carbonic acid, the absorption of sulphuric ether, as shown in Table XXXV., is between 6 and 7 per cent. in excess of that of formic ether; while for the radiation from hot aqueous vapour, the absorption of formic ether, as shown in Table XXXVI., is 7 per cent. in excess of that of sulphuric. That the periods of aqueous vapour, as compared with those of carbonic acid, are slow, and that it is the aqueous vapour, and not the carbonic acid, of the flame of Bunsen's burner which causes the reversal noticed in Table XXIII., may therefore be inferred from these experiments.

§ 18.

Radiation of Hydrogen Flame, and of Platinum Spiral plunged in Hydrogen Flame, through Liquids.—Conversion of Long Periods into Short ones.

Water at moderate thickness is a very transparent substance; that is to say, the periods of its molecules are in discord with those of the visible spectrum. It is also highly transparent to the ultra-violet rays; so that we may safely infer from the deportment of this substance its incompetence to enter into rapid molecular vibration. When, however, we once quit the visible spectrum for the rays beyond the red, the opacity of the substance begins to show itself: for such rays, indeed, its absorbent power is unequalled. The synchronism of the periods of the water-molecules with those of the ultra-red waves is thus demonstrated.

The vibrating-period of a molecule is, no doubt, determined by the elastic forces which separate it from other molecules, and it is worth inquiring how these forces are affected when a change so great as that of the passage of a vapour to a liquid occurs. The fact established in the earlier sections of this paper, that the order of absorption for liquids and their vapours is the same, renders it extremely probable that the period of vibration is not materially affected by the change from vapour to liquid; for, if changed, it would probably be changed in different degrees for the different liquids, and the order of absorption would be thereby disturbed.* The following table will throw additional light upon this question:—

TABLE XXXVII.—*Radiation through Liquids.*

Source of heat: *Hydrogen Flame.*

Thickness of liquid layer, 0·07 of an inch.

Name of liquid	Absorption per 100	Transmission
Bisulphide of carbon	27·7	72·3
Chloroform	49·3	50·7
Iodide of ethyl	75·6	24·4
Benzol	82·3	17·7
Amylene	87·9	12·1
Sulphuric ether	92·6	7·4
Formic ether	93·5	6·5
Acetic ether	93·9	6·1
Water	100	

Through a layer of water 9·21 millimètres thick, Melloni found a transmission of 11 per cent. of the heat of a Locatelli lamp.

* The general agreement in point of colour between a liquid and its vapour favours the idea that the period, at all events in the great majority of cases, remains constant when the state of aggregation is changed.

Here we employ a source of heat of higher temperature, and a layer of water only one-fifth of the thickness used by Melloni, and still we find the whole of the heat intercepted.* A layer of water, 0·07 of an inch in thickness, is sensibly opaque to the radiation from a hydrogen flame. Hence we may infer the coincidence in period between cold water and aqueous vapour heated to a temperature of 3259° C.; and inasmuch as the period of the water-molecules has been proved to be ultra-red, the period of the vapour-molecules in the hydrogen flame must be ultra-red also.

Another point of considerable interest may here be adverted to. Professor Stokes has demonstrated that a change of period is possible to those rays which belong to the violet and ultra-violet end of the spectrum, the change showing itself by a degradation of the refrangibility. That is to say, vibrations of a rapid period are absorbed, and the absorbing substance has become the source of vibrations of a longer period. Efforts, I believe, have been made to obtain an analogous result at the red end of the spectrum, but hitherto without result; and it has been considered improbable that a change of period can occur which should *raise* the refrangibility of the light or heat.

Such a change, I believe, occurs when we plunge a platinum wire into a hydrogen flame. The platinum is rendered white by the collision of molecules whose periods of oscillation are incompetent to excite vision. There is, therefore, in this common experiment an actual breaking up of the long periods into short ones—a true rendering of unvisual periods visual. The change of refrangibility differs from that of Professor Stokes, first, in its being in the opposite direction—that is, from low to high; and secondly, in the circumstance that the platinum is heated by the collision of the molecules of aqueous vapour, and before their heat has assumed the *radiant form*. But *it cannot be doubted that the same effect would be produced by radiant heat of the same period, provided the motion of the æther could be raised to a sufficient intensity*. The effect in principle is

* From the opacity of water to the radiation from aqueous vapour, we may infer the opacity of aqueous vapour to the radiation from water, and hence conclude that the very act of nocturnal refrigeration which causes the condensation of water on the earth's surface gives to terrestrial radiation that particular character which renders it most liable to be intercepted by the aqueous vapour of the air.

the same, whether we consider the platinum wire to be struck by a particle of aqueous vapour oscillating at a certain rate, or by a particle of æther oscillating at the same rate. And thus, I imagine, by a chain of rigid reasoning, we arrive at the conclusion that a degree of incandescence, equal to that of the sun itself, might be produced by the impact of waves of themselves incompetent to excite vision.

The change of quality produced in the radiation by the introduction of a platinum spiral into a hydrogen flame is illustrated by a series of experiments, executed for me by my assistant, Mr. Barrett, and inserted subsequently to the presentation of this memoir.

TABLE XXXVIII.—*Radiation through Liquids.*

Sources of heat: 1. *Hydrogen Flame;*

2. *Hydrogen Flame and Platinum Spiral.*

Name of liquid	Transmission: Thickness of Liquid 0·04 inch, Flame only	Flame and spiral	Thickness of Liquid 0·07 inch, Flame only	Flame and spiral
Bisulphide of carbon . .	77·7	87·2	70·4	86
Chloroform . . .	54	72·8	50·7	69
Iodide of methyl . .	31·6	42·4	26·2	36·2
Iodide of ethyl. . .	30·3	36·8	24·2	32·6
Benzol	24·1	32·6	17·9	28·8
Amylene	14·9	25·8	12·4	24·3
Sulphuric ether . .	13·1	22·6	8·1	22
Acetic ether . . .	10·1	18·3	6·6	18·5
Alcohol	9·4	14·7	5·8	12·3
Water	3·2	7·5	2	6·4

Here the introduction of the platinum spiral changed the periods of the flame into others more in discord with the periods of the liquid molecules, and hence the more copious transmission when the spiral was employed. It will be seen that a transmission of 2 per cent. is here obtained through a layer of water 0·07 of an inch in thickness, while in Table XXVII. all was absorbed.

To test this point further, another series of experiments was executed, and gave the following results for the radiation of a hydrogen flame through layers of water of five different thicknesses:—

Radiation through Water.

Source of heat: *Hydrogen Flame.*

	Thickness of liquid				
	0·02 inch	0·04 inch	0·07 inch	0·14 inch	0·27 inch
Transmission per 100 . .	5·8	2·8	1·1	0·5	0·0

§ 19.

Radiation of Small Gas Flame compared with that of Hydrogen Flame.—Further Changes of Diathermic Position.

Wishing to compare the radiation from a flame of ordinary coal-gas with that of our hydrogen flame, I reduced the former to the dimensions of the latter. The flame thus diminished had a blue base and bright top, and the whole of it was permitted to radiate through our series of liquids. The following results were obtained:—

TABLE XXXIX.—*Radiation through Liquids.*

Source of heat: *Small Gas Flame.*

Thickness of liquid, layer 0·07 of an inch.

Name of Liquid	Deflection	Absorption per 100	From Table XXXVII.
Chloroform . . .	28°·7	39·8	49·3
Bisulphide of carbon .	36	53·2	27·7
Iodide of ethyl . .	41·7	72·3	75·6
Benzol	43·4	79·4	82·3
Amylene	45	86·1	87·9
Sulphuric ether . .	46·6	93·3	92·6
Formic ether . . .	46·6	93·3	93·5
Alcohol	46·8	94·1	
Acetic ether . . .	46·9	94·4	93·9
Water	47·4	97·1	100
Total heat . . .	48	100	

I have placed the results obtained with the hydrogen flame in the third column of figures. It will be observed that the absorption of the heat issuing from the small gas flame is, in some cases, nearly the same as that of the heat issuing from the flame of hydrogen. A very remarkable difference, however, shows itself in the deportment of bisulphide of carbon as compared with that of chloroform. For the small gas flame chloroform is the most transparent body in the list; it is markedly more transparent than the bisulphide of carbon, while for the hydrogen flame the bisulphide greatly excels the chloroform in transparency. The large luminous gas flame previously experimented with differs also from the small one here employed. With the large flame, the absorption by the bisulphide is to that by the chloroform as

100 : 121,

while with the small flame the absorptions of the same two substances stand to each other in the ratio of

100 : 76.

Numerous experiments were subsequently made, with a view of testing this result, but in all cases the bisulphide was found more opaque than the chloroform to the radiation of the small gas flame. The same result was obtained when a very small oil flame was employed; and it came out in a very decided manner when the source of heat was a flame of bisulphide of carbon.

It was found, moreover, that, whenever two liquids underwent a change of position of this kind, the vapours of the liquids underwent a similar change; in its finest gradations the deportment of the liquid was imitated by that of its vapour.

§ 20.

Explanation of Certain Results of Melloni *and* Knoblauch.

And here we find ourselves in a position to offer solutions of various facts which have hitherto stood as enigmas in researches upon radiant heat. It was for a long time supposed that the power of heat to penetrate diathermic substances augmented with the temperature of the source of heat, and from the exceptional penetrative power of solar heat inferences were drawn as to the enormous temperature of the sun. Knoblauch contended against this notion, showing that the heat emitted by a platinum wire plunged into an alcohol flame was less absorbed by certain diathermic screens than the heat of the flame itself, and justly arguing that the temperature of the spiral could not be higher than that of the body from which it derived its heat. A plate of glass being introduced between his source of heat and his thermo-electric pile, the deflection of his needle fell from 35° to 19° when the source of heat was the platinum spiral; while, when the source of heat was the flame of alcohol, the introduction of the same glass caused the deflection to fall from 35° to 16°, proving that the radiation from the flame was intercepted more powerfully than that from the spiral —showing, in other words, that the heat emanating from the body of highest temperature possessed the least penetrative power. Melloni afterwards corroborated this experiment.

Transparent glass allows the rays of the visible spectrum to pass freely through it; but it is well known to be highly opaque to the radiation from obscure sources of heat—in other words, to waves of long period. A plate 2·6 millimètres thick intercepts all the rays from a source of heat of 100° C., and transmits only 6 per cent. of the heat emitted by copper raised to 400° C.* Now the products of the combustion of alcohol are aqueous vapour and carbonic acid, whose waves have just been proved to be of slow period, or of the particular character most powerfully intercepted by glass. But by plunging a platinum wire into such a flame, we virtually convert its heat into heat of higher refrangibility; we break up the long periods into shorter ones, and thus establish the discord between the periods of the source of heat and the periods of the diathermic glass, which, as before defined, is the physical cause of transparency. On purely *à priori* grounds, therefore, we might infer that the introduction of the platinum spiral would augment the penetrative power of the heat through the glass. With two plates of glass, of different thicknesses, Melloni found the following transmissions for the flame and the spiral:—

For the flame	For the platinum
41·2	52·8
5·7	26·2

The same remarks apply to the transparent selenite examined by Melloni. This substance is highly opaque to the ultra-red undulations; but the radiation from an alcohol flame is almost wholly of this character, and hence the opacity of the selenite to this radiation. The introduction of the platinum spiral shortens the periods and increases the transmission. Thus, with two specimens of selenite, of different thicknesses, Melloni found the transmissions to be as follows:—

Flame	Platinum
4·4	19·5
1·7	3·5

* Melloni.

So far the results of Melloni correspond with those of Knoblauch; but the Italian philosopher pursues the matter further, and shows that Knoblauch's results, though true for the particular substances examined by him, are far from being applicable to diathermic media generally. In the case of *black* glass and *black* mica, a striking inversion of the effect is observed; by these substances the radiation from the flame is more copiously transmitted than the radiation from the platinum. For two pieces of black glass, of different thicknesses, Melloni found the following transmissions:—

From the flame	From the platinum
52·6	42·8
29·9	27·1

And for two plates of black mica the following transmissions:—

From the flame	From the platinum
62·8	52·5
43·3	28·9

These results were left unexplained by Melloni; but the solution is now easy. The black glass and the black mica owe their blackness to the carbon incorporated in them, and the blackness of this substance, as already remarked, proves the accord of its vibrating-periods with those of the visible spectrum. But it has been proved that carbon is in a considerable degree pervious to the waves of long period—that is to say, to those emitted by a flame of alcohol. The case of the carbon is therefore precisely antithetical to that of the transparent glass—the former transmitting the heat of long period and the latter the heat of short period most freely. Hence it follows that the introduction of the platinum wire, by converting the long periods of the flame into short ones, augments the transmission through the transparent glass and selenite, and diminishes it through the black glass and the black mica.

§ 21.

Radiation of Hydrogen Flame through Lampblack, Iodine, and Rock-salt.—Diathermancy of Rock-salt examined.

Lampblack, as already stated, is in accord with the undulations of the visible spectrum; it absorbs them all; but it is partially transparent to the waves of slow period. As, therefore, the waves issuing from a flame of hydrogen have been proved to be of slow period, we may with probability infer that its radiation will penetrate the lampblack. A plate of rock-salt was placed over an oil-lamp until the layer of soot deposited on it was sufficient to intercept the light of the brightest gas-flame. The smoked plate was introduced in the path of the rays from the hydrogen flame, and its absorption was measured; the plate was then cleansed, and its absorption again determined. The difference of both gave the absorption of the layer of lampblack. The results were as follows:—

TABLE XL.

	Deflection	Absorption per 100
Smoked rock-salt	44°·2	82·7
Unsmoked plate	15·8	24

The difference between these gives us the absorption of the lampblack; it is 58·7 per cent.; and this corresponds to a transmission of

41·3 per cent.

of the radiation from the hydrogen flame.

Iodine, in a solution sufficiently opaque to cut off the light of our most brilliant lamps, transmitted of the heat of the hydrogen flame

99 per cent.

In experimenting on liquids with heat of slow period, it was noticed that the introduction of the empty rock-salt cell caused the needle to move through a much larger arc than when the source of heat was a luminous one. This suggested that a greater proportion of the heat of slow period was absorbed by the rock-salt. A few experiments were made to test the diathermancy of the salt, with the following results:—

For the heat of a hydrogen flame, the transmission through a perfectly transparent plate of rock-salt was

82·3 per cent.

For a spiral of platinum wire heated to whiteness by an electric current, the transmission was

87 per cent.

For the same spiral lowered to bright redness, the transmission was

84·4 per cent.

For the same spiral lowered to moderate redness, the transmission was

83·6 per cent.

Nothing was changed in these experiments but the heat of the spiral; the direction of the rays, and the size of the radiating body, remained throughout the same; still we find a gradually augmenting opacity on the part of the rock-salt as the temperature of the source of heat is lowered. There cannot, I think, be a doubt that MM. De la Provostaye and Desains are right in their conclusion that rock-salt acts differently on different calorific rays, and is not, as Melloni supposed, equally transparent to all. For the heat of the hydrogen flame, moreover, it is more opaque than for that of the moderately red spiral.

§ 22.

Physical Connexion between Radiation and Conduction.

This memoir ought perhaps to end here. I would, however, ask permission to make a few additional remarks on a subject which was briefly touched upon towards the conclusion of the first of this series of memoirs. These remarks are made with diffidence, for I have reason to know that authorities worthy of the highest respect do not share my views regarding the connexion which subsists between the radiation and conduction of heat.

Let us suppose heat to be communicated to a superficial stratum of the molecules of any body; say, those at the extremity of a metal bar. They vibrate, and the motion communicated by them to the *external* æther is despatched in waves through space. But motion must also be imparted to the æther *within* the body, and a portion of this motion will be transferred to the adjacent stratum of molecules, heat as a consequence appearing to penetrate the mass. But irrespective of the æther, the molecules occupy positions determined by their own attractive and repulsive forces; so that if any one molecule be disturbed, it will of necessity disturb its neighbours.

In an aggregate of molecules so related, motion would be transmitted independently of the æther. If, indeed, we could imagine the æther entirely away, the motion that we call heat would still be propagated from molecule to molecule. In other words, *conduction* would manifest itself, while radiation would be absent through want of a medium.

In matter, however, as we know it, molecular motion is only in part transmitted *immediately* from molecule to molecule, being more or less transmitted mediately by the æther. Now in the case just supposed, the quantity of motion transmitted by the internal æther to our second stratum of molecules cannot be the *whole* of that imparted to it by the superficial stratum. The æther must, to some extent, squander externally the internal molecular motion; so that were the medium absent—were the cushion removed which interferes with the direct propagation of motion from molecule to molecule—conduction would be freer than at present; the heat, suffering no lateral loss, would penetrate further into the mass than when the æther intervenes.

This reasoning leads to the inference that those molecules which yield their motion most freely to the æther must on that account be the most wasteful as regards conduction; in other words, that the best radiators ought to prove themselves the worst conductors.

A broad consideration of the subject shows this conclusion to be in general harmony with observed facts. Organic substances are all exceedingly imperfect conductors of heat, and they are all excellent radiators. The moment, moreover, we pass from the metals to their compounds we pass from good conductors to bad ones, and from bad radiators to good ones.*

* And we also pass, as a general rule, from a series of bodies which vibrate in accord with the visible spectrum to a series which vibrate in discord with the spectrum. The lowering of the rate of vibration is a consequence of chemical union. The comparative incompetence of *compound* bodies to oscillate in visual periods has incessantly declared itself in these researches. I would here refer to a most interesting illustration of the same kind, derived from the experiments of MM. De la Provostaye and Desains. These distinguished experimenters were the first to record the important fact that the qualities of heat emitted by bodies at the same temperature may be very unlike. Two experiments illustrate this. The first is recorded in the *Comptes Rendus*, vol. xxxiv. p. 951. One half of a cube was coated with lampblack, and the other half with cinnabar. The cube being filled with oil at a temperature of 173° C., it was found that the emission from the cinnabar was more copiously absorbed by a

From the earlier memoirs of MM. de la Provostaye and Desains,* and in that of MM. Wiedemann and Franz, I cull the following facts:—The radiative power of platinum is five times that of silver; its conductive power is one-tenth that of silver. Platinum has more than twice the radiative power of gold; it has only one-seventh of the conducting power. Zinc and tin are almost equal as conductors, and they are also nearly equal as radiators. Silver has about six times the conductive power of zinc and tin; it has only one-fourth of their radiative power. Brass possesses but one-half the radiative energy of platinum; it possesses more than twice its conductivity. Other experiments of MM. de la Provostaye and Desains† confirm those first referred to. Taking the absorbent power, as determined by these excellent experimenters, to express the radiating power which will be allowed, and multiplying their results by a common factor to facilitate comparison with those of MM. Wiedemann and Franz on conduction, we obtain the following table:—

TABLE XLII.—*Comparison of Conduction and Radiation.*

Name of metal	Conduction	Radiation
Silver	100	11
Gold	53	27
Brass	24	42
Tin	15	90
Platinum	8	100

We here find that, as the power of conduction diminishes, the power of radiation augments—a result, I think, completely

plate of glass than that from the lampblack. In the second experiment, they found that, while 39 per cent. of the radiation from a bright surface of platinum was transmitted by a plate of glass, only 29 per cent. of the radiation from the opposite surface of the same plate, which was coated with borate of lead, was transmitted. These results are quite in harmony with the views which I have ventured to enunciate. We may infer from them that the heat emitted by the respective *compounds*—the cinnabar and the borate of lead—is of slower period than that emitted by the *elements*; for experiment proves that as the periods are quickened the glass becomes more transparent. At a temperature of 100° C., moreover, the emission from borate of lead was found equal to that from lampblack (*Comptes Rendus*, vol. xxxviii. p. 442), while at a temperature of 550° C. it had only three-fourths of the emissive power of the lampblack. With reference to the theoretic views which these researches are intended to foreshadow, the results of MM. De la Provostaye and Desains are of the highest interest.

* *Comptes Rendus*, 1846, vol. xxii. p. 1139.

† *Annales de Chimie*, 1850, vol. xxx. p. 442.

in harmony with that to which a consideration of the molecular mechanism leads us.

There is but one serious exception known to me to the law here indicated; this is copper, which MM. de la Provostaye and Desains place higher than gold as a radiator, though it is also higher as a conductor. When, however, the immense change in radiative power which the slightest film of an oxide can produce, and the liability of heated copper to contract such a film, are taken into account, the apparent exception will not have too much weight ascribed to it. I have had a cube of brass coated electrolytically with copper, silver, and gold; and, of all its faces, that coated with copper has the least emissive power. This is probably due to some slight impurity contracted by the silver. What we know of the deportment of minerals also illustrates the law. Rock-salt I find to be a far better conductor than glass, while MM. de la Provostaye and Desains find the relative emissive powers of the two substances to be as 17 to 6. So also with regard to alum: as a conductor it is immensely behind rock-salt; as a radiator it is immensely in advance of it.

Royal Institution, March 1864.

VII.

ON LUMINOUS AND OBSCURE RADIATION.

ANALYSIS OF MEMOIR VII.

In the foregoing investigation, the conclusion had been *reasoned out* that the quality of the heat radiated by a flame of hydrogen was almost exclusively ultra-red, and the change produced by the plunging of solid bodies into the flame was pronounced to be a virtual exaltation of refrangibility. These conclusions it was important to verify, and accordingly, the necessary rock-salt lenses and prisms having been secured, the emission from the hydrogen flame was subjected in 1864 to strict analysis.

By direct experiment the reasoning was verified, and the emission was proved to be sensibly ultra-red. The rays of greatest energy of the hydrogen flame were proved to be of precisely the same refrangibility as the rays of greatest energy from a luminous gas-flame.

The other conclusions enunciated in Memoir VI. regarding the raising of solid bodies to incandescence by a hydrogen or an oxyhydrogen flame were also experimentally established.

Intent on clearly bringing out the differences between elementary and compound bodies in relation to radiant heat, I tried at an early period of these researches to extend the experiments to solids and liquids. From the physico-chemical point of view, the deportment of lampblack already revealed by Melloni was to me of peculiar interest and significance. But the interest was greatly augmented by the deportment of bromine and iodine. With various sources of heat the diathermancy of these two substances was illustrated. Leslie's cubes containing boiling water, copper balls heated to various degrees of incandescence, gas and candle flames, were respectively examined, the surprising transparency of bromine and iodine to the calorific rays being in all cases demonstrated.

The step from these experiments to sifting or filtering the radiation from luminous sources, by quenching the light and permitting the heat to pass, was inevitable and indeed immediate. After numerous experiments in the laboratory of the Royal Institution, the filtering of the electric lamp and the formation of powerful dark foci by the heat-rays emitted from the carbon-points, were publicly illustrated in the theatre of the Institution on March 27, 1862. The experiments are referred to in my 'Notes on Heat,' published at the time. They are also mentioned in a foot-note bearing date June 13, 1862, at the bottom of page 79 of this collection.

The discussion with Professor Magnus being, as I thought, finally closed by the experimental evidence brought forward in Memoir VI., and I being still further assured by that investigation of the surprising diathermancy of iodine, the filtration of the emission from incandescent bodies became the subject of special investigation.

The augmentation of the energy of the invisible heat-rays by the increase of temperature necessary to produce the visible ones, is determined; in the first instance, by placing the pile in the ultra-red emission from a platinum spiral, as it rose gradually from a dark heat to an intense white one, a rock-salt prism being used to decompose the beam; in a second instance by causing the spiral to pass through the same range of temperature, and cutting off its luminous rays by the iodine filter.

The experiments are then extended to flames of coal-gas and to the electric-light.

It is thus found that of the radiation from platinum heated to whiteness, one twenty-fourth only consists of luminous rays.

Of the emission from the most brilliant portion of a gas flame, one twenty-fifth only consists of luminous rays.

Of the emission from a dazzling electric-light, one-tenth only consists of luminous rays.

Iodine is found to be perfectly transparent to the emission of bodies at all temperatures under incandescence.

With the rock-salt lens and the iodine filter, the invisible rays of the electric-light are afterwards so concentrated as to ignite combustible bodies placed at the focus.

The eye is proved capable of bearing without inconvenience the heat of a focus where paper and other combustible bodies are ignited and gunpowder was exploded.

Employing greater battery power, precisely the same effects are produced with the glass lenses used in 1862 to concentrate the invisible rays.

From experiments on water, and on the vitreous humour of an ox, it is concluded that nearly two-thirds of the rays from the electric-light, which actually reach the retina, are obscure.

It is further shown that the visible radiation from a red-hot platinum spiral is incapable of thermal measurement.

The paper winds up with some remarks on the relation of light to heat, and on the application of radiant heat to fog-signalling.

VII.

ON LUMINOUS AND OBSCURE RADIATION.*

§ 1.

Spectrum of Hydrogen Flame.

SIR William Herschel discovered the obscure rays of the sun, and proved the position of maximum heat to be beyond the red of the solar spectrum.† Forty years subsequently Sir John Herschel succeeded in obtaining a thermograph of the calorific spectrum, and in giving striking visible evidence of its extension beyond the red.‡ Melloni proved that an exceedingly large proportion of the emission from a flame of oil, of alcohol, and from incandescent platinum heated by a flame of alcohol, is obscure.§ Dr. Miller inferred from its paucity of luminous rays evident to the eye, and a like paucity of ultra-violet rays, that the radiation from a flame of hydrogen must be mainly ultra-red; and he concluded from this that the glowing of a platinum wire in a hydrogen flame, as also the brightness of the Drummond light in the oxyhydrogen flame, are produced by a change in the period of vibration.|| By a different mode of reasoning I arrived at the same conclusion myself, and published the conclusion subsequently.¶

A direct experimental demonstration of the character of the radiation from a hydrogen flame was, however, wanting, and this want I have sought to supply. I had constructed for me, by Mr. Becker, a complete rock-salt train of a size sufficient to permit of its being substituted for the ordinary glass train of a

* *Philosophical Magazine* for November 1864.

† *Phil. Trans.* 1800.

‡ *Phil. Trans.* 1840. I hope very soon to be able to turn my attention to the remarkable results described in note III. of Sir J. Herschel's paper.

§ *La Thermochrose*, p. 304.

|| *Phil. Trans.* vol. cliv. p. 327.

¶ *Report of the British Association*, 1863.

Duboscq's electric lamp. A double rock-salt lens placed in the camera rendered the rays parallel; the rays passed through a slit, and a second rock-salt lens placed without the camera produced, at an appropriate distance, an image of this slit. Behind this lens was placed a rock-salt prism, while laterally stood a thermo-electric pile intended to examine the spectrum produced by the prism. Within the camera of the electric lamp was placed a small burner, so that the flame issuing from it occupied the position usually taken up by the coal points. This burner was connected with a T-piece, from which two pieces of india-rubber tubing were carried, the one to a large hydrogen-holder, the other to the gas-pipe of the laboratory. It was thus in my power to have, at will, either the gas flame or the hydrogen flame. When the former was employed, it produced a visible spectrum, which enabled me to fix the thermo-electric pile in its proper position. To obtain the hydrogen flame, it was only necessary to turn on the hydrogen until it reached the gas flame and was ignited; then to turn off the gas and leave the hydrogen flame behind. In this way the one flame could be substituted for the other without opening the door of the camera, or producing any change in the positions of the source of heat, the lenses, the prism, and the pile.

The thermo-electric pile employed is a beautiful instrument constructed by Ruhmkorff. It belongs to my friend Mr. Gassiot, and consists of a single row of elements properly mounted and attached to a double brass screen. It has in front two silvered edges, which, by means of a screw, can be caused to close upon the pile so as to render its face as narrow as desirable, reducing it to the width of the finest hair, or, indeed, shutting it off altogether. By means of a small handle and long screw, the plate of brass and the pile attached to it can be moved gently to and fro, and thus the vertical slit of the pile can be caused to traverse the entire spectrum, or to pass beyond it in both directions. The width of the spectrum was in each case equal to the length of the face of the pile, which was connected with an extremely delicate galvanometer.

I began with a luminous gas flame. The spectrum being cast upon the brass screen (which, to render the colours more visible, was covered with tinfoil), the pile was gradually moved in the direction from blue to red, until the deflection of the galvanometer became a maximum. To reach this it was necessary to pass entirely through the spectrum and a little way beyond the red; the deflection then observed was

30°.

When the pile was moved *in either direction* from this position, the deflection diminished.

The hydrogen flame was now substituted for the gas flame; the visible spectrum disappeared, and the deflection fell to

12°.

Hence, as regards rays of this particular refrangibility, the emission from the luminous gas flame was two and a-half times that from the hydrogen flame.

The pile was now moved to and fro, and the movement in both directions was accompanied by a diminished deflection. Twelve degrees, therefore, was the maximum deflection for the hydrogen flame; and the position of the pile, determined previously by means of the luminous flame, proves that this deflection was produced by ultra-red undulations. I moved the pile a little forwards, so as to reduce the deflection from 12° to 4°, and then, in order to ascertain the refrangibility of the rays which produced this small deflection, I relighted the gas. The rectilinear face of the pile was found invading the red. When the pile was caused to pass successively through positions corresponding to the various colours of the spectrum, and to its ultra-violet rays, no measurable deflection was produced by the hydrogen flame.

I next placed the pile at some distance from the invisible spectrum of the flame of hydrogen, and *felt* for the spectrum by moving the pile to and fro. Having found it, the place of maximum heating was without difficulty ascertained. Changing nothing else, the luminous flame was substituted for the non-luminous one; the position of the pile when thus revealed was beyond the red.

The action was still very sensible when the distance of the pile from the red end of the spectrum on the one side was as great as that of the violet rays on the other, the heat-spectrum thus proving itself to be at least as long as the light-spectrum.

It is thus *proved experimentally* that the radiation from a hydrogen flame is sensibly ultra-red. The other constituents of the radiation are so feeble as to be thermally insensible. Hence, *when a body is raised to incandescence by a hydrogen flame, the vibrating periods of its atoms must be shorter than those to which the radiation of the flame itself is due.*

§ 2.

Influence of Solid Particles.

The falling of the deflection from 30° to 12° when the hydrogen flame was substituted for the gas flame is doubtless due to the absence of all solid matter in the former. We may, however, introduce such matter, and thus make a radiation originating in the hydrogen flame much greater than that of the gas flame. A spiral of platinum wire plunged in the former gave a maximum deflection of

52°

at a time when the maximum deflection of the gas-flame was only

33°.

It is mainly *by convection* that the hydrogen flame disperses its heat: though its temperature is higher, its sparsely-scattered molecules are not able to cope, in radiant energy, with the solid carbon of the luminous flame. The same is true for the flame of a Bunsen's burner; the moment the air (which destroys the solid carbon particles) mingles with the gas flame, the radiation falls considerably. Conversely, a gush of radiant heat accompanies the shutting out of the air which deprives the gas flame of its luminosity. *When, therefore, we introduce a platinum wire into a hydrogen flame, or carbon particles into a Bunsen's flame, we obtain not only waves of a new period, but also convert a large portion of the heat of convection into the heat of radiation.*

§ 3.

Persistence and Strengthening of Obscure Rays by Augmentation of Temperature.

Bunsen and Kirchhoff have proved that, for incandescent metallic vapours, the period of vibration is, within wide limits, independent of temperature. My own experiments with flames of hydrogen and carbonic oxide as sources of heat, and with cold aqueous vapour and cold carbonic acid as absorbing media, point to the same conclusion.* But in *solid* metals augmented temperature introduces waves of shorter periods into the radiation. It may be asked, 'What becomes of the long obscure periods when we heighten the temperature? Are they broken up or changed into shorter ones, or do they maintain themselves side by side with the new vibrations?' The question is worth an experimental answer.

A spiral of platinum wire, suitably supported, was placed within the camera of the electric lamp at the place usually occupied by the carbon points. This spiral was connected with a voltaic battery; and by varying the resistance it was possible to raise it gradually from a state of darkness to an intense white heat. Raising it to a white heat in the first instance, the rock-salt train was placed in the path of its rays, and a brilliant spectrum was obtained. A thermo-pile was then moved into the region of obscure rays beyond the red of the spectrum. Altering nothing but the strength of the current, the spiral was reduced to darkness, and lowered in temperature till the deflection of the galvanometer fell to 1°. Our question is, 'What becomes of the waves which produce this deflection when new ones are introduced by augmenting the temperature of the spiral?'

Causing the spiral to pass from this state of darkness through various degrees of incandescence, the following deflections were obtained:—

TABLE I.

Appearance of Spiral	Deflection by obscure rays	Appearance of Spiral	Deflection by obscure rays
	°		°
Dark	1	Full red	27
Dark	6	Bright red	44·4
Faint red	10·4	Nearly white	54·3
Dull red	12·5	Full white	60
Red	18		

The deflection of 60° here obtained is equivalent to 122 of the first degrees of the galvanometer. Hence the intensity of

* See Sections 13 and 14 of Memoir VI.

the obscure rays in the case of the full white heat is 122 times that of the rays of the same refrangibility emitted by the dark spiral used at the commencement. Or, as the intensity is proportional to the square of the amplitude, this, in the case of the last deflection, was eleven times that of the waves which produced the first. The *wave-length*, of course, remained the same throughout.

The experimental answer, therefore, to the question above proposed is, that the amplitude of the old waves is augmented by the same accession of temperature that gives birth to the new ones. The case of the obscure rays is, in fact, that of the luminous ones (of the red of the spectrum, for example), which glow with augmented intensity as the temperature of the radiant source of heat is heightened.

§ 4.

Persistence and Strengthening of Rays illustrated by means of a Ray-filter of Iodine and Bisulphide of Carbon.

In my last memoir * the wonderful transparency of the element iodine to the ultra-red undulations was demonstrated. It was there shown that a quantity of iodine sufficient to quench the light of our most brilliant flames transmitted 99 per cent. of the radiation from a flame of hydrogen.

Fifty experiments on the radiant heat of a hydrogen flame, recently executed, make the transmission of its rays, through a quantity of iodine which is perfectly opaque to light,

100 per cent.

To the radiation from a hydrogen flame the dissolved iodine is therefore, according to these experiments, *perfectly transparent.*

It is also sensibly transparent to the radiation from solid bodies heated under incandescence.

It is also sensibly transparent to all the obscure heat-rays emitted by luminous bodies.

To the mixed radiation which issues from solid bodies at a very high temperature, the pure bisulphide of carbon is eminently

* Section 22, Memoir VI.

transparent. Hence, as the bisulphide of carbon interferes but slightly with the obscure rays issuing from a highly luminous source, and as the dissolved iodine seems not at all to interfere with them, we have in a combination of both substances a means of almost entirely detaching the purely thermal rays from the luminous ones.

If vibrations of a long period, established when the radiating body is at a low temperature, maintain themselves, as just indicated, side by side with the new periods which augmented temperature introduces, it would follow that a body once pervious to the radiation from any source must always remain pervious to it. We cannot so alter the character of the radiation that a body once in any measure transparent to it shall become quite opaque to it. We may, by augmenting the temperature, diminish the percentage of the total radiation transmitted by the body; but inasmuch as the old vibrations have their amplitudes enlarged by the very accession of temperature which produces the new ones, the total quantity of heat of any given refrangibility transmitted by the body must increase with increase of temperature.

This conclusion is thus experimentally illustrated. A cell with parallel sides of polished rock-salt was filled with the solution of iodine, and placed in front of the camera within which was the platinum spiral. Behind the rock-salt cell was placed a thermo-electric pile, to receive such rays as had passed through the solution. The rock-salt lens was in the camera in front, but a small sheaf only of the parallel beam emergent from the lamp was employed. Commencing at a very low dark heat, the temperature was gradually augmented to full incandescence with the following results :—

Table II.

Appearance of Spiral	Deflection	Appearance of Spiral	Deflection
	°		°
Dark	1	Full red	45
Dark but hotter	3	Bright red	53
Dark but still hotter	5	Very bright red	63
Dark but still hotter	10	Nearly white	69
Feeble red	19	White	75
Dull red	25	Intense white	80
Red	35		

To the luminous rays from the intensely white spiral the solution was perfectly opaque; but though by the introduction of such rays the transmission, *as expressed in parts of the total radiation,* was diminished, the quantity *absolutely transmitted* was enormously increased. The value of the last deflection is 440 times that of the first; by raising therefore the platinum spiral from darkness to whiteness, we augment the intensity of the obscure rays which it emits in the ratio of 1 : 440.

A rock-salt cell filled with the *transparent* bisulphide of carbon was placed in front of the camera which contained the platinum spiral raised to a dazzling white heat. The transparent liquid was then drawn off and its place supplied by the solution of iodine. The deflections observed in the respective cases are as follows:—

Radiation from White-hot Platinum.

Through Transparent CS^2	Through Opaque Solution
73°·9	73°
73·8	72·9

All the luminous rays passed through the transparent bisulphide; *none* of them passed through the solution of iodine. Still we see what a small difference is produced by their withdrawal. The actual proportion of luminous to obscure, as calculated from the above observations, may be thus expressed:—

Dividing the radiation from a platinum wire raised to a dazzling whiteness by an electric current into twenty-four equal parts, one of these parts is luminous and twenty-three obscure.

A bright gas flame was substituted for the platinum spiral, the top and bottom of the flame were shut off, and its most brilliant portion chosen as the source of rays. The result of forty experiments with this source may be thus expressed:—

Dividing the radiation from the most brilliant portion of a flame of coal-gas into twenty-five equal parts, one of those parts is luminous and twenty-four obscure.

I next examined the ratio of obscure to luminous rays in the electric light. A battery of fifty cells was employed, and the rock-salt lens was used to render the rays from the coal points parallel. To prevent the deflection from reaching an inconvenient magnitude, the parallel rays were caused to issue from a circular aperture 0·1 of an inch in diameter, and were sent alternately through the transparent bisulphide and through the opaque solution. It is not easy to obtain perfect steadiness on the part of the electric-light; but three experiments carefully executed gave the following deflections:—

Radiation from Electric-light.—Experiment No. I.

Through Transparent CS^2	Through Opaque Solution
72°	70°

Experiment No. II.

76·5°	75°

Experiment No. III.

77·5°	76·5°

Calculating from these measurements the proportion of luminous to obscure heat, the result may be thus expressed:—

Dividing the radiation from the electric-light, generated by a Grove's battery of fifty cells, into ten equal parts, one of those parts is luminous and nine obscure.

The results may be thus presented in a tabular form:—

TABLE III.—*Radiation through dissolved Iodine.*

Source of heat	Absorption per 100	Transmission
Dark spiral	0	100
Lampblack at 212° Fahr .	0	100
Red-hot spiral	0	100
Hydrogen flame	0	100
Oil flame	3	97
Gas flame	4	96
White-hot spiral	4·6	95·4
Electric-light.	10	90

Repeated experiments may slightly alter these results, but they are extremely near the truth.

§ 5.

Combustion by Invisible Rays.

Having thus in the solution of iodine found a means of almost perfectly detaching the obscure from the luminous heat-rays of any source, we are able to operate at will upon the former. Here are some illustrations:—The rock-salt lens was so placed in the camera that the coal points themselves and their image beyond the lens were equally distant from the latter. A battery of forty cells being employed, the track of the cone of rays emergent from the lamp was plainly seen in the air, their point of convergence being therefore easily fixed. The cell containing the opaque solution was now placed in front of the lamp. The luminous cone was thereby entirely cut off, but the intolerable temperature of the focus, when the hand was placed there, showed that the calorific rays were still transmitted. Thin plates of tin and zinc were placed successively in the dark focus and speedily fused; matches were ignited, gun-cotton exploded, and brown paper set on fire. Employing the iodine solution and a battery of sixty of Grove's cells, all these results were readily obtained with the ordinary glass lenses attached to Duboscq's electric lamp. They cannot, I think, fail to give pleasure to those who repeat the experiments. It is extremely interesting to observe in the middle of the air of a perfectly dark room a piece of black paper suddenly pierced by the invisible rays, and the burning ring expanding on all sides from the centre of ignition.

On the 15th of this month I made a few experiments on solar light. The heavens were not free from clouds, nor the London atmosphere from smoke, and at best only a portion of the action which a clear day would have given, was obtained. I happened to possess a hollow lens, which I filled with the concentrated solution of iodine. Placed in the path of the solar rays, a faint red ring was imprinted on a sheet of white paper held behind the lens, the ring contracting to a faint red spot when the focus of the lens was reached. It was immediately found that this ring was produced by the light which had penetrated the thin rim of the liquid lens. Pasting a zone of black paper round the rim, the ring was entirely cut off and no visible trace of solar light crossed the lens. At the focus, whatever light passed would be intensified nine-hundredfold; still even here no light was visible.

Not so, however, with the sun's obscure rays; the focus was burning hot. A piece of black paper placed there was instantly pierced and set on fire; and by shifting the paper, aperture after aperture was formed in quick succession. Gunpowder was also exploded. In fact we had in the focus of the sun's dark rays heat decidedly more powerful than that of the electric-light similarly condensed, and all the effects obtained with the latter could be obtained in an increased degree with the former.

A plano-convex lens of glass, larger than the opaque lens just referred to, was introduced into the path of the sun's rays. The focus on white paper was of dazzling brilliancy; and in this focus the results already described were obtained. A cell containing a solution of alum was then introduced in front of the focus. The intensity of the *light* at the focus was not sensibly changed; still these almost intolerable visual rays, aided as they were by a considerable quantity of invisible rays which had passed through the alum, were incompetent to produce effects which were obtained with ease in the perfectly dark focus of the opaque lens.

To show that this reduction of power was not due to the withdrawal of heat by reflection from the sides of the glass cell, I put in its place a rock-salt cell filled with the opaque solution. Behind this cell the rays manifested the power which they exhibited in the focus of the opaque lens.

§ 6.

Melloni's Method of determining the Ratio of Visible to Invisible Rays.—Diathermancy of Alum and of the Humours of the Eye.

Melloni's experiments led him to conclude that rock-salt transmits obscure and luminous rays equally well, and that a solution of alum of moderate thickness entirely intercepts the invisible rays, while it allows all the luminous ones to pass. Hence the difference between the transmissions of rock-salt and alum ought to give the obscure radiation. In this way Melloni found that 10 per cent. only of the radiation from an oil flame consists of luminous rays. The method above employed proves that the proportion of luminous heat to obscure, in the case of an oil flame, is probably not more than one-third of what Melloni made it.

In fact this distinguished man clearly saw the possible inaccuracy of the conclusion that none but luminous rays are transmitted by alum; and the following experiments justify the clauses of limitation which he attached to his conclusion:—

The solution of iodine was placed in front of the electric lamp, the luminous rays being thereby intercepted. Behind the rock-salt cell containing the opaque solution was placed a glass cell, empty in the first instance. The deflection produced by the obscure rays which passed through both produced a deflection of

80°.

The glass cell was now filled with a concentrated solution of alum; the deflection produced by the obscure rays passing through both solutions was

50°.

Calculating from the values of these deflections, it was found *that of the obscure heat emergent from the solution of iodine, and from the side of the glass cell,* 20 *per cent. was transmitted by the alum.*

A point of very considerable importance forces itself upon our attention here—namely, the vast practical difference which may exist between the two phrases, 'obscure rays,' and 'rays from an obscure source of heat.' Many writers seem to regard these phrases as equivalent to each other, and are thus led into grave errors. A stratum of alum solution $\frac{1}{25}$th of an inch in thickness is, according to Melloni, entirely opaque to the radiation from all bodies heated under incandescence. In the foregoing experiments the layer of alum solution traversed by the obscure rays of our luminous source of heat was thirty times the thickness of the layer which Melloni found sufficient to quench all rays emanating from obscure sources of heat.

There cannot be a doubt that the invisible rays which have shown themselves competent to traverse such a thickness of the most powerful adiathermic liquid yet discovered are also able to pass through the humours of the eye. The very careful and interesting experiments of M. Janssen,* prove that the humours of the eye absorb an amount of radiant heat exactly equal to that absorbed by a layer of water of the same thickness, and in our solution the power of alum is added to that of water. Direct experiments on the vitreous humour of an ox lead me to conclude that one-fifth of the obscure rays emitted by an intense electric-light reaches the retina; and inasmuch

* *Annales de Chimie et de Physique*, tom. lx. p. 71.

as in every ten equal parts of the radiation from the carbon points nine consist of obscure rays, it follows that nearly two-thirds of the whole radiant energy which actually reaches the retina is incompetent to excite vision. With a white-hot platinum spiral as source of heat, the mean of four good experiments gave a transmission of 11·7 per cent. of the obscure heat of the spiral through a layer of distilled water 1·2 inch in thickness. A larger proportion no doubt reaches the retina.*

Converging the beam from the electric lamp by a glass lens, I placed the opaque solution of iodine before my open eye, and brought the eye into the focus of obscure rays; the heat was immediately unbearable. But the unpleasant effect seemed to be mainly due to the action of the obscure rays upon the eyelids and other opaque parts round the eye. Through an aperture in a card, somewhat larger than the pupil, the concentrated calorific beam was subsequently permitted to enter the eye. The sense of heat entirely disappeared. Not only were the rays received by the retina incompetent to excite vision, but the optic nerve seemed unconscious of their existence even as heat.

On a tolerably clear night a candle flame can be readily seen at the distance of a mile. The intensity of the electric-light used by me is 650 times that of a good composite candle, and as the non-luminous radiation from the coal points which reaches the retina is equal in energy to twice the luminous, it follows that at a common distance of a foot, the energy of the invisible rays of the electric-light which reach the optic nerve, but are incompetent to provoke vision, is 1,300 times that of the light of a candle. But the intensity of the candle's light at the distance of a mile is less than the twenty-millionth of its intensity at the distance of a foot, hence the energy which renders the candle perfectly visible a mile off would have to be multiplied by 1,300 × 20,000,000, or by twenty-six thousand millions, to bring it up to the energy sent to a retina placed at a foot distance from the electric-light, but which, notwithstanding its enormous relative magnitude, is utterly incompetent to excite vision. Nothing, I think, could more forcibly illustrate the special relationship which subsists between the optic nerve and the oscillating periods of luminous bodies. The nerve may be compared to a musical string, which responds

* M. Franz has shown that a portion of the sun's obscure rays reach the retina.

to periods with which it is in accordance, while it refuses to be excited by others of vastly greater energy which are not in unison with its own.

By means of the opaque solution of iodine, I have already shown that the quantity of luminous heat emitted by a bright red platinum spiral is immeasurably small.* Here are some determinations since made with the same source of heat and a solution of iodine in iodide of ethyl, the strength and thickness of the solution being such as entirely to intercept the luminous rays :—

Radiation from Red-hot Platinum Spiral.

Through Transparent Liquid	Through Opaque Solution
43°·7	43°·7
43·7	43·7

These experiments were made with exceeding care, and all the conditions were favourable to the detection of the slightest difference in the amount of heat reaching the galvanometer; still the quantity of heat transmitted by the opaque solution was found to be the same as that transmitted by the transparent one. In other words, the luminous radiation intercepted by the former, though competent to excite vividly the sense of vision, was, when expressed in terms of actual energy, absolutely incapable of measurement.

And here we have the solution of various difficulties which from time to time have perplexed experimenters. When we see a vivid light incompetent to affect our most delicate thermoscopic apparatus, the idea naturally presents itself that light and heat must be totally different things. The pure light emerging from a combination of water and green glass, even when rendered intense by concentration, has, according to Melloni, no sensible heating power.† The light of the moon is also a case in point. Concentrated by a polyzonal lens more than a yard in diameter upon the face of his pile, it required all Melloni's acuteness to *nurse* the calorific action up to a measurable quantity. Such experiments, however, demonstrate, not that the two agents are dissimilar, but that the sense of vision can be excited by an amount of power almost infinitely small.

* Section 7, Memoir VI.

† Taylor's *Scientific Memoirs*, vol. i. p. 392.

Here also we are able to offer a remark as to the applicability of radiant heat to fog-signalling.* The proposition, in the abstract, is a philosophical one; for were our fogs of a physical character similar to that of the iodine held in solution by the bisulphide of carbon, or to that of iodine or bromine vapour, it would be possible to transmit through them powerful beams of radiant heat, even after the entire stoppage of the light from our signal lamps. But our fogs are not of this character. They are unfortunately so constituted as to act very destructively upon the purely calorific rays; and this fact, taken in conjunction with the marvellous sensitiveness of the eye, leads to the conclusion that, long before the *light* of our signals ceases to be visible, their radiant heat has lost the power of affecting, in any sensible degree, the most delicate thermoscopic apparatus that we could apply to their detection.

Royal Institution, October 1864.

* Which had been proposed a short time prior to the writing of this paper.

VIII.

ON CALORESCENCE, OR THE TRANSMUTATION OF HEAT-RAYS.

ANALYSIS OF MEMOIR VIII.

The separate memoirs of this collection are seen to be so many links in a connected chain of investigation, each new inquiry arising out of some observation or suggestion made in those preceding it.

Thus the diathermancy of the elementary gases led to the discovery of the diathermancy of elementary liquids, embracing solutions of sulphur and iodine. This again led to the experiments made in 1862 on the formation of invisible foci by the filtered beam of the electric-light. The possession of rock-salt lenses in 1864 enabled me to ignite combustible substances at the dark focus; and having obtained this mastery over the subject, it was found possible, by adopting the precise arrangement employed in March 1862, to produce combustion.

The present memoir is a further step in this direction. Reference is first made to the discovery of the sun's invisible rays by Sir William Herschel. The results of Professor Müller are also referred to, and experiments are described in which the distribution of heat in the spectrum of the electric-light is strictly determined by measurement, and represented graphically.

It is thus proved by experiments involving prismatic analysis that the thermal energy of the invisible radiation of a very powerful electric-light is eight times that of the visible.

The same result is deduced from experiments with the iodine-filter.

Various efforts made to intensify the dark foci of the electric-light; the introduction and improvement of small concave mirrors with a view to concentration, and the greatly augmented effects of combustion at the dark foci thus intensified, are fully described and illustrated.

In Memoir VI. a vibrating molecule of aqueous vapour is compared with a vibrating particle of the luminiferous æther, and it is contended that, as regards the change of period produced by rendering refractory bodies incandescent, the principle is the same. This conclusion is confirmed. For it is proved that a sheet of platinum, which when plunged into a flame of hydrogen becomes white-hot, is also heated to whiteness at the perfectly dark focus of the filtered electric beam.

From the dark rays thus transmuted by the platinum, a spectrum may be obtained embracing all the visual vays from red to blue.

A perfectly invisible image of the carbon points is shown to be formed by the heat-rays, the image being changed to a bright incandescent one when the invisible rays fall on platinised platinum.

It is shown that the eye can be placed, without inconvenience, at a dark focus sufficiently intense to heat platinum to redness.

This change of vibration from slow to quick, and from unvisual to visual periods, is called *calorescence.*

The precautions needed in dealing with the inflammable bisulphide of carbon are fully dwelt upon, and various methods of handling the dark rays with safety and certainty are described.

Experiments on the calorescence of the sun's obscure rays and those of the lime-light are also described.

Experiments on the relation of colour to combustion by the dark rays are also recorded.

Calorescence with other filters than the iodine one is also shown to be possible. The memoir winds up by some remarks on the defects of the black-bulb thermometer.

VIII.

ON CALORESCENCE, OR THE TRANSMUTATION OF HEAT RAYS.*

Forsitan et roseâ sol altè lampade lucens
Possideat multum cæcis fervoribus ignem
Circum se, nullo qui sit fulgore notatus,
Æstiferum ut tantum radiorum exaugeat ictum.
LUCRET. v. 610.†

§ 1.

General Statement of the Nature of this Inquiry.

In the year 1800, and in the same volume of the 'Philosophical Transactions' that contains Volta's celebrated letter to Sir Joseph Banks on the Electricity of Contact,‡ Sir William Herschel published his discovery of the invisible rays of the sun. Causing thermometers to pass through the various colours of the solar spectrum, he determined their heating-power, and found that this power, so far from ending at the red extremity of the spectrum, rose to a maximum at some distance beyond the red. The experiment proved that, besides its luminous rays, the sun emitted others of low refrangibility, possessing great calorific power, but incompetent to excite vision.

Drawing a datum-line to represent the length of the spectrum, and erecting at various points of this line perpendiculars to represent the calorific intensity existing at those points, on uniting the ends of the perpendiculars Sir William Herschel

* Received October 20th, and read before the Royal Society, November 23, 1865; *Philosophical Transactions* for 1866, p. 1; *Philosophical Magazine* for May and June 1866. The phrase 'transmutation of rays' is, I believe due to Professor Challis.

† I am indebted to my excellent friend Sir Edmund Head for this extract.

‡ Vol. lxx.

obtained the subjoined curve (fig. 19), which shows the distribution of heat in the solar spectrum, according to his observations. The space A B D represents the invisible, and B D E the visible radiation of the sun.

Fig. 19.

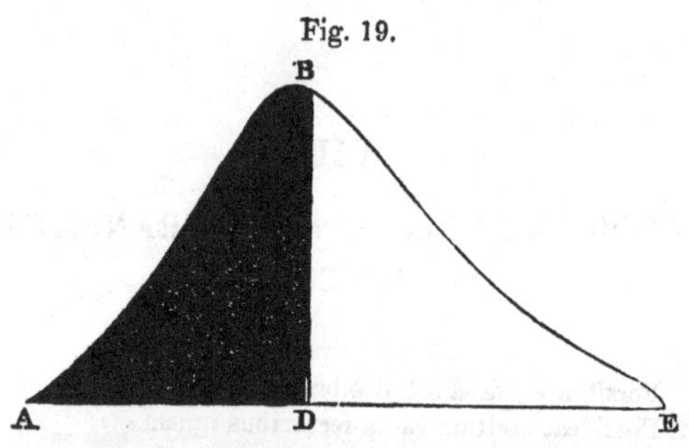

SPECTRUM OF SUN (HERSCHEL) REDUCED.

With the more perfect apparatus subsequently devised, Professor Müller of Freiburg examined the distribution of heat in the spectrum,* and the results of his observations are rendered graphically in fig. 20. Here the area A B C D represents the invisible, while C D E represents the visible radiation.

Fig. 20.

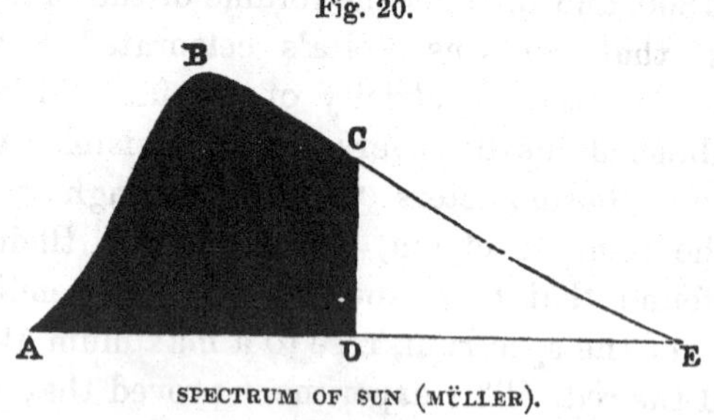

SPECTRUM OF SUN (MÜLLER).

With regard to terrestrial sources of heat, it may be stated that all such sources hitherto examined emit those obscure rays. Melloni found that 90 per cent. of the emission from an oil flame, 98 per cent. of the emission from incandescent platinum, and 99 per cent. of the emission from an alcohol flame consists of obscure rays.† The visible radiation from a hydrogen flame is, according to my own experiments, too small to admit of measurement. With regard to solid bodies, it may be stated generally that, when they are raised from a state of obscurity to vivid incandescence, the invisible rays emitted in the first

* *Philosophical Magazine*, S. 4. vol. xvii. p. 242.

† *La Thermochrose*, p. 304.

instance continue to be emitted with augmented power when the body glows. For example, with a current of feeble power the carbons of the electric lamp may be warmed and caused to emit invisible rays. But the intensity of these same rays may be augmented a thousandfold by raising the carbons to the temperature necessary for the electric-light. Here, in fact, the luminous and non-luminous emission augment together, the maximum of brightness of the visible rays occurring simultaneously with the maximum calorific power of the invisible ones.*

At frequent intervals during the past ten or twelve years I have had occasion to experiment on the invisible rays of the electric-light, and the discovery of the iodine-filter enables me now to make them the subject of special investigation. I endeavour, in the first place, to compare the luminous with the non-luminous radiation of the electric-light, and to determine their relative energy; then a method is pointed out of detaching the luminous from the non-luminous rays, and of concentrating the latter in intense invisible foci. Various experiments illustrative of the calorific power of the invisible rays, and of their transmutation into visible ones are afterwards described.

§ 2.

Source of Rays.—Employment of Rock-salt Train.

Through the kindness of my friend Mr. Gassiot, a very beautiful linear thermo-electric pile, constructed by Ruhmkorff, has remained in my possession for several years, and been frequently employed in my researches. It consists of a double metallic screen, with a rectangular aperture in the centre, a single row of thermo-electric elements 1·2 inch in length being fixed to the screen behind the aperture. Connected with the latter are two moveable side pieces, which can be caused to approach or recede so as to vary the width of the exposed face of the pile from zero to $\frac{1}{10}$th of an inch. The instrument is mounted on a slider, which, by turning a handle, is gradually moved along a massive metal stand. A spectrum of a width equal to the length of the thermo-electric pile being cast at the proper

* On this point see the *Rede Lecture* for 1865, p. 33 (Longmans). Reprinted in *Fragments of Science*, 1871 (Longmans).

elevation on the screen, by turning the handle of the slider the vertical face of the pile can be caused to traverse the colours, and also the spaces right and left of them.

To produce a steady spectrum of the electric-light, I employed the regulator devised by M. Foucault and constructed by Duboscq, the constancy of which is admirable. A complete rock-salt train of high transparency, constructed for me by Mr. Becker, was arranged in the following manner:—In the camera was placed a rock-salt lens, which reduced to parallelism the divergent rays proceeding from the carbon points. The parallel beam was permitted to pass through a narrow vertical slit. In front of this was another rock-salt lens, which produced a sharply-defined image* of the slit at a distance beyond it equal to that at which the spectrum was to be formed. Immediately behind this lens was placed a pure rock-salt prism with its axis vertical—sometimes a pair of prisms. The beam was thus decomposed, a brilliant horizontal spectrum being cast upon the screen which bore the thermo-electric pile. By turning the handle already referred to, the face of the pile could be caused to traverse the spectrum, an extremely narrow band of light or radiant heat falling upon it at each point of its march.† The pile was connected with an exceedingly sensitive galvanometer, by which the heating-power of every part of the spectrum, visible and invisible, was determined.

§ 3.

Methods of Experiments and Tabulated Results.

Two modes of moving the instrument were practised. In the first the face of the pile was brought up to the violet end of the spectrum, where the heat was insensible, and then moved through the colours to the red, then past the red up to the position of maximum heat, and afterwards beyond this position until the heat of the invisible spectrum gradually faded away. The following table contains a series of measurements executed in this manner. The motion of the pile is measured by turns of its handle, every turn corresponding to the shifting of the face of the instrument through a space of one

* The width of the image was about 0·1 of an inch.

† The width of the linear pile was 0·03 of an inch.

millimètre, or $\frac{1}{25}$th of an inch. At the beginning, where the increment of heat was slow and gradual, the readings were taken at every two turns of the handle; on quitting the red, where the heat suddenly increases, the intervals were only half a turn, while near the maximum, where the changes were most sudden, the intervals were reduced to a quarter of a turn, which corresponded to a translation of the pile through $\frac{1}{100}$th of an inch. Intervals of one and of two turns were afterwards resumed until the heating-power ceased to be distinct. At every halting-place the deflection of the needle was noted, and its value ascertained from the table of calibration.

It was found convenient to call the maximum effect in each series of experiments 100. The first column of figures in the table gives the values of the deflections, expressed in terms of the lowest degree of the galvanometer; the second column, obtained by multiplying the first by the constant factor 1·37, expresses the heat of all the parts of the spectrum with reference to the maximum of 100.

TABLE I.—*Distribution of Heat in Spectrum of Electric Light.*

Movement of Pile	Value of deflection	Calorific intensity, in 100ths of the maximum
Before starting (pile in the blue).	0	0
Two turns forward (green entered)	1·5	2
„	3·5	4·8
„	5·5	7·5
„ (red entered) .	15·5	21
„ (extreme red) .	32·6	44·6
Half turn forward . . .	44	60
„	54	74
„	62	85
„	70	95·8
„	72·5	99
Quarter turn forward, *maximum* .	73	100
„	70·8	97
Half turn forward . . .	57	78
„	45·5	62
„	32·6	44·5
„	26	35·6
Two turns forward . . .	10·5	14·4
„	6·5	9
„	5	6·8
„	3·5	5
„	2·5	3·4
„	1·7	2·3
. . . . .	1·3	1·8

Here, as before stated, we begin in the blue, and pass first through the whole visible spectrum. Quitting this at the place marked 'extreme red,' we enter the invisible calorific spectrum and reach the position of maximum heat, from which, onwards, the thermal power falls till it practically disappears.

In other observations the pile was first brought up to the position of maximum heat, and moved thence to the extremity of the spectrum in one direction. It was then brought back to the maximum, and moved to the extremity in the other direction. There was generally a small difference between the two maxima, arising, no doubt, from some slight alteration of the electric-light during the period which intervened between the two observations. The following table contains the record of a series of such measurements. As in the last case, the motion of the pile is measured by turns of the handle, and the values of the deflections are given with reference to a maximum of 100.

TABLE II.—*Distribution of Heat in Spectrum of Electric Light.*

Movement of Pile	Calorific intensity, in 100ths of the maximum
Maximum	100
One turn *towards* visible spectrum	94·4
,, ,, ...	65·5
,, ,, ...	42·6
,, ,, (extreme red) .	28·3
, ,, ...	20
,, ,, ...	14·8
,, ,, ...	11·1
Two turns in the same direction (green entered) .	7·4
,, ,, ...	4·6
,, ,, ...	2
,, ,, (pile in blue) .	0·9

Pile brought back to maximum.

Movement of Pile	Calorific intensity, in 100ths of the maximum
Maximum	100
One turn *from* visible spectrum	67·1
,, ,, ...	41
,, ,, ...	23
,, ,, ...	13
,, ,, ...	9·4
Two turns	5
,, ...	3·4
,, ...	0

§ 4.

Graphic Representation of Results.—Curve of the Electric Spectrum. —Deviations from Solar Spectrum.

More than a dozen series of such measurements were executed, and afterwards plotted as ordinates from a datum-line representing the length of the spectrum. Uniting the ends of these ordinates a number of curves were obtained, each of which represented the distribution of heat in the spectrum, as shown by the corresponding series of observations. On superposing, by means of tracing-paper, the different curves, a very close agreement was found to exist between them. The annexed diagram (fig. 21), which is the mean of several, expresses, with a close approximation to accuracy, the distribution of heat in the spectrum of the electric-light from fifty cells of Grove. The space A B C D represents the invisible, while C D E represents the visible radiation. We observe the gradual augmentation of thermal power, from the blue end of the spectrum to the red. But in the region of dark rays beyond the red the curve shoots suddenly upwards in a steep and massive peak, which quite dwarfs by its magnitude the portion of the diagram representing the visible radiation.*

The sun's rays before reaching the earth have to pass through

* How are we to picture the vibrating atoms which produce the different wave-lengths of the spectrum? Does the infinity of the latter, between the extreme ends of the spectrum, answer to an infinity of atoms each oscillating at a single rate? or are we not to figure the atoms as virtually capable of oscillating at different rates at the same time? When a sound and its octave are propagated through the same mass of air, the resultant motion of the air is the algebraic sum of the two separate motions impressed upon it. The ear decomposes this motion into its two components (Helmholtz, *Ton-Empfindungen*, p. 54); still we cannot here figure certain particles of the air occupied in the propagation of the one sound, and certain other particles in the propagation of the other. May not what is true of the air be true of the æther? and may not, further, a single atom, controlled and jostled as it is in *solid* bodies by its neighbours, be able to impress upon the æther a motion equivalent to the sum of the motions of several atoms each oscillating at one rate?

It is perhaps worthy of remark, that there appears to be a definite rate of vibration for all solid bodies having the same temperature, at which the *vis viva* of their atoms is a maximum. If, instead of the electric-light, we examine the lime-light, or a platinum wire raised to incandescence by an electric current, we find the apex of the curve of distribution (B, fig. 21) corresponding throughout to very nearly, if not exactly, the same refrangibility. There seems, therefore, to exist one special rate at which the atoms of heated solids oscillate with greater energy than at any other rate—a non-visual period, which lies about as far from the extreme red of the spectrum on the invisible side as the commencement of the green on the visible one.

our atmosphere, where they encounter the atmospheric aqueous vapour, which exercises a powerful absorption on the invisible

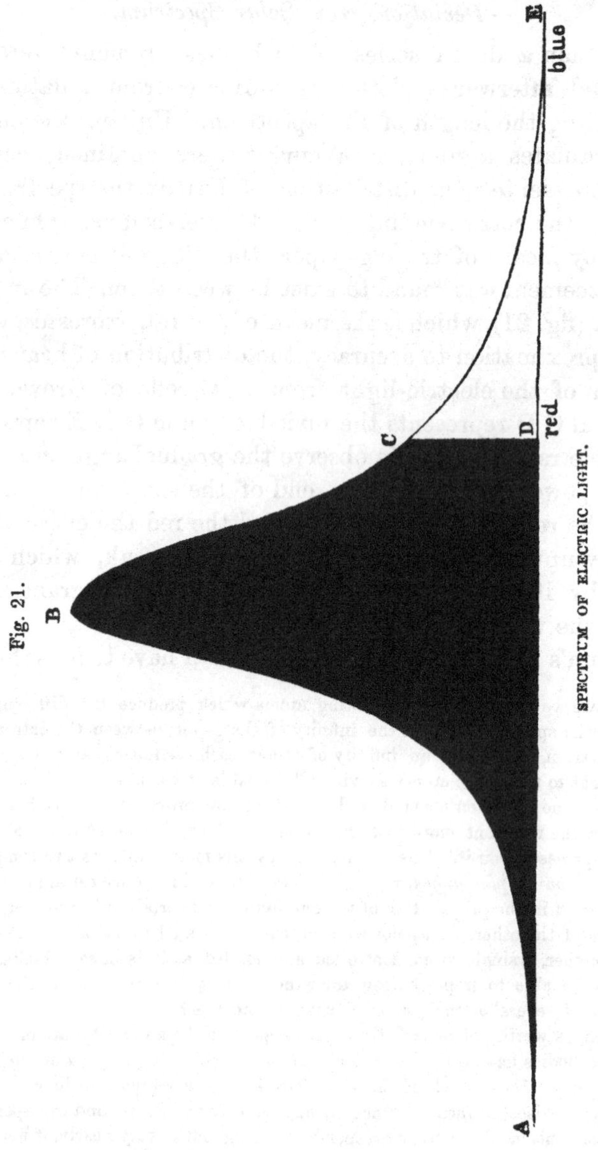

calorific rays. From this, apart from other considerations, it would follow that the ratio of the invisible to the visible radiation in the case of the sun must be less than in the case of the electric-light. Experiment, we see, justifies this conclusion; for whereas fig. 20 shows the invisible radiation of the sun to be about twice the visible, fig. 21 shows the invisible radiation of the electric-light to be nearly eight times the visible. If we cause the beam from the electric lamp to pass through a layer of water of suitable thickness, we place its radiation in approximately the same condition as that of the sun; and on decomposing the beam after it has been thus sifted, we obtain a distribution of heat closely resembling that observed in the solar spectrum.

The curve representing the distribution of heat in the electric spectrum falls most steeply on that side of the maximum which is most distant from the red. On both sides, however, we have a *continuous* falling-off. I have had numerous experiments made to ascertain whether there is any interruption of continuity in the calorific spectrum; but all the measurements hitherto executed with artificial sources reveal a gradual and continuous augmentation of heat from the point where it first becomes sensible up to the maximum. Sir John Herschel has shown that this is not the case with the radiation from the sun when analysed by a flint-glass prism. Permitting the solar spectrum to fall upon a sheet of blackened paper, over which had been spread a wash of alcohol, this eminent philosopher determined by its drying-power the heating-power of the spectrum. He found that the wet surface dried in a series of spots representing thermal maxima separated from each other by spaces of comparatively feeble calorific intensity. No such maxima and minima were observed in the spectrum of the electric-light, nor in the spectrum of a platinum wire raised to a white heat by a voltaic current. Prisms and lenses of rock-salt, of crown glass, and of flint glass were employed in these cases. In subsequent experiments the beam intended for analysis was caused to pass through layers of water and other liquids of various thicknesses. Gases and vapours of various kinds were also introduced into the path of the beam. In all cases there was a general lowering of the calorific power, but the descent of the curve on both sides of the maximum was unbroken.*

* At a future day I hope to subject this question to a more severe examination.

§ 5.

Rays from Obscure Sources of Heat contrasted with Obscure Rays from Luminous Sources of Heat.—Further Observations on the Construction of a Ray-filter.

The rays from an obscure source of heat cannot compete in point of intensity with the obscure rays of a luminous source of heat. No body heated under incandescence could emit rays of an intensity comparable to those of the maximum region of the electric spectrum. If therefore we wish to produce intense calorific effects by invisible rays, we must choose those emitted by an intensely luminous source of heat. The question then arises, how are the invisible calorific rays to be isolated from the visible ones? The interposition of an opaque screen suffices to cut off the visible spectrum of the electric-light, and leaves us the invisible calorific rays to operate upon at our pleasure. Sir William Herschel experimented thus when he sought, by concentrating them, to render the invisible rays of the sun visible. But to form a spectrum in which the invisible rays shall be completely separated from the visible ones, a narrow slit or a small aperture is necessary; and this circumstance renders the amount of heat separable by prismatic analysis very limited. If we wish to ascertain what the intensely concentrated invisible rays can accomplish, we must devise some other mode of detaching them from their visible companions. We must, in fact, discover a substance which shall *filter* the composite radiation of a luminous source of heat by stopping the visible rays and allowing the invisible ones free transmission.

Could we obtain a *black* elementary body thoroughly homogeneous, and with all its parts in perfect optical contact, experiments already published would lead me to expect that such a body would form an effectual filter for the radiation of the sun or of the electric-light. While cutting off the visible radiation, the black element would, I imagine, allow the invisible to pass. Carbon in the state of soot is black, but its parts are not optically continuous. In black glass the continuity is far more perfect, and hence the result established by Melloni, that black glass possesses a considerable power of transmission. Gold in ruby glass, or in the state of jelly prepared by Mr. Faraday, I find to be exceedingly transparent to the invisible calorific rays, but it is not black enough to quench the visible ones. The densely brown liquid bromine is better suited to our purpose; for, in thicknesses sufficient to quench the light of our brightest flames, this element displays extraordinary diathermancy. Iodine cannot be applied in the solid condition, but it dissolves freely in various liquids, the solution in some cases being intensely dark. Here, however, the action of the element may be masked by that of its solvent. Iodine, for example, dissolves freely in alcohol; but then alcohol is so destructive of the ultra-red rays that it would be entirely unfit for experiments the object of which is to retain these rays while quenching the visible ones. The same remark applies in a greater or less degree to most other solvents of iodine.

The deportment of bisulphide of carbon, both as a vapour and a liquid, suggests the thought that it would form a most suitable solvent. It is extremely diathermic, and there is hardly another substance able to hold so large a quantity of iodine in solution. Experiments already recorded prove that, of the rays emitted by a red-hot platinum spiral, 94·5 per cent. is transmitted by a layer of the liquid 0·02 of an inch in thickness, the transmission through layers 0·07 and 0·27 of an inch thick being 87·5 and 82·5 respectively.* The following experiment with a layer of far greater thickness exhibits the deportment of the transparent bisulphide towards the more intense radiation of the electric-light. A cylindrical cell, 2 inches in length and 2·8 inches in diameter, with its ends stopped by plates of perfectly transparent rock-salt, was placed empty in front of an electric lamp; the radiation from the lamp, after having crossed the cell, fell upon a thermo-electric pile, and produced a deflection of

73°.

Leaving the cell undisturbed, the transparent bisulphide of carbon was poured into it: the deflection fell to

72°.

A repetition of the experiment gave the following results:—

* *Philosophical Transactions*, vol. cliv. p. 333; *Philosophical Magazine*, S. 4, vol. xxviii. p. 446.

	Deflection
	°
Through empty cell	74
Through bisulphide	73

Taking the values of these deflections from a table of calibration and calculating the transmission, that through the empty cell being 100, we obtain the following results :—

	Transmission
For the first experiment . . .	94·9
For the second experiment . . .	94·6
Mean . . .	94·8

Hence the introduction of the bisulphide lowers the transmission only from 100 to 94·8.*

The vehicle which holds the iodine in solution would, if perfect for our purpose, be perfectly transparent to the *total* radiation; and the bisulphide of carbon is shown by the foregoing experiment to approach tolerably near perfection. We have in it a body capable of transmitting with little loss the entire radiation of the electric-light. Our object is now to filter this total, by the introduction into the bisulphide of a substance competent to quench the visible and transmit the invisible rays. Iodine does this with marvellous sharpness. In a short paper 'On Luminous and Obscure Radiation,' published in the 'Philosophical Magazine' for November 1864,† the diathermancy of this substance is illustrated by the following table :—

TABLE III.—*Radiation through dissolved Iodine.*

Source	Transmission
Dark spiral of platinum wire . .	100
Lampblack at 212° Fahr. . . .	100
Red-hot platinum spiral . . .	100
Hydrogen flame	100
Oil flame	97
Gas flame	96
White-hot spiral	95·4
Electric-light, battery of 50 cells .	90

These experiments were made in the following way :—A rock-salt cell was first filled with the transparent bisulphide, and the quantity of heat transmitted by the pure liquid to the pile was determined. The same cell was afterwards filled with the opaque

* The diminution of the reflexion from the sides of the cell by the introduction of the bisulphide is not here taken into account.

† Being Memoir VII. of this volume.

solution, the transmission through which was also determined. Calling the transmission through the transparent liquid 100, the foregoing table gives the transmission through the opaque. The results, it is plain, refer solely to the iodine dissolved in the bisulphide,—the transmission 100, for example, indicating, not that the solution itself, but that the dissolved iodine is, within the limits of observation, *perfectly diathermic* to the radiation from the first four sources of heat.

The layer of liquid employed in these experiments was not sufficiently thick to quench utterly the luminous radiation from the electric lamp. A cell was therefore constructed whose parallel faces were 2·3 inches apart, and which, when filled with the solution of iodine, allowed no trace of the most highly concentrated luminous beam to pass through it. Five pairs of experiments executed with this cell yielded the following results:—

Radiation from Electric Light; battery 40 cells.

	Deflection	
	°	°
{ Through transparent bisulphide	47	46
{ Through opaque solution	42·3	43·5
{ Through transparent bisulphide	44	43·7
{ Through opaque solution	41·2	40
Through transparent bisulphide	42	43

Calling the transmission through the transparent liquid 100, and taking the mean of all these determinations, the transmission through the opaque solution is found by calculation to be 86·8. An absorption of 13·2 per cent. is therefore to be set down to the iodine. This was the result with a battery of forty cells; subsequent experiments with a battery of fifty cells made the transmission 89, and the absorption 11 per cent.

Considering the transparency of the iodine for heat emitted by all sources heated up to incandescence, as exhibited in Table III., it may be inferred that the above absorption of 11 per cent. represents the calorific intensity of the *luminous rays* alone. By the method of filtering, therefore, we make the invisible radiation of the electric-light eight times the visible. Computing, by means of a proper scale, the area of the spaces A B C D, C D E (fig. 21), the former, which represents the invisible emission, is found to be 7·7 times the latter. *Prismatic analysis, therefore, and the method of filtering yield almost exactly the same result.*

§ 6.

Invisible Foci of Electric Light.—Efforts to intensify their Heat.—Danger of Bisulphide of Carbon, and trial of other substances.—Final Precautions.

In the combination of bisulphide of carbon and iodine we find a means of filtering the composite radiation from any luminous source. The solvent is practically transparent, while the dissolved iodine cuts off every visible ray, its absorptive power ceasing with extraordinary suddenness at the extreme red of the spectrum. Doubtless the absorption extends a little way beyond the red, and with a very great thickness of solution the absorption of the ultra-red rays might become very sensible. But the solution may be employed in layers which, while competent to intercept every trace of light, allow the invisible calorific rays to pass with scarcely sensible diminution.

The *ray-filter* here described was first publicly employed in the early part of 1862.* Concentrating by large glass lenses the radiation of the electric lamp, I cut off the visible portion of the radiation by the solution of iodine, and thus formed invisible foci of an intensity at that time unparalleled. In the autumn of 1864 similar experiments were executed with rock-salt lenses and with mirrors. The paper 'On Luminous and Obscure Radiation,' already referred to, contains an account of various effects of combustion and fusion which were then obtained with the invisible rays of the electric-light and of the sun.†

* *Philosophical Transactions*, 1862, p. 67, note.

† To the experiments there described the following may be added, as made at the time:—A glass, globe 3¾ inches in diameter, was filled with the opaque solution, and placed in front of the electric-light. An intense focus of invisible rays was formed immediately beyond the globe. Black paper held in this focus was pierced, a burning ring being produced. A second spherical flask, 9 inches in diameter, was filled with the solution and employed as a lens. The effects, however, were less powerful than those obtained with the smaller flask.

Two plano-convex lenses of rock-salt, 3 inches in diameter, were placed with their plane surfaces opposed, but separated from each other by a brass ring ⅜ths of an inch thick. The space between the plates was filled with the solution, an opaque lens being thus formed. Paper was fired by this lens. In none of these cases, however, could the paper be caused to blaze. Hollow plano-convex lenses filled with the solution were not effective, the focal length of those at my disposal being too great.

Mr. Mayall was so extremely obliging as to transfer his great photographic camera

From the setting of paper on fire, and the fusion of non-refractory metals, to the rendering of refractory bodies incandescent, the step was immediate. To avoid waste by conduction, it was necessary to employ the metals in plates as thin as possible. A few preliminary experiments with platinum-foil, which resulted in failure, raised the question whether, even with the *total radiation* of the electric-light, it would be possible to obtain incandescence without combustion. Abandoning the use of lenses altogether, I caused a thin leaf of platinum to approach the ignited coal points. It was observed by myself from behind, while my assistant stood beside the lamp, and looking through a dark glass, observed the distance between the platinum-foil and the electric-light. At half an inch from the carbon points the metal became red-hot. The problem now was *to obtain, at a greater distance, a focus which should possess a heating-power equal to that of the direct rays at a distance of half an inch.*

In the first attempt the direct rays were utilized as much as possible. A piece of platinum-foil was placed an inch distant from the carbon points, there receiving the direct radiation. The rays emitted *backwards* from the points were at the same time converged upon the foil by a small mirror, and were found more than sufficient to compensate for the diminution of intensity due to the withdrawal of the foil to the distance of an inch. By the same method incandescence was subsequently obtained when the foil was removed two, and even three, inches from the carbon points.

The last-mentioned distance allowed me to introduce between the focus and the source of rays a cell containing the solution of iodine. The transmitted invisible rays were found of sufficient power to inflame paper, and to raise platinum-foil to incandescence.

These experiments, however, were not unattended with danger. The bisulphide of carbon is an extremely inflammable substance; and on the 2nd of November, while employing a very powerful battery and intensely heated carbon points, the

from Brighton to London, for the purpose of enabling me to operate with the fine glass lens, 20 inches in diameter, which belonged to it: the result was not successful. It will, however, be subsequently shown that both the hollow lens and the glass lens are effective when, instead of the divergent rays of the electric lamp, we employ the parallel rays of the sun.

solution took fire, and instantly enveloped the electric lamp and all its appurtenances in flame. The precaution, however, had been taken of placing the entire apparatus in a flat vessel containing water, into which the flaming mass was summarily turned. The bisulphide of carbon being heavier than the water, sank to the bottom, so that the flames were speedily extinguished. Similar accidents occurred twice subsequently.

Such occurrences caused me to seek earnestly for a substitute for the bisulphide. Pure chloroform, though not so diathermic, transmits the obscure rays pretty copiously, and it freely dissolves iodine. In layers of the thickness employed, however, the solution was not sufficently opaque; and in consequence of its absorptive power, comparatively feeble effects only were obtained with it. The same remark applies to the iodides of methyl and ethyl, to benzol, acetic ether, and other substances. They all dissolve iodine, but they enfeeble the results by their action on the ultra-red rays.

I had special cells constructed for bromine and chloride of sulphur: neither of these substances is inflammable; but they are both intensely corrosive, and their action upon the lungs and eyes was so irritating as to render their employment impracticable. With both of these liquids powerful effects were obtained; still their diathermancy, though very high, did not come up to that of the dissolved iodine. Bichloride of carbon would be invaluable if its solvent power were equal to that of the bisulphide. It is not at all inflammable, and its own diathermacy appears to excel that of the bisulphide. But in reasonable thicknesses the quantity of iodine which it can dissolve is not sufficient to render the solution perfectly opaque. The solution forms a purple colour of indescribable beauty. Though unsuited to strict crucial experiments on dark rays, this filter may be employed with good effect in lecture experiments.

Thus foiled in my attempts to obtain a solvent equally good and less dangerous than the bisulphide of carbon, I sought to reduce to a minimum the danger of employing this substance. At an earlier period of the investigation a tin camera was constructed, within which were placed both the lamp and its mirror. Through an aperture in front, 2¾ inches wide, the cone of reflected rays issued, forming a focus outside the camera. Underneath this aperture was riveted a stage, on which the solution of iodine rested, covering the aperture and cutting off all the light. In the first experiments nothing intervened between the cell and the carbon points; but the peril of thus exposing the bisulphide caused me to make the following improvements:—First, a perfectly transparent plate of rock-salt, secured in a proper cap, was employed to close the aperture; and by it all direct communication between the

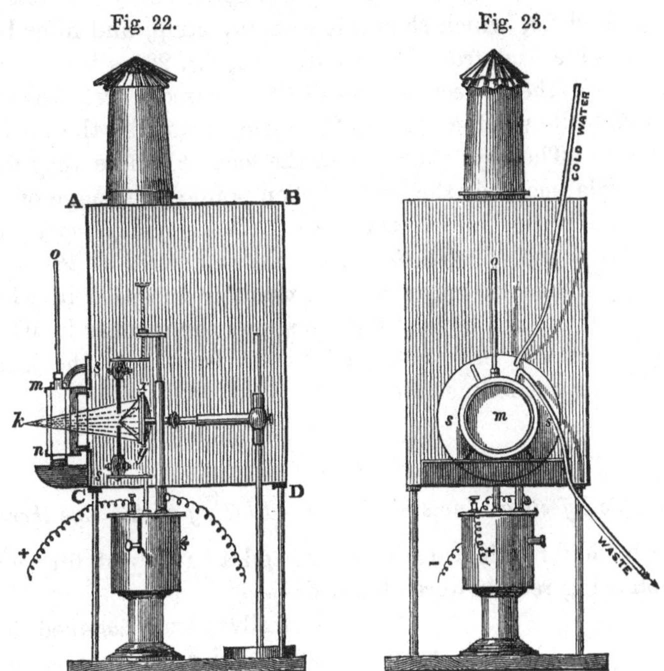

Fig. 22. Fig. 23.

solution and the incandescent carbons was cut off. The camera itself, however, became quickly heated by the intense radiation falling upon it, and the cell containing the solution was liable to be warmed both by the camera and by the luminous heat which it absorbed. The aperture above referred to was therefore surrounded by an annular space, about $2\frac{1}{2}$ inches wide and a quarter of an inch deep, through which cold water was caused to circulate. The cell containing the solution was moreover surrounded by a jacket, and the current, having completed its course round the aperture, passed round the solution.

Thus the apparatus was kept cold. The neck of the cell was stopped by a closely-fitting cork; through this passed a piece of glass tubing, which, when the cell was placed upon its stage, ended at a considerable distance from the focus of the mirror. Experiments on combustion might therefore be carried on at the focus without fear of igniting the small amount of vapour which even under the improved conditions might escape from the bisulphide of carbon.

The arrangement will be at once understood by reference to figs. 22 and 23, which show the camera, lamp, and filter both from the side and from the front. *x y*, fig. 22, is the mirror, from which the reflected cone of rays passes, first, through the rock-salt window and afterwards through the iodine filter *m n*. The rays converge to the focus *k*, where they form an invisible image of the lower carbon point; the image of the upper one being thrown below *k*. *Both images spring vividly forth when a leaf of platinized platinum is exposed at the focus.* At *s s*, fig. 22, is shown, in section, the annular space in which the cold water circulates. Fig. 23 shows the manner in which the water enters this space and passes from it to the jacket surrounding the iodine-cell *m*.

§ 7.

Calorific Effects at Invisible Focus.—Placing of the Eye there.

With the foregoing arrangement, and a battery of fifty cells, the following results were obtained:—

Fig. 24.

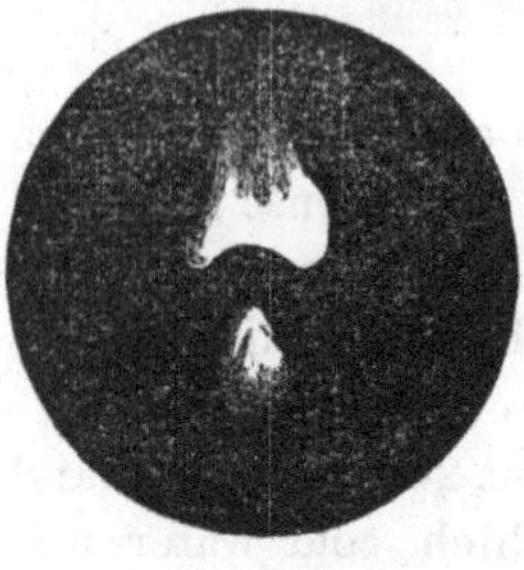

A piece of silver-leaf, fastened to a wire ring and tarnished by exposure to the fumes of sulphide of ammonium, being held in the dark focus, the film flashed out occasionally into vivid redness.

Copper-leaf tarnished in a similar manner, when placed at the focus, was heated to redness.

A piece of platinized platinum-foil was supported in an exhausted receiver, the vessel being so placed that the focus fell upon the platinum. The heat of the focus was instantly converted into light, a clearly-defined and inverted image of the points being stamped upon the metal. Fig. 24 represents the thermograph of the carbons.

Blackened paper was now substituted for the platinum in the exhausted receiver. Placed at the focus of invisible rays, the paper was instantly pierced, a cloud of smoke was poured through the opening, and fell like a cascade to the bottom of the receiver. The paper seemed to burn without incandescence. Here also a thermograph of the coal points was stamped out. When black paper is placed at the focus, where the thermal image is well defined, it is always pierced in two points, answering to the images of the two carbons. The superior heat of the positive carbon is shown by the fact that its image *first* pierces the paper; it burns out a large space, and shows its peculiar crater-like top, while the negative carbon usually pierces but a small aperture.

Paper reddened by the iodide of mercury had its colour discharged at the places on which the invisible image of the coal points fell upon it.

Disks of paper reduced to carbon by different processes were raised to brilliant incandescence, both in the air and in the exhausted receiver.

In these earlier experiments I turned to account apparatus which had been constructed for other purposes. The mirror employed, for example, was detached from a Duboscq's camera, the ordinary silvering at the back being first made use of; the mirror being subsequently improved by silvering in front. The cell employed for the iodine solution was also that which usually accompanies Duboscq's lamp, being intended by its maker for a solution of alum. Its sides are of good white glass, the width from side to side being 1·2 inch.

A point of considerable theoretic importance was involved in these experiments. In his excellent researches on fluorescence, Professor Stokes had invariably found the refrangibility of the incident light to be *lowered*. This rule was so constant as almost to enforce the conviction that it was a law of nature. But if the rays which in the foregoing experiments raised platinum and gold and silver to a red heat were wholly ultra-red, the rendering visible of the metallic films would be an instance of *raised* refrangibility.

And here I thought it desirable to make sure that no trace of visible radiation passed through the solution, and also that the invisible radiation was exclusively ultra-red.

This latter condition might seem to be unnecessary, because the calorific action of the ultra-violet rays is so exceedingly feeble (in fact so immeasurably small) that, even supposing them to reach the platinum, their heating power would be an utterly vanishing quantity. Still there were considerations which rendered the exclusion of *all* rays, of a higher refrangibility than those generated at the focus, necessary to the rigid solution of the problem. Hence, though the iodine employed in the foregoing experiments was sufficient to cut off the light of the sun at noon, I wished to submit its opacity to a severer test. The following experiments were accordingly executed:—

The iodine cell being placed in position, a piece of thick black paper, mounted on a retort-ring, was caused gradually to approach the focus of obscure rays. The position of the focus was announced by the piercing of the paper; the combustion being quenched, the retort-ring was moved slightly nearer to the lamp, so that the converged beam passed through the burnt aperture, the focus falling about half an inch beyond it. A bit of blackened platinum held immediately behind the aperture was raised to redness over a considerable space. The platinum was then moved to and fro until the maximum degree of incandescence was obtained, the point where this occurred being accurately marked. A second cell containing a solution of alum, was then placed between the diaphragm of black paper and the iodine-cell. The alum solution diminished materially the invisible radiation, but it was without sensible influence on such *visible* rays as the concentrated beam contained.

The two cells being in position, all stray light issuing from the crevices in the lamp was cut off; the daylight was also excluded from the room, and the eye being brought to a level with the aperture was slowly moved towards it, until the point which marked the focus was reached. A singular appearance presented itself. The incandescent coke points were seen perfectly black, projected on a deep-red ground. When the points were moved up and down, their black images moved also. When brought into contact, a white space was seen at the extremities of the points, appearing to separate them. The points were seen erect. By careful observation the whole of the carbon-rods could be seen, and even the holders which supported them. The darkness of the incandescent portion of the carbon-rods could of course only be *relative*; they, in fact, intercepted more of the light reflected from the mirror behind than they could make good by their direct emission, hence their apparent blackness.

The solution of iodine, 1·2 inch in thickness, proving unequal to the severe test here applied to it, I had two other cells constructed—the one with transparent rock-salt sides, the other with glass ones. The width of the former was 2 inches, that of the latter nearly 2½ inches. Filled with the solution of iodine, these cells were placed in succession in front of the camera, and the concentrated beam was sent through them. Determining the focus as before, and afterwards introducing the alum-cell, the eye on being brought up to the focus received no impression of light. The alum-cell was then abandoned, and the undefended eye was caused to approach the focus: the heat was intolerable, but it seemed to affect the eyelids and not the retina. An aperture somewhat larger than the pupil being made in a metal screen, the eye was placed behind it, and brought slowly and cautiously up to the focus. Nearly the whole concentrated beam here entered the pupil; but no impression of light was produced, nor was the retina sensibly affected by the heat. The eye was then withdrawn, and a plate of platinized platinum was placed in the position occupied by the retina a moment before. *It instantly rose to vivid redness.**

The rays which produced this incandescence were certainly invisible ones, and the subsequent failure to obtain, with the most sensitive media, and in the darkest room, the slightest evidence of fluorescence at the obscure focus, proved the invisible rays to be exclusively ultra-red.

§ 8.

Improvement of Mirrors.—Exalted Effects of Combustion at Dark Focus.

When intense effects are sought after, the problem is to collect as many of the invisible rays as possible, and to concentrate

* I do not recommend the repetition of these experiments.

them on the smallest possible space. The nearer the mirror is to the source of rays, the more of these rays will it intercept and reflect; and the nearer the focus is to the same source, the smaller, and more intense, will the image be. To secure proximity both of focus and mirror, the latter must be of short focal length. If a mirror of long focal length be employed, its distance from the source of rays must be considerable to bring the focus near the source; but when placed thus at a distance, a great number of rays escape the mirror altogether. If, on the other hand, the mirror be too deep, spherical aberration comes into play; and, though a vast quantity of rays may be collected, their convergence is imperfect.

To determine the best form of mirror, I had three constructed: the first is 4·1 inches in diameter, and of 1·4 inch focal length; the second 7·9 inches in diameter, and of 3 inches focal length; the third 9 inches in diameter, with a focal length of 6 inches. Fractures caused by imperfect annealing repeatedly occurred; but at length I was fortunate enough to obtain the three mirrors, each without a flaw. The mirrors were all silvered in front, and thus the absorption due to the transmission of the heat through glass was avoided. The most convenient distance of the focus from the source I find to be about five inches; and the position of the mirror ought to be arranged accordingly. This distance permits of the introduction of an iodine-cell of sufficient depth, while the heat at the focus is exceedingly powerful.

The isolation of the luminiferous æther from the air is strikingly illustrated by these experiments. The air at the focus may be of a freezing temperature, while the æther possesses an amount of heat competent, if absorbed, to impart to that air the temperature of flame. *An air-thermometer is unaffected where platinum is raised to a white heat.* Numerous experiments will suggest themselves to every one who wishes to operate upon the invisible heat-rays. Dense volumes of smoke rise from a blackened block of wood when it is placed in the dark focus: matches are of course at once ignited, and gunpowder instantly exploded. Dry black paper held there bursts into flame. Chips of wood are also inflamed, the dry wood of a hat-box being very suitable for this experiment. When a sheet of brown paper is placed a little beyond the focus, it is first brought to vivid incandescence over a large space; the paper then yields, and the combustion prapagates itself as a burning ring round the centre of ignition. Charcoal is reduced to an ember at the focus, and disks of charred paper glow with extreme vividness. Sheet-lead and tin, if blackened, may be fused, while a thick cake of fusible metal is quickly pierced and melted. Blackened zinc-foil placed at the focus bursts into flame; and by drawing the foil slowly through the focus, its ignition may be kept up till the whole of the foil is consumed. Magnesium wire, flattened at the end and blackened, also bursts into vivid combustion. A cigar or a tobacco-pipe may of course be instantly lighted at the dark focus. The bodies experimented on may be enclosed in glass receivers; the concentrated rays will still burn them, after having crossed the glass. A small chip of wood in a jar of oxygen bursts suddenly into flame: charcoal burns, while charcoal-bark throws out suddenly showers of scintillations.

§ 9.

Transmutation of Heat Rays.—Calorescence.

In all these cases the body exposed to the action of the invisible rays was more or less combustible. But it required to be heated to initiate the attack of the atmospheric oxygen. Its vividness was in great part due to combustion, and does not furnish a conclusive proof that the refrangibility of the incident rays was elevated. This, which is the result of greatest theoretic import, is effected by exposing non-combustible bodies at the focus, or by enclosing combustible ones in a space devoid of oxygen. Both in air and *in vacuo* platinised platinum-foil has been repeatedly raised to a white heat. The same result has been obtained with a sheet of charcoal or coke suspended *in vacuo*. On looking at the white-hot platinum through a prism of bisulphide of carbon, a rich and complete spectrum was obtained. All the colours, from red to violet, glowed with extreme vividness. The waves from which these colours were primarily extracted had neither the visible nor the ultra-violet rays commingled with them; they were exclusively ultra-red. The action of the atoms of platinum, copper, silver, and carbon upon these rays transmutes them from heat rays into light rays. They impinge upon the platinum at a certain rate; they return from it at a quicker rate. Their refrangibility is thus raised, the invisible being rendered visible.

To express this transmutation of heat rays into others of higher refrangibility, I would propose the term *calorescence.* It harmonises well with the term 'fluorescence' introduced by Professor Stokes, and is also suggestive of the character of the effects to which it is applied. The phrase 'transmutation of rays,' introduced by Professor Challis,* covers both classes of effects.

§ 10.

Various Modes of obtaining with the Electric-light Invisible Foci for Combustion and Calorescence.

In the foregoing section arrangements are described which were made with a view of avoiding the danger incidental to the use of so inflammable a substance as the bisulphide of carbon. I have since thought of accomplishing this end in a simpler way, and thus facilitating the repetition of the experiments. The arrangement shown in Figure 25 may be adopted with safety.

A B C D is an outline of the camera;

x y the silvered mirror within it;

c the carbon points of the electric-light;

o p the aperture in front of the camera, through which issues the beam reflected by the mirror *x y*.

Let the distance of the mirror from the carbon points be such as to render the reflected beam slightly convergent.

Fill a round glass flask with the solution of iodine, and place the flask in the path of the reflected beam at a safe distance from the lamp. The flask acts as a lens and as a filter at the same time, the bright rays are intercepted, and the dark ones are powerfully converged. F, fig. 25, represents such a flask; and at the focus formed a little beyond it combustion and calorescence may be produced.

The following results have been obtained with a series of flasks of different dimensions, at a distance of 3½ feet from the carbon points:—

* *Philosophical Magazine*, S. 4, vol. xii. 521.

1. With a spherical flask, 6¾ inches in diameter: platinum was raised to redness at the focus, and black paper inflamed.

2. Ordinary Florence flask, 3¼ inches in diameter: platinum raised to bright redness over a large irregular space. Near the lamp, the effects obtained with this flask were very striking.

Fig. 25.

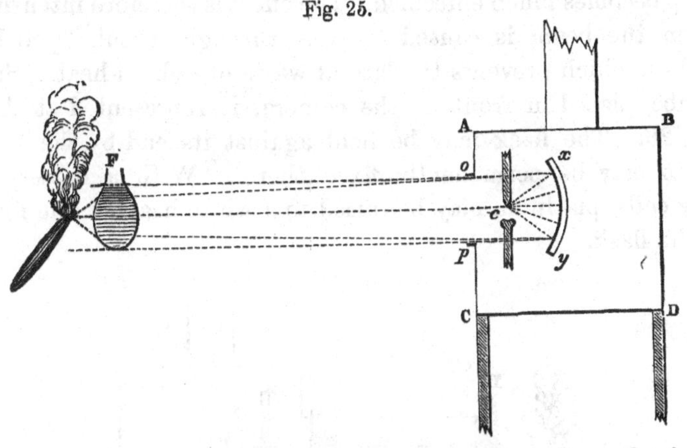

3. Small flask, 1·8 inch in diameter, not quite spherical: platinum rendered white-hot; paper immediately inflamed.

4. A still smaller flask, 1·5 inch in diameter: effects very good; about the same as the last.

Fig. 26.

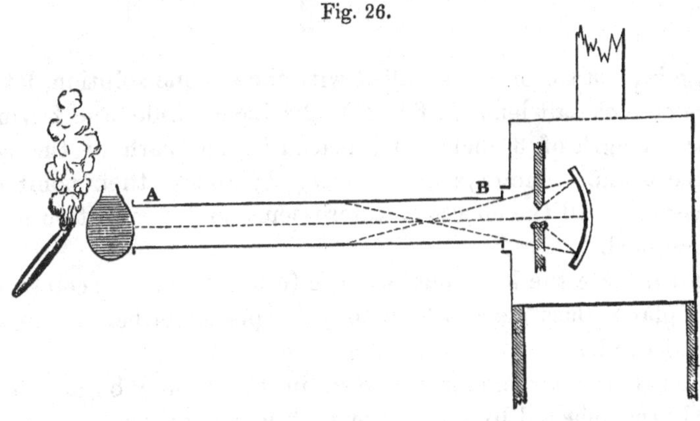

5. The bulb of a pipette: effects striking, but not quite so brilliant as with the less regularly shaped small flasks.

It follows, as a matter of course, that where platinum is heated to whiteness, the combustion of wood, charcoal, zinc, and magnesium may also be effected.

By the arrangement here described, platinum has been heated to redness at a distance of 22 feet from the source of the rays.

The best of mirrors, however, scatters the rays more or less; and, by this scattering, the beam at a great distance from the lamp becomes much enfeebled. The effect is therefore intensified when the beam is caused to pass through a tube polished within, which prevents the lateral waste of radiant heat. Such a tube, placed in front of the camera, is represented at A B, fig. 26. The flask may be held against its end by the hand, or it may be permanently fixed there. With a battery of fifty cells, platinum may be raised to a white heat at the focus of the flask.

Fig. 27.

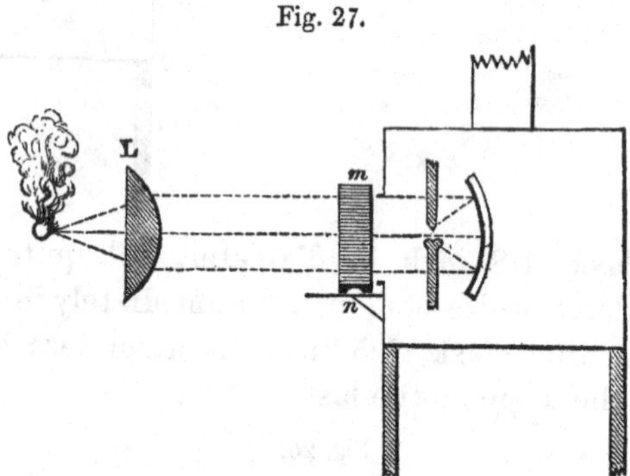

Again, instead of a flask filled with the opaque solution, let a glass or rock salt lens (L, fig. 27), 2·5 inches wide, and having a focal length of 3 inches, be placed in the path of the reflected beam. The rays are converged; and at their point of convergence all the effects of calorescence and combustion may be obtained.

In this case the luminous rays are to be cut off by a cell (*m n*) with plane glass sides, which may be placed either before or behind the lens.

Finally, the arrangement shown in fig. 28 may be adopted. The beam reflected by the mirror within the camera is received and converged by a second mirror, *x′ y′*. At the point of convergence, which may be several feet from the camera, all the effects hitherto described may be obtained. The light of the beam may be cut off at any convenient point of its course; but in ordinary cases the experiment is best made by employing the bichloride instead of the bisulphide of carbon, and placing the cell (*m n*) containing the opaque solution close to the camera. The moment the coal points are ignited, explosion, combustion, or calorescence, as the case may be, occurs at the focus.

The ordinary lamp and camera of Duboscq may be employed in these experiments. With proper mirrors, which are easily procured, a series of effects which, I venture to affirm, will interest everybody who witnesses them, may with the greatest facility be obtained.

It is also manifest that, save for experiments made in darkness, the camera is not necessary. The mirrors and filter may be associated with the naked lamp.

Fig. 28.

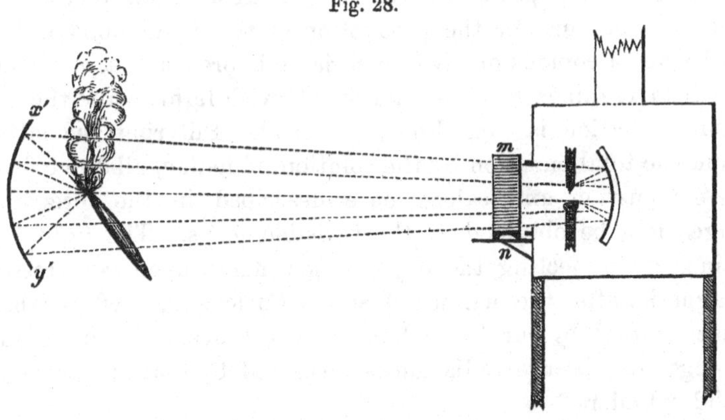

I have sought several times to *fuse* platinum with the invisible rays of the electric-light, but hitherto without success. In some experiments a large model of Foucault's lamp was employed, and a battery of 100 cells. In others I employed two batteries, one of 100 cells and one of 70, making use of two lamps, two mirrors, and two ray-filters, and converging the heat of both lamps upon opposite sides of the same film of platinum placed between them. The platinum was heated thereby to dazzling whiteness. I am persuaded that the metal could be fused, if the platinum-black could be retained upon its surface. But this being dissipated by the intense heat, the reflecting-power of the metal comes into play, and lowers the absorption so much as to prevent fusion. By coating the platinum with lampblack it has been brought to the verge of melting, the incipient yielding of the mass being perfectly apparent after it had cooled. Here, however, as in the case of the platinized platinum, the absorbing substance disappears too quickly. Copper and aluminium, however, when thus treated, are speedily burnt up.

§ 11.

Invisible Foci of the Lime-light and the Sun.

Thus far I have dealt exclusively with the invisible radiation of the electric-light; but all solid bodies raised to incandescence emit these invisible calorific rays. The denser the incandescent body, moreover, the more powerful is its obscure radiation. We possess at the Royal Institution very dense cylinders of lime for the production of the Drummond light; and when a copious oxyhydrogen flame is projected against one of them it shines with an intense yellowish light, while the obscure radiation is exceedingly powerful. Filtering the latter from the total emission by the solution of iodine, all the effects of combustion and calorescence described in the foregoing pages may be obtained at the invisible focus. The light obtained by projecting the oxyhydrogen flame upon compressed magnesia, after the manner of Signor Carlevaris, is whiter than that emitted by our lime; but the substance being light and spongy, its obscure radiation is surpassed by that of our more solid cylinders.*

The invisible rays of the sun have also been transmuted. A concave mirror, 3 feet in diameter, was mounted on the roof of the Royal School of Mines in Jermyn Street. The focus was formed in a darkened chamber, in which the platinized platinum-foil was exposed. Cutting off the visible rays by the solution of

* The discovery of fluorescence by Professor Stokes naturally excited speculation as to the possibility of a change of refrangibility in the opposite direction. Mr. Grove, I believe, made various experiments with a view to effect such a change; but very soon after the publication of Professor Stokes' memoir, Dr. Miller pointed to the lime-light itself as an instance of raised refrangibility. From its inability to penetrate glass screens, he inferred that the radiation of the oxyhydrogen flame was almost wholly ultra-red, an inference the truth of which has been since established by direct prismatic analysis. See Memoir VII. The intense light produced by the oxyhydrogen flame when projected upon lime must, he concluded, involve a change of period from slow to quick, or, in other words, a virtual elevation of refrangibility.—*Elements of Chemistry*, 1855, p. 210.

iodine, feeble but distinct incandescence was produced by the invisible rays.

In a blackened tin tube (A B, fig. 29), with square cross section and open at the end, B, a plane mirror (*x y*) was fixed, at an angle of 45° with the axis of the tube. A lateral aperture (*x o*), about 2 inches square, was cut out of the tube, and over the aperture was placed a leaf of platinized platinum. Turning the leaf towards the concave mirror, the concentrated sunbeams were permitted to fall upon the platinum. In the glare of daylight it was quite impossible to see whether the platinum was incandescent or not; but on placing the eye at B, the glow of the metal could be seen by reflexion from the plane mirror. Incandescence was thus obtained at the focus of the large mirror, X Y, after the removal of the visible rays by the iodine solution, *m n*.

Fig. 29.

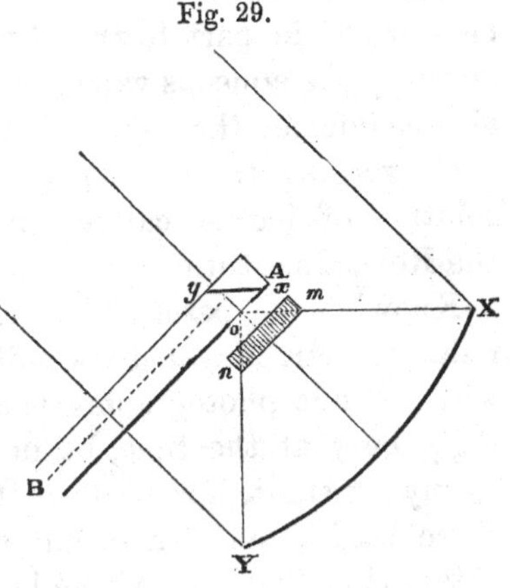

To obtain a clearer sky, I had this mirror transferred to the garden of my friend, Mr. Lubbock, near Chislehurst. The effects obtained with the total solar radiation were extraordinary. Large spaces of the platinum-leaf, and even thick foil, when exposed at the focus, disappeared as if vaporized.* The handle of a pitchfork, similarly exposed, was soon burnt quite across. Paper placed at the focus burst into flame with almost explosive suddenness. The high ratio which the visible radiation of the sun bears to the invisible was strikingly manifested in these experiments. With a *total* radiation vastly inferior, the invisible rays of the electric-light, or of the lime-light, are competent to heat platinum to whiteness, while the most that could be obtained from the dark rays of the sun was a bright red calorescence. The heat of the luminous rays, moreover, is so great

* Concentrating the solar rays with a mirror 9 inches in diameter and of 6 inches focal length upon a leaf of platinized platinum, the metal was instantly pierced. Causing the focus to pass *along* the leaf, it was cut by the sunbeam as if a sharp instrument had been drawn along it.

as to render it exceedingly difficult to work with the solution of iodine. It boiled up incessantly, exposure for two or three seconds being sufficient to raise it to ebullition. The high ratio of the luminous to the non-luminous radiation is doubtless to be ascribed in part to the absorption of a large portion of the latter by the aqueous vapour of the air. From it, however, may also be inferred the enormous temperature of the sun.

Converging the sun's rays with a hollow lens filled with the solution of iodine, calorescence was obtained on the roof of the Royal Institution.

Knowing the permeability of good glass to the solar rays, I requested Mr. Mayall to permit me to make a few experiments with his fine photographic lens at Brighton. Though exceedingly busy at the time, he in the kindest manner abandoned to my assistant, Mr. Barrett, the use of his apparatus for the three best hours of a bright summer's day. A red heat was obtained at the focus of the lens after the complete withdrawal of the luminous portion of the radiation.

§ 12.

Relation of Colour to Combustion by Dark Rays.

Black paper has been very frequently employed in the foregoing experiments, the action of the invisible rays upon it being most energetic. This suggests that the absorption of those rays is not independent of colour. A red powder is red because of the entrance and extinction of the luminous rays of higher refrangibility than the red, and the ejection of the unabsorbed red light by reflexion at the limiting surfaces of the particles of the red body. This feeble absorption of the red extends to the rays of greater length beyond the red; and the consequence is that red paper when exposed at the focus of invisible rays is scarcely charred, while black paper bursts in a moment into flame. The following table exhibits the condition of paper of various kinds when exposed at the dark focus of an electric-light of moderate intensity:—

Paper	Condition.
Glazed orange-coloured paper .	Barely charred.
„ red „ „ .	Scarcely tinged; less than the orange.
„ green „ „ .	Pierced with a small burning ring.
„ blue „ „ .	The same as the last.

Paper.	Condition.
Glazed black paper . . .	Pierced; and immediately set ablaze.
„ white „ . . .	Charred; not pierced.
Thin foreign-post . . .	Barely charred; less than the white.
Foolscap	Still less charred; about the same as the orange.
Thin white blotting-paper . .	Scarcely tinged.
„ whitey-brown „ . .	The same; a good deal of heat seems to get through these last two papers.
Ordinary brown „ . .	Pierced immediately; a burning ring expanding on all sides.
Thick brown „ . .	Pierced; not so good as the last.
Thick white sand-paper . .	Pierced with a burning ring.
Brown emery „ . .	The same as the last.
Dead-black „ . .	Pierced; and immediately set ablaze.

We have here an almost total absence of absorption on the part of the red paper. Even white absorbs more, and is consequently more easily charred.

Rubbing the red iodide of mercury over paper, and exposing the reddened surface at the focus, a thermograph of the coal points is obtained, which discharges the colour at the place on which the invisible image falls. Expecting that this change of colour would be immediate, I was at first surprised at the time necessary to produce it. We are here reminded of Franklin's experiments on cloths of different colours, and his conclusion that dark colours are the best absorbers. This conclusion might readily be pushed too far. Franklin's colours were of a special kind, and their deportment by no means warrants a general conclusion. The invisible rays of the sun possess, according to Müller, twice the energy of the visible ones. A white substance may absorb the former, while a dark substance—dark because of its absorption of the feeblest portion of the radiation—may not do so. The white powder of alum and the dark powder of iodine, exposed to the action of a source in which the invisible rays greatly surpass the visible in calorific power, exhibit a deportment at direct variance with the popular notion that dark colours are the best absorbers.

§ 13.

Calorescence through Ray-filters of Glass.—Remarks on the Black-bulb Thermometer.

In conclusion, I would briefly refer to a few experiments made to determine the calorescence obtainable through glasses of various colours. In the first column of the subjoined table the colour of the glass is given; in the second column the effect observed when a brilliant spectrum was regarded through the glass; and in the third column the appearance of a leaf of platinized platinum when placed at the focus, after the converging beam had passed through the glass:—

Colour of Glass	Prismatic Examination	Calorescence
Dark red	Red only transmitted	Dull white heat.
Mean red	Red only transmitted	White heat.
Light red	Yellow intercepted with greatest power	Bright white.
Yellow	All the blue end absorbed	Vivid red with bright yellow in centre.
Green	Besides the green, a dull red fringe and a blue band were transmitted	No incandescence.
Dark purple	Extreme blue and red transmitted	Vivid orange.
Mean purple	Central portion of spectrum cut out	Vivid orange.
Light purple	Dims the whole spectrum, but chiefly absorbs the green	Vivid orange.
Dark blue	Transmits the blue, a green band, and a band in the extreme red	Red heat.
Mean blue	Blue; a yellowish-green band, and the extreme red transmitted	Reddish-pink heat.
Light blue	Transmits a series of bands—blue and green, a red band next orange, then a dark-red band, and finally extreme red	Pink heat, passing into red.
Another blue glass	. . . . . .	Pink heat.
Black glass No. 1	Dims all the spectrum: white light transmitted	Barely visible red.
Black glass No. 2	Whitish-green light transmitted	Dull red.
Black glass No. 3	Deep-red light transmitted	Bright red, orange in the middle.

The extremely remarkable fact here reveals itself, that when the beam of the electric lamp is sifted by certain blue glasses, the platinum at the focus glows with a distinct pink colour. Every care was taken to avoid subjective illusion here. The pink colour was also obtained at the focus of invisible rays. Withdrawing all the glasses, and filtering the beam by a solution of iodine alone, platinum was heated nearly to whiteness at the focus. On introducing the pale blue glass between the iodine-cell and the focus, the calorescence of the platinum was greatly enfeebled—so much so that a darkened room was necessary to bring it out in full distinctness; when seen, however, the thermograph was pink. A disk of carbonized paper being exposed at the obscure focus, rose at once to vivid whiteness when the blue glass was absent; but when present, the colour of the light emitted by the carbon was first a distinct pink; the attack of the atmospheric oxygen soon changes this colour, the combustion of the carbon extending on all sides as a white-hot circle. If subsequent experiments should confirm this result, it would follow that there is a gap in the calorescence, the atoms of the platinum vibrating in red and blue periods, and not in intermediate ones. But I wish here to say that further experiments, which I hope shortly to make, are necessary to satisfy my own mind as to the cause of this phenomenon.

The incandescent thermograph of the coal points being obtained on platinum, a very light-red glass introduced between the opaque solution and the platinum reduced the thermograph both in size and brilliancy. A second red glass, of deeper colour, rendered the thermograph still smaller and feebler. A dark-red glass reduced it still more—the visible surface being in this case extremely minute, and the heat a dull red merely. When, instead of the coloured glass, a sheet of pure white glass was introduced, the image of the coal points stamped upon the platinum-foil was scarcely diminished at all in brilliancy. A thick piece of glass of deep ruby-red proved equally transparent; its introduction scarcely changed the vividness of the thermograph. The colouring matter in this instance was the *element* gold, while the colour in the other red glasses was due to the compound suboxide of copper. Many specimens of gold-jelly, prepared by Mr. Faraday for his investigation of the colours of gold, though of a depth approaching to absolute blackness, showed themselves eminently transparent to the obscure heat-rays; their introduction scarcely dimmed the brilliancy of the thermograph. *Hence it would appear that even the metals themselves, in certain states of aggregation, share that high diathermic power which the elementary metalloids have been found to display.*

I have just said that a sheet of pure-white glass, when interposed in the path of the condensed invisible beam, scarcely dimmed the brilliancy of the thermograph. The intense calorific rays of the electric-light pass through such glass with freedom. We here come to a point of considerable practical importance to meteorologists. When such pure-white glass in a molten condition has carbon mixed with it, the resulting black glass is still eminently transparent to those invisible heat-rays which constitute the greater part of the sun's radiation. I have pieces of glass, to all appearance black, which transmit 63 per cent. of the total heat of the electric light; and there is not the slightest doubt that, in thicknesses sufficient to quench entirely the light of the sun, such glasses would transmit a large portion of the invisible heat-rays. This is the glass often, if not uniformly, employed in the construction of our black-bulb thermometers, under the impression that the blackening secures the entire absorption of the solar rays. This conclusion is fallacious, and the instruments are correspondingly defective. *A large portion of the sun's rays pass through such black glass, impinge upon the mercury within the bulb, and are ejected by reflexion.* Such rays contribute nothing to the heating of the thermometer.

A sheet of common window-glass, apparently transparent, was placed between the iodine solution and the platinum-leaf at the focus; the thermograph was more dimmed than by the black glass last referred to. The window-glass here employed, when looked at edgeways, was *green*; and this experiment proves how powerfully this green colouring-matter, even in infinitesimal quantity, absorbs the invisible heat-rays. Perfect imperviousness might doubtless be secured by augmenting the quantity of green colouring-matter. *It is with glass of this description that the carbon should be mixed in the construction of black-bulb thermometers*; on entering such glass the solar rays would be entirely absorbed, and greater differences than those now observed would probably be found to exist between the black-bulb and the ordinary thermometer.

IX.

ON THE INFLUENCE OF COLOUR AND MECHANICAL CONDITION ON RADIANT HEAT.

ANALYSIS OF MEMOIR IX.

Franklin's conclusion, that dark bodies absorb heat more greedily than light ones, glanced at in the last memoir (§ 12), is proved at the outset of this one to be by no means general. The dark powder of iodine is contrasted with the white powder of alum, and it is shown that the white body is by far the most powerful absorber.

The fact of its being an element, and its proved diathermancy in solution, suggested iodine as a fitting substance to test the law of Franklin.

The deportment of other elementary bodies is then glanced at, and two kinds of opacity are distinguished from each other—the one caused by a true absorption, the other by the multiplied reflexions at the surfaces of particles not in optical contact.

Thus ordinary sulphur stops the calorific rays, but not by absorption, and it is difficult therefore to burn it at the most intense focus. Were all its parts in optical contact, free transmission would follow. In pure crystallised sulphur this contact is attained, and the diathermancy of the substance is then remarkable.

Black amorphous phosphorus, notwithstanding its inflammability, bears intense radiant heat for some time without ignition; ordinary phosphorus does the same. And this substance, which is dissolved in large quantities by bisulphide of carbon, proves, when thus dissolved, almost perfectly diathermic.

That substances which absorb radiant heat are those only which can be burnt by radiant heat, is further illustrated by pounded loaf-sugar when compared with table-salt. The one is immediately fused and ignited at the dark focus, the other is scarcely warmed.

It is also shown that a beam sifted by water, though its heat may be very great, has no power to melt the most delicate hoar-frost, and from this it is inferred that the melting of the Alpine snows and ice is the work not of the visible but of the invisible rays of the sun.

Alcohol is shown to boil almost instantaneously at a focus where bisulphide of carbon is scarcely warmed. The errors on these points which have crept into physical treatises are shown to be due to the fact that a large portion of the sun's emission consists of rays regarding which neither colour nor optical transparency gives us any information.

Throughout these memoirs radiation and absorption have been looked upon as the acts of *atoms* and *molecules*; chemical constitution, rather than physical condition, being regarded as the really potent agency. Still the numerous experiments of Masson and Courtépée seemed clearly to prove that all bodies, however different chemically, in a fine state of division possessed the same power of radiation and absorption. Melloni had also compared white and black substances, and found no difference in their radiative powers.

It is shown, however, that in all those experiments it was not the powders, but the *varnish* in which the powders were imbedded, that was really the subject of experiment. Hence the observed equality. When the varnish is abandoned, differences in radiative and absorptive power immediately appear.

The powders here employed as radiators were first attached to a hot surface by sulphur cement; they were afterwards held there by electric attraction. Both methods showed the diversity existing among chemical precipitates as regards the radiation and absorption of heat.

As regards colour, no general conclusion is possible: black, in some cases, transcends white; white, in other cases, excels black; while two blacks, or two whites, may differ greatly from each other. The use of glue, or varnish, to attach the powders to the hot surface abolished these differences.

Thirty-two different powders held by sulphur cement are thus examined, and found to vary in radiating power from 35·3 in the case of rock-salt, to 84 in the case of lampblack. Twenty-six powders held by electric attraction are examined, and found to vary from 24·5, in the case of rock-salt, to 65·8, in the case of black oxide of iron.

The transmission by rock-salt of the heat emitted from twenty-four different substances is determined; and found to vary from 62·8, where rock-salt is itself the source of heat, to 89 where black platinum is the source. Looking at the absorptions by rock-salt instead of the transmissions, they are found to vary from 3·7 with the platinum as source of heat to 29·9 with rock-salt itself as a source. These experiments leave no doubt upon the mind that the investigators were correct who affirmed rock-salt to be *not* equally diathermic to all kinds of heat.

Experiments then follow illustrating the reciprocity of radiation and absorption.

IX.

ON THE INFLUENCE OF COLOUR AND MECHANICAL CONDITION ON RADIANT HEAT.*

§ 1.

Proof that White Bodies sometimes absorb Heat more copiously than Dark ones.—Explanation.

FRANKLIN placed cloths of various colours upon snow and allowed the sun to shine upon them. They absorbed the solar rays in different degrees, became differently heated, and sank therefore to different depths in the snow beneath them. His conclusion was that dark colours were the best absorbers, and light colours the worst; and to this hour we appear to have been content to accept Franklin's generalization without qualification. In my last memoir I briefly pointed out its probable defects. Did the emission from luminous sources of heat consist exclusively of visible rays, we might fairly infer from the colour of a substance its capacity to absorb the heat of such sources. But the emission from luminous sources of heat is by no means all visible. In terrestrial sources of heat by far the greater part, and in the sun a very great part, of the emission consists of invisible rays, regarding which colour teaches us nothing.

It remained therefore to examine whether the results of Franklin were the expression of a law of nature. Two cards were taken of the same size and texture; over one of them was shaken the white powder of alum, and over the other the dark powder of iodine. Placed before a glowing fire and permitted to assume the maximum temperature due to their position, it was found that the card bearing the alum became extremely

* Received December 21st, 1865; and read before the Royal Society, January 18th, 1866. *Philosophical Transactions* for 1866, page 83; *Philosophical Magazine*, October 1866.

hot, while that bearing the iodine remained cool. No thermometer was necessary to demonstrate this difference. Placing, for example, the back of the iodine card against the forehead or cheek, no inconvenience was experienced; while the back of the alum card similarly placed proved intolerably hot.

This result was corroborated by the following experiments:— One bulb of a differential thermometer was covered with iodine, and the other with alum powder. A red-hot spatula being placed midway between both, the liquid column associated with the alum-covered bulb was immediately forced down, and maintained in an inferior position. Again, two delicate mercurial thermometers had their bulbs coated, the one with iodine, the other with alum. On exposing them at the same distance to the radiation from a gas flame, the mercury of the alum-covered thermometer rose nearly twice as high as that of its neighbour. Two sheets of tin were coated, the one with alum, and the other with iodine powder. The sheets were placed parallel to each other, and about 10 inches asunder; at the back of each was soldered a little bar of bismuth, which with the tin plate to which it was attached constituted a thermo-electric couple. The two plates were connected together by a wire, and the free ends of the bismuth bars were connected with a galvanometer. Placing a red-hot ball midway between both, the calorific rays fell with the same intensity on the two sheets of tin, but the galvanometer immediately declared that the sheet which bore the alum was the most highly heated.

In some of the foregoing cases the iodine was simply shaken through a muslin sieve; in other cases it was mixed with bisulphide of carbon and applied with a camel's-hair brush. When dried afterwards it was almost as black as soot; but as an absorber of radiant heat it was no match for the perfectly white powder of alum.

The difficulty of warming iodine by radiant heat is evidently due to the diathermic property which it manifests so strikingly when dissolved in bisulphide of carbon. The heat enters the powder, is reflected at the limiting surfaces of the particles, but it does not lodge itself among the atoms of the iodine. When shaken in sufficient quantity on a plate of rock-salt and placed in the path of a calorific beam, iodine cuts the latter off. But its opacity is mainly that of a white powder to light; it is impervious, *not through absorption, but through repeated internal reflexion.* Ordinary roll sulphur, even in thin cakes, allows no radiant heat to pass through it; but its opacity is also due to repeated internal reflexion. The temperature of ignition of sulphur is about 244° C.; but on placing a small piece of the substance at the focus of the electric lamp, where the temperature was sufficient to heat platinum-foil in a moment to whiteness, it required exposure for a considerable time to fuse and ignite the sulphur. Though impervious to the heat, it was not adiathermic. The milk of sulphur was also ignited with some difficulty. Sugar is a much less inflammable substance than sulphur, but it is a far better absorber; exposed at the focus, it is speedily fused and burnt up. The heat, moreover, which is competent to inflame sugar is scarcely competent to warm table-salt.

A fragment of almost black amorphous phosphorus was exposed at the dark focus of the electric lamp, but refused to be ignited. A still more remarkable result was obtained with ordinary phosphorus. A small fragment of this exceedingly inflammable substance could be exposed for twenty seconds without ignition at a focus where platinum was almost instantaneously raised to a white heat. Placing a morsel of phosphorus on a plate of rock-salt and holding it before a glowing fire, it bears, as proved by my assistant, Mr. Barrett, an intense radiation without ignition; but laid upon a plate of glass and similarly exposed, the phosphorus soon fuses and ignites; its ignition, however, is not entirely due to radiant heat, but mainly to the heat imparted to it by the glass.*

The fusing-point of phosphorus is about 44° C., that of sugar is 160°; still at the focus of the electric lamp the sugar fuses before the phosphorus. All this is due to the diathermancy of the phosphorus: a thin disk of the substance placed between two plates of rock-salt permits of a copious transmission. *This substance therefore takes its place with other elementary bodies as regards deportment towards radiant heat.*

The more diathermic a body is, the less it is warmed by radiant heat, and no *perfectly* transparent body could be warmed by purely luminous heat. The surface of a vessel covered with

* I believe this deportment of phosphorus towards radiant heat is not unknown to chemists.

a thick fur of hoar frost was exposed to the beam of the electric lamp condensed by a powerful mirror, the beam having been previously sent through a cell containing water; *the sifted beam was powerless to remove the frost, though it was competent to set wood on fire.* We may largely apply this result. It is not, for example, the luminous rays but the dark rays of the sun which sweep the snows of winter from the slopes of the Alps. Every glacier-stream that rushes through the Alpine valleys is almost wholly the product of invisible radiation. It is also the invisible solar rays which lift the glaciers from the sea-level to the summits of the mountains; for the luminous rays penetrate the tropical ocean to great depths, while the non-luminous ones are absorbed close to the surface, and become the main agents in evaporation.

It is often stated, without limitation, in chemical treatises that sulphuric ether may be exposed at the focus of a concave mirror without being sensibly heated; but this can only be true of a sifted beam. At the focus of the electric lamp, not only ether, but alcohol and water are speedily caused to boil, while bisulphide of carbon, whose boiling-point is only 48° C., cannot be raised to ebullition. *In fact exposure for a period sufficient to boil alcohol or water is scarcely sufficient to render bisulphide of carbon sensibly warm.*

§ 2.

Melloni on Colours, and Masson and Courtépée on Powders, in relation to Radiant Heat.

If any one point came out with more clearness than any other in my experiments on gases, liquids, and vapours, it was the paramount influence which chemical constitution exerted upon the phenomena of radiation and absorption. And, seeing how little the character of the radiation was affected by the change of a body from the state of vapour to the state of liquid, I held it to be exceedingly probable that even in the solid state chemical constitution would exert its power. But opposed to this conclusion we had the experiments of Melloni on chalk and lampblack, and the far more extensive ones of Masson and Courtépée on powders, which seemed clearly to show that in a state of extremely fine division, as in chemical precipitates, the radiant and absorbent powers of all bodies are the same. From these experiments it was inferred that the influence of physical condition was so predominant as to cause that of chemical constitution to disappear.*

A serious oversight, however, seems to have connected itself with all the experiments of these distinguished men. Melloni mixed his lampblack and powdered chalk with gum or glue, and applied them by means of a camel's-hair brush on the surfaces of his radiating tube. Masson and Courtépée did the same. Melloni, it is true, thus compared a black surface with a white one; but the surfaces were seen to be white and black *through* the transparent gum, which in both cases was the real radiator. The same remark applies to Masson and Courtépée. Every particle of the precipitates they employed was *a varnished particle*; and the constancy they observed was, I imagine, due to the fact that the main radiant in all their experiments was the substance employed to make their powders cling to the surfaces of their cubes.

§ 3.

New Experiments on Chemical Precipitates.—Influence of Colour and Chemical Constitution.—Sulphur Cement.

Gum or glue is a powerful radiator—in fact, equal to lampblack; and it is a correspondingly powerful absorber; the particles surrounded by it had therefore but small chance of radiating through it. I sought to remedy this by the employment of a diathermic cement. Sulphur is highly diathermic; it dissolves freely in bisulphide of carbon; and at the suggestion of a chemical friend it was employed to fix the powders. The cube was laid upon its side, the surface to be coated being horizontal, and the bisulphide, containing the sulphur in solution, was poured over the surface. Before the liquid film had time to evaporate, the powder was shaken upon it through a muslin sieve. The bisulphide passed rapidly away in vapour, leaving the powder behind imbedded in the sulphur cement. Each powder, moreover, was laid on sufficiently thick to prevent the sulphur from surrounding its particles. This, though not per-

* Masson and Courtépée, *Comptes Rendus*, vol. xxv. p. 938; Jamin, *Cours de Physique*, vol. ii. p. 289.

haps a perfect way of determining the radiation of powders, was at all events an improvement on former methods, and yielded different results.

Ten or twelve cubes of tin were employed in the investigation. One side of each of them was coated with milk of sulphur, and the fact that this substance was used throughout the entire series of cubes enabled me to connect all the results together. The cubes were heated with boiling water, and placed in succession at the same fixed distance in front of the thermo-electric pile, which as usual was well defended from air-currents and other extraneous sources of disturbance. Before giving the complete table of results, I will adduce a few of them, which show in a conclusive manner that, *in solid bodies* also, radiation is molecular rather than mechanical.

The biniodide of mercury and the red oxide of lead resemble each other physically, both of them being of a brilliant red. Chemically, however, they are very different. Examined in the way indicated, their relative powers as radiators were found to be as follows:—

Name	Chemical Formula	Radiation
Biniodide of mercury	$(Hg\ I^2)$	39·7
Red oxide of lead	$(2\ Pb\ O,\ Pb\ O^2)$	74·1

Mixed with gum and applied with a camel's-hair brush to the surfaces of the cube, the radiation from these two substances fell out thus:—

Name	Radiation
Biniodide of mercury	80
Red oxide of lead	80

Here the influence of the gum entirely masks the difference due to molecular constitution.

The effect of atomic complexity upon the radiation is well illustrated by the deportment of these two substances. It is further illustrated by the deportment of two different iodides of mercury:—

	Radiation
Biniodide of mercury $(Hg\ I^2)$	39·7
Iodide of mercury $(Hg^2\ I^2)$	46·6

Here the addition of a second atom of mercury to the molecule of the biniodide raises the radiation 7 per cent. The experiment furnishes a kind of physical justification of the practice of chemists in regarding the molecule of yellow iodide of mercury to be $Hg^2 I^2$, and not Hg I.

The peroxide and protoxide of iron gave the following results:—

	Radiation
Peroxide of iron	78·4
Protoxide of iron	81·3

I did not expect this, the protoxide being a less complex molecule than the peroxide. On examination, however, the protoxide was found to be in part the magnetic oxide. The formulæ of the two substances are $Fe^2 O^3$ and $Fe O, Fe^2 O^3$, and the anomaly therefore disappears.

Amorphous phosphorus and sulphide of iron gave the following results:—

	Radiation
Amorphous phosphorus	63·6
Sulphide of iron	81·7

Sugar and salt were reduced in a mortar to the state of exceedingly fine powders. In point of cohesion and physical aspect these substances closely resemble each other; their radiative powers, however, are as follows:—

	Radiation
Salt	35·3
Sugar	70

In his last interesting paper on emission at a red heat,* M. Desains mentions oxide of zinc as a body which at 100° C. has the same emissive power as lampblack. This is nearly true for the hydrated oxide; with the calcined oxide the following is the relation:—

	Radiation
Lampblack	84
Hydrated oxide of zinc	80·4
Calcined „ „	53·2

Two red powders have been already compared together. With black platinum and black oxide of iron the following results were obtained:—

	Radiation
Black platinum (electrolytic)	59
Black oxide of iron	81·3

* *Comptes Rendus*, July 3, 1865; *Philosophical Magazine*, August 1865.

The black platinum here employed was obtained by electrolysis, a sheet of platinum-foil being coated with the substance.

Chloride of silver and carbonate of zinc, two white powders, gave the following results:—

	Radiation
Chloride of silver	32·5
Carbonate of zinc	77·7

When held upon its cube by the sulphur cement, the chloride of silver soon darkens in the diffuse light of the laboratory. It first becomes lavender, and passes through various stages of brown to black. During these changes, which may be associated with a chemical reaction between the chloride of silver and the sulphur in which it is embedded, the radiation steadily augments. Beginning in one instance with a radiation of 25, the chloride ended with a radiation of 60.

We have thus far compared two red surfaces, two black surfaces, and two white surfaces together. The comparison of a black and white surface gave the following result:—

	Radiation
Black platinum	59
White hydrated oxide of zinc	80·4

Here the radiation of the white body far transcends that from the black one.

Comparing black and white a second time, we have the following result:—

	Radiation
Oxide of cobalt	76·5
Carbonate of zinc	77·7

Here the black radiation is sensibly equal to the white one.

Comparing black and white a third time, we have the following result:—

	Radiation
Lampblack	84
Chloride of lead	55·4

Here the radiation of the black body far transcends that of the white one.

We have thus compared red powders with red, black with black, white with white, and black with white; and the conclusion to be drawn is, I think, that chemical constitution, so far from being of vanishing value, is a really potent influence in the experiments. Combined with previous ones, they show that the influence of chemical constitution makes itself felt in all states of aggregation—gaseous, liquid, and solid—whether the solid be a chemical precipitate or a coherent mass.

Were the radiative power of these substances determined by the state of division, it would probably make itself sensible even in a case where the division is effected by the pestle and mortar; but I do not find this to be the case. A plate of glass was fixed against the polished surface of a Leslie's cube, and on the plate the powder of glass, rendered as fine as the pestle and mortar could make it, was strewn. It was caused to adhere without cement of any kind. The cube was filled with boiling water and presented to the thermo-electric pile until a permanent deflection was obtained. The cube being permitted to remain in its position, the powder was simply removed with a camel's-hair brush. The increase of radiation was only such as might be expected from the slight difference of temperature between the surface of the glass plate and the powder strewn upon it. Similar experiments, with precisely similar results, were made with a plate of rock-salt, on which the finely divided powder of rock-salt was shaken.

Still the same substance in different states of molecular aggregation may produce very different effects, both as regards radiation and absorption. We have already had an instance of this in the case of ozone as compared with ordinary oxygen. The following instance may also be cited:—

One side of a Leslie's cube was covered with a sheet of bright platinum-foil, and a second face by a similar sheet on which black platinum had been deposited by electrolysis. As radiators these two sheets of foil behaved in the following manner:—

	Radiation
Bright platinum-foil	6
Platinized platinum	45·2

Here the radiation of the black platinum is nearly eight times that of the bright substance.

§ 4.

Tabulation of the Radiant Powers of Powders.—Employment of Electric Attraction instead of Sulphur Cement.

For the sake of reference, I will here tabulate the results obtained with a considerable number of precipitates when subjected to the described conditions of experiment.

TABLE I.—*Radiation from Powders imbedded in Sulphur Cement.*

Name of Substance	Radiation of Substance named	Name of Substance	Radiation of Substance named
Rock-salt	35·3	Sulphide of molybdenum	71·3
Biniodide of mercury	39·7	Sulphate of baryta	71·6
Milk of sulphur	40·6	Chromate of lead	74·1
Common salt	41·3	Red oxide of lead	74·2
Yellow iodide of mercury	46·6	Sulphide of cadmium	76·3
Sulphide of mercury	46·6	Subchloride of copper	76·5
Iodide of lead	47·3	Oxide of cobalt	76·7
Chloride of lead	55·4	Sulphate of lime	77·7
Chloride of cadmium	56·5	Carbonate of zinc	77·7
Chloride of barium	58·2	Red oxide of iron	78·4
Chloride of silver (dark)	58·6	Sulphide of copper	79
Fluor-spar	68·4	Hydrated oxide of zinc	80·4
Tersulphide of antimony	69·4	Black oxide of iron	81·3
Carbonate of lime	70·2	Sulphate of iron	81·7
Oxysulphide of antimony	70·5	Iodide of copper	82
Sulphide of calcium	71	Lampblack	84

I subsequently endeavoured to get rid of the sulphur cement, and to make the powders adhere by wetting them with pure bisulphide of carbon, applying them to the cubes while wet. Some of the powders clung, others did not. My ingenious friend Mr. Duppa suggested to me that the powders might be held on by electrifying the cubes. I tried this plan, and found it simple and practicable. It was, however, aided by a circumstance which we did not anticipate. The cube being placed upon an insulating stand with its side horizontal, the powder was shaken over it, and electrified by a few turns of a machine. It was found that the cube might then be discharged and set upright, the powders clinging to it in this position. The results obtained with this arrangement are recorded in the following table :—

TABLE II.—*Radiation from Powders held by Electricity.*

Substance	Radiation	Substance	Radiation
Rock-salt . . .	24·5	Sulphide of calcium .	49·1
Chloride of silver (white)	25	Sulphate of baryta . .	51·3
Milk of sulphur . .	25·8	Sugar	52·1
Biniodide of mercury .	26	Red oxide of lead . .	56·5
Iodide of lead. . .	36	Sulphide of cadmium .	56·9
Sulphide of mercury .	30·6	Sulphate of lime . .	59·3
Spongy platinum . .	31·5	Chloride of silver (black)	60
Washed sulphur (flowers)	32·3	Carbonate of zinc . .	62
Sulphide of zinc . .	36·1	Oxide of cobalt . .	62·5
Amorphous phosphorus .	38	Iodide of copper . .	63
Chloride of lead . .	39	Red oxide of iron . .	63·8
Chloride of cadmium .	40	Sulphide of iron . .	65·5
Fluor-spar . . .	48·6	Black oxide of iron. .	65·8

The agreement as regards relative radiative power between this and the former table is as good as could under the circumstances be expected. The experiments have been several times repeated; and the table contains the means of the results, which were never widely different from each other.

§ 5.

Qualitative Experiments.—Radiation of various Bodies through Rock-salt.—Unequal Diathermancy of the Substance.

The *quantity* of radiant heat emitted by bodies in all states of aggregation having been thus conclusively shown to depend mainly upon their molecular character, the question as to the *quality* of the heat emitted next arises. In examining this point, I contented myself with testing the heat radiated from various substances by its transmission through rock-salt. The choice of this substance involved the solution of the still disputed question whether rock-salt is equally pervious to all kinds of rays.* For if it absorbed the radiation from two different bodies in different degrees, it would not only show a difference of quality in the radiations, but also demonstrate the incapacity of rock-salt to transmit equally rays of all descriptions.

* The last publication on this subject is from the pen of that extremely able experimenter Professor Knoblauch. After discussing the results of De la Provostaye and Desains, and of Mr. Balfour Stewart, he arrives at a different conclusion—namely, that pure rock-salt is equally pervious to all kinds of heat.—Poggendorff's *Annalen*, 1863, vol. cxx. p. 177.

The plate of salt chosen for this purpose was a very perfect one. I have never seen one more pellucid. The thickness was 0·8 of an inch, and its size, compared with the aperture in front of which it was placed, was such as to prevent any part of the rays reflected from its lateral boundaries from mingling with the direct radiation. M. Knoblauch has clearly shown how the absence of caution in this particular may lead to error. The mode of experiment was that usually followed: the source of heat was first permitted to radiate against the pile, the deflection produced by the total radiation being noted. The plate of rock-salt was then interposed; the deflection sank, and from its new value the transmission through the rock-salt was calculated and expressed in hundredths of the total radiation.

TABLE III.—*Transmission through Rock-salt of Heat radiated by the following Substances heated to* 100° C.

Substance	Transmission in 100ths of the Total Radiation	Total Radiation
Rock-salt	67·2	35·3
Biniodide of mercury	76·3*	39·7
Milk of sulphur	76·9*	40·6
Common salt	70·8	41·3
Yellow iodide of mercury	79*	46·6
Sulphide of mercury	73·1	46·6
Iodide of lead	73·8	47·3
Chloride of lead	73·1	55·4
Chloride of cadmium	73·2	56·5
Chloride of barium	70·7*	58·2
Chloride of silver (dark)	74·2	58·6
Fluor-spar	70·5*	68·4
Tersulphide of antimony	77·1	69·4
Carbonate of lime	77·6	70·2
Oxysulphide of antimony	77·6	70·5
Sulphide of molybdenum	78·4	71·3
Sulphate of baryta	71·3	78·4
Chromate of lead	71·6	79·2
Red oxide of lead	74·1	79·2
Subchloride of copper	76·3	78·6
Oxide of cobalt	76·5	79·7
Red oxide of iron	78·4	81
Sulphide of copper	79	82·3
Black oxide of iron	81·3	82·7
Sulphide of iron	81·7	83·3
Lampblack	84	83·3

Here we have a transmission varying from 67 per cent. in the case of powdered rock-salt to 84 per cent. in the case of lamp-black. The second column of figures will be referred to immediately.

The powders here employed were fixed by the sulphur cement. The same powders held by electricity, and permitted to radiate through the rock-salt, gave the following transmissions :—

TABLE IV.

Substance	Transmission	Substance	Transmission
Rock-salt . . .	62·8	Carbonate of zinc .	74·8
Chloride of silver (white)	69·7	Sulphate of baryta .	75
Fluor-spar . . .	70·7	Common sugar .	75·4
Sulphide of mercury .	71	Sulphide of copper .	76·5
Sulphide of calcium .	72·5	Iodide of copper .	76·5
Milk of sulphur . .	72·8	Red oxide of iron .	76·8
Sulphide of cadmium .	73·3	Chloride of silver (black)	77·3
Biniodide of mercury .	73·7	Amorphous phosphorus .	78
Washed sulphur . .	74	Oxide of cobalt .	78·2
Iodide of lead . .	74·1	Sulphide of iron .	78·5
Sulphate of lime . .	74·2	Black oxide of iron .	79·7
Sulphide of zinc . .	74·4	Black platinum .	89

The transmissions here are lower than when the sulphur cement was employed. I do not, however, think that the differences are due to the employment of the cement, but to a slight source of disturbance, which was removed in the later experiments.

It will be remarked that, as a general rule, powerful radiators have their heat more copiously transmitted by the rock-salt than feeble ones. To render this clear, in Table III., appended to the transmission, is the corresponding total radiation. The only striking exceptions to the rule are marked with asterisks. This result, I think, is what might fairly be expected; for the peculiarity which enables one molecule to radiate more heat than another, may also be expected to introduce dissonance between their rates of oscillation. The probability is therefore that greater dissonance will exist between the vibrating periods of good radiators and bad radiators than between the periods of the members of either class. But the greater the dissonance the less will be the absorption; hence, as regards transmission through rock-salt, we have reason to expect that powerful radiators will find a more open door to their emission than feeble ones. This is, as I have said, in general the case. But the rule is not without its exceptions; and the most striking of these is the case of black platinum, which, though but a moderate radiator, sends a greater proportion of heat through rock-salt than any other known substance.

In his latest investigation, Knoblauch examined at great length the diathermancy of rock-salt. With his usual acuteness, he points out several possible sources of error, and with his customary skill he neutralizes these sources. His conclusion is the same as that of Melloni, namely, that rock-salt transmits in the same proportion all sorts of rays. On the opposite side we find the experiments of MM. De la Provostaye and Desains, and those of Mr. Balfour Stewart,* both of which are discussed by Knoblauch. He differs from those experimenters, while my results bear them out. Considering the slow augmentation of transmission which the foregoing tables reveal, and the considerable number of bodies whose heat is transmitted in almost the same proportion by rock-salt, it is easy to see that, where the number of radiants is restricted, such a uniformity of transmission might manifest itself as would lead to the conclusion of Melloni and Knoblauch. It was only by the deliberate selection and extension of the substances chosen as radiators that the differences were brought out with the fulness and distinctness recorded in the foregoing tables.

The differences in point of quality and the absence of perfect diathermancy in rock-salt appear more striking when instead of the transmissions we take the absorptions. In the case of the radiation from powdered rock-salt, for example, 37·2 per cent. of the whole radiation is intercepted by the rock-salt plate. According to Melloni, between 7 and 8 per cent. of this is lost by reflexion at the two surfaces of the salt. This would leave in round numbers a true absorption of 30 per cent. by the plate of rock-salt. In the case of black platinum, the absorption similarly deduced amounts to only 4 per cent. of the total radiation. Instead, therefore, of the radiation from those two sources of heat being absorbed in the same proportion, the ratio in the one case is more than seven times that in the other. For the sake

* I think the important experiment first executed by Mr. Balfour Stewart, of rock-salt radiating through rock-salt, is of itself sufficient to demonstrate in the most unequivocal manner that this substance is not equally pervious to all kinds of rays.

of illustration, I here introduce a few of the absorptions determined in this way:—

TABLE V.—*Radiation through Rock-salt.*

Source of heat	Absorption per 100
Black platinum	3·7
Black oxide of iron	13
Red oxide of iron	15·9
Sugar	17·3
Chloride of silver	22·6
Rock-salt	29·9

These differences are so great as to enable every experimenter to satisfy himself with the utmost ease as to the unequal permeability of rock-salt; and this facility of demonstration will, I trust, contribute to make inquirers unanimous on this important point.

§ 6.

Radiation of Powders.—Reciprocity of Radiation and Absorption.

Theory alone would lead us to the conclusion that the absorptive power of the substances mentioned in Table I. is proportional to their radiative power; nevertheless a few actual experiments on absorption will serve as a check upon those recorded in the table. These were conducted in the following manner:—A B (fig. 30) is a sheet of common block tin, 5 inches high and 4 wide, fixed upon a suitable stand. At the back of A B is soldered one end of the small bar of bismuth *b*, the remainder of the bar, to its free end, being kept out of contact with the tin by a bit of cardboard. To the free end of *b* is soldered a wire which can be connected with a galvanometer. A′ B′ is a second plate of tin furnished also with its bismuth bar, and in every respect similar to A B. From one plate to the other stretches the wire W. C is a cube containing boiling water, placed midway between the two plates of metal.

Fig. 30.

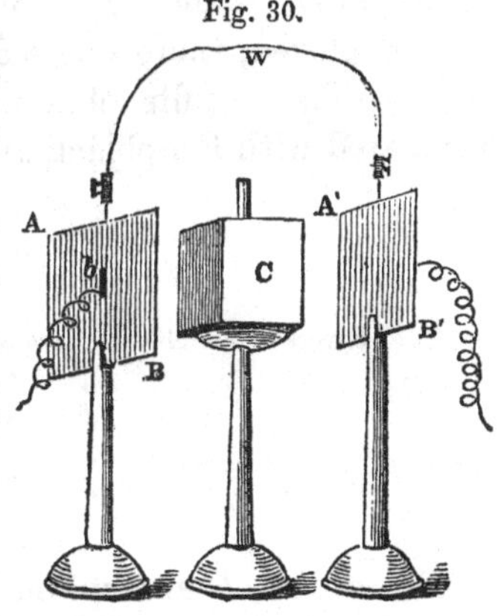

The tin plates were in the first instance coated uniformly with lampblack, and the two surfaces of the cube which radiated against the plates were similarly coated. The rays from C being emitted equally right and left, and absorbed equally by the two coated plates A B and A′ B′, warmed these plates to the same degree. It is manifest from the arrangement that, if the thermo-electric junctions were equally sensitive, the current generated at the one ought exactly to neutralize the current from the other. This was found to be very nearly the case. It is difficult to make both junctions of absolutely the same sensitiveness; but the moving of the feebler plate a hair's breadth nearer to the cube C enabled it to neutralize exactly the radiation from its opposite neighbour. My object now was to compare the lampblack coating of the plate A B with a series of other coatings, placed in succession on the other plate. These latter coatings were the powders already employed, and they were held upon A′ B′ by their own adhesion.

When A B was coated with lampblack and A′ B′ with rock-salt powder, the equilibrium observed when both the plates were coated with lampblack did not exist. The lampblack, by its greater absorption, heated its bismuth junction most, and a permanent deflection of 59° in favour of the lampblack was obtained. Other powders were then substituted for the rock-salt, and the difference between them and the lampblack was determined in the same way. When, for example, sulphide of iron was employed, there was a deflection of 30° in favour of lampblack. The results obtained with six different powders thus compared with lampblack are given in the following table :—

Table VI.

Excess of lampblack above	rock-salt . . .	59° = 112	units.
" "	fluor-spar . . .	46 = 68	"
" "	red lead . . .	40 = 45	"
" "	oxide of cobalt . .	37 = 42	"
" "	sulphide of iron .	30 = 30	"

The order of absorption here coincides with the order of radiation of the same substances shown in Table III.

But we can go further than the mere order of absorption. Removing the opposing plate, connecting the wire W direct with the galvanometer, and allowing the standard lampblack to exert its full action, the deflection observed was

$$65^\circ = 163 \text{ units.}$$

The numbers in Table VI. show us the excess of the lampblack over the substances there employed,—its excess in the case of rock-salt, a bad absorber, is 112, its excess in the case of sulphide of iron being only 30. Deducting, therefore, the numbers given in Table VI. from 163, the total absorption of lampblack, we obtain a series of numbers which expresses the absorption of the other substances. This series stands as follows:—

TABLE VII.

Substance	Relative Absorptions		Radiation from Table II.
Rock-salt	51	25·5	25
Fluor-spar	95	47·5	49
Red lead	118	59	57
Oxide of cobalt	121	60·5	63
Sulphide of iron	133	66·5	66

The first column of figures expresses the relative absorptions; for the sake of comparison with the corresponding radiations, I have placed the halves of these numbers in the second column of figures, and in the third column the radiations obtained from Table II. The approximation of the figures in the second and third columns is close enough to establish the accurate proportionality of radiation and absorption.

Throughout some of these investigations I have been efficiently assisted by Mr. Robert Chapman, and throughout others, with great skill and assiduity, by Mr. W. F. Barrett.

X.

ON THE ACTION OF RAYS OF HIGH REFRANGIBILITY UPON GASEOUS MATTER.

ANALYSIS OF MEMOIR X.

In the researches thus far placed before the reader rays of low refrangibility were invoked as the explorers of molecular condition. The present memoir deals with the interaction of rays of high refrangibility and gaseous matter.

Theoretic considerations regarding atoms and molecules, and their relation to the waves of æther, preface the memoir, and the special origin of the inquiry is indicated. The obstacles to be overcome are referred to, one of these being the floating matter of the air.

This was found competent to pass through tubes containing caustic potash and sulphuric acid, thus showing the insecurity of experiments based on the assumption that these substances destroy the floating matter of the air.

The vapours of various substances inclosed in a glass experimental tube are subjected to the action of a concentrated beam of light. The vapours are decomposed; non-volatile products are formed which are precipitated as actinic clouds in the experimental tube.

The nitrite of amyl, the iodide of allyl, and the iodide of isopropyl are adduced as examples of this new action of light upon vapours.

In Memoir VI. it was shown that the order of absorption in vapours and their liquids is the same. So constant was this relation that in the various singular shiftings of diathermic position, revealed in the memoir referred to, the vapours followed with undeviating precision the fluctuations of their liquids.

In the present memoir a similar relation is shown to hold good with rays of high refrangibility. The *chemical penetrability* of nitrite of amyl, both in the liquid and vaporous condition, is contrasted with that of the iodide of allyl. A layer one-eighth of an inch thick of the liquid nitrite suffices to remove the rays which act upon the nitrite vapour; while a foot or so of the nitrate vapour almost accomplishes the same thing. On the other hand, a whole inch of the liquid iodide of allyl is found incompetent to stop the rays which act upon the iodide, and six or seven feet of the iodide vapour are also found incompetent to accomplish this.

The influence of a second body on the decomposition of vapours by light is illustrated; and it is compared with the influence of chlorophyl on the decomposition of carbonic acid by the rays of the sun.

The character of the actinic clouds produced, when the quantities of vapour acted on are very small, is described and illustrated. In all cases the cloud commences with a beautiful azure, which discharges perfectly polarized light at right angles to the direction of the beam.

The identity of deportment of this azure with the blue of the firmament justifies the conclusion that the physical origins of both are identical: that in the experimental tube we have, to all intents and purposes, an artificial sky. Section 9 contains a condensed description of this portion of the inquiry; and

Section 10 embraces additional illustrations of the production of the firmamental blue.

Section 11 describes the difficulty which beset the earlier stages of the inquiry, and the extreme liability to error arising out of the action of infinitesimal residues of active vapours. It is shown that an extraordinary amount of light may be scattered by cloudy matter of almost infinite tenuity, the phenomena of comets' tails being thus rendered the subjects of experimental illustration.

I may add here that *cometary envelopes* may be imitated with perfect accuracy by these actinic clouds. Motions of different angular magnitudes round the axis of the experimental tube always set in, and by the sliding of the layers of cloud over each other envelopes are formed, the interior ones being plainly seen through the exterior ones. As many as five such envelopes are frequently produced. May not the envelopes of comets be formed in a similar way? Differential motions started by differences of temperature would be undoubtedly competent to distribute the cometary matter in the layers or envelopes which observation reveals. Cometary envelopes according to this hypothesis would be the result of *convexion currents*; and these again would intimate that the *visible* matter floats in an *invisible* gas or vapour.

Eight different vapours mixed in various proportions with air and hydrochloric acid and nitric acid, are subjected to examination, the resulting phenomena being described in detail.

Finally, the vapours are subjected to the powerful dark foci obtained by the use of an iodine-filter; but the ultra-red rays are found totally incompetent to produce decomposition. In these cases, therefore, as in so many others, the rays of high refrangibility are the chemical rays.

X.

ON THE ACTION OF RAYS OF HIGH REFRANGIBILITY UPON GASEOUS MATTER.*

§ 1.

Introduction.

WITHIN the last ten years I have had the honour of submitting to the Royal Society a series of investigations the principal aim of which was to render the less refrangible rays of the spectrum interpreters and expositors of the molecular condition of matter.

Unlike the beautiful researches of Melloni and Knoblauch, these inquiries made radiant heat a means to an end. My thoughts were fixed on it in relation to the matter through which it passed. Placing before my mind such images of molecules and their constituents as modern science justifies or renders probable, such images of the luminiferous æther and its motions as the undulatory theory enables us to form, I endeavoured to fashion and execute experiments founded upon these conceptions which should give us a surer hold upon molecular constitution.

Thus definite physical ideas have accompanied and guided the whole course of these researches. That matter is constituted of atoms and molecules has been accepted as a verity throughout. The phenomena under examination rendered it impossible for me to halt at the law of multiple proportions, which so many chemists of the present day appear inclined to make their intellectual bourne. In following up a train of æther waves, in idea, to their source, I could not place at that source a multiple proportion; for the waves could not be connected physically with such a multiple. I was forced to put there a bit of

* Received December 4, 1869; read before the Royal Society, January 27, 1870. *Philosophical Transactions* for 1870, p. 333.

matter—in other words, a *molecule*—bearing the same relation to the æther as a vibrating string does to the air, which accepts its motion and transmits them as waves of sound.

One result among many others which these researches established will probably play an important part in the chemistry of the future. I refer to the proved change of relation between the luminiferous æther and ordinary matter which accompanies the act of chemical combination. Here, without any alteration in the quantity, or in the ultimate quality of the medium traversed by the æthereal waves, vast changes occur in the amount of wave-motion intercepted. Let pure nitrogen and ordinary oxygen be mixed mechanically together in the proportion by weight of 14 : 8. Radiant heat, it is now known, will pass through the mixture as through a vacuum. No doubt a certain amount of heat is intercepted; but it is so small an amount as to be practically insensible. At all events it is multiplied by hundreds, if not by thousands, the moment the oxygen and nitrogen combine to form nitrous oxide. Or let nitrogen and hydrogen be mixed mechanically together in the proportion of 14 : 3; the amount of radiant heat which they then absorb is augmented more than a thousandfold* the moment they build themselves together into the molecules of ammonia. Neither the quantity nor the ultimate quality of the matter is here changed; the act of chemical union is the sole cause of the enormous alteration in the amount of heat intercepted. The converse of these statements is of course also true; dissolve the chemical bond, either of the nitrous oxide or of the ammonia, and you instantly destroy the absorption. As a proof that our atmosphere is a mixture, and not a compound, no experiment with which I am acquainted matches in point of conclusiveness that which demonstrates the deportment of dry air to radiant heat.

But the molecules which can thus intercept the waves of æther must be shaken by those waves, possibly shaken asunder. That ordinary thermometric heat provokes chemical actions is one of the commonest facts of observation. These actions, considered from a physical point of view, are changes of molecular position and arrangement consequent on the acceptance of

* It may be a millionfold; for we do not yet know how small the absorption of the absolutely pure mixture really is.

motion from the source of heat. Radiant heat also, if sufficiently intense, and if absorbed with sufficient avidity, could produce all the effects of ordinary thermometric heat. The dark rays, for example, which can make platinum white hot, could also, if absorbed, produce the chemical effects of white-hot platinum. They could decompose water, as now in a moment they can boil water. But the decomposition in this case would be effected by the virtual conversion of the radiant heat into thermometric heat. There would be nothing in the act characteristic of *radiation*, or demanding it as an essential element in the decomposition.

The dark calorific rays are powerfully absorbed by various bodies, but, as a general rule, they do not appear competent to set up that particular motion among the constituents of a molecule which breaks the tie of chemical affinity. All the rays of the spectrum exercise no doubt chemical powers. We should have scant vegetation upon the earth's surface if the red and ultra-red rays of the sun were abolished. But the chemical actions in which the *radiant form* comes into play are mainly produced by the least energetic rays of the spectrum. The photographer has his heat focus in advance of the chemical focus; which latter, though potent for his special purposes, possesses almost infinitely less mechanical energy than its neighbour. Some special relation must, therefore, as a general rule, subsist between chemical molecules and the more refrangible rays; we arrive at the conclusion, that chemical decomposition by *rays*, to keep to the ordinary term, is less a matter of *amplitude* on the part of the vibrating æther particles than of *time of vibration.*

The decomposition of a molecule must result from the internal strain of its atoms; to the atoms, therefore, and not to the molecule as a whole, the vibrations which produce chemical change must be imparted. The question remains an outstanding one in molecular physics, why it is that the longer and more powerful æther waves are generally incompetent to set up the motion which results in decomposition. The influence of synchronism here suggests itself. These shorter waves are effectual because their motion is *stored up*. Their infinitesimal impulses, because imparted at the proper intervals, *accumulate* and finally become intense enough to jerk asunder the atoms with whose periods they are in accordance.

§ 2.

Theoretic Notions: Formation of Actinic Clouds through the Decomposition of Vapours by Light.

The present investigation is in a certain sense complementary to those referred to at the outset of this paper. It deals with the relations of gaseous matter to the most refrangible rays of the spectrum. It treats of the chemical energies of such rays as exerted upon such matter. If we except the combination of chlorine and hydrogen by light, and the decomposition of carbonic acid by the solar rays in the leaves of plants, which latter, however, may not be the decomposition of *a gas*, no fact, I believe, has hitherto been known to exist in which light, or heat in the radiant form, acts chemically upon a gas or vapour.* By this inquiry the range of radiant energy as a chemical agent is, therefore, considerably extended; the phenomena resulting from that energy are exhibited in a new and exceedingly impressive form, and they prompt reflexions regarding the possible influence of solar radiation on the gases, vapours, and effluvia of our atmosphere which could not previously be entertained.

The inquiry was started thus:—It is known to the Society that the experiments on radiant heat already referred to, were for the most part performed in tubes of brass or glass, called, for the sake of distinction, 'experimental tubes.' It is also known that a difference exists between my eminent friend, Professor Magnus,† and myself, with regard to the deportment of aqueous vapour towards radiant heat. Last autumn, and in reference to the reasons assigned by him for this difference, I scrutinised, by means of a powerful beam of light, the appearance of my experimental tubes during the entrance into them of various gases and vapours. The vapours were carried into the tubes by dry air, which had been permitted to bubble through their liquids. I watched carefully, and with the aid of magnifying-lenses, for any signs of the precipitation of moisture either upon the surface of the experimental tube itself, or upon the plates of rock-salt employed to close it, keeping at

* Professor Stokes reminds me that Phosgen gas derives its name from its formation under the influence of light.—[J. T., July 1870.]

† Unhappily lost to science since these words were written.—[J. T., July 1870.]

the same time my mind open to any other action which the intensely concentrated luminous beam might reveal.

On October 9, 1868, while thus engaged upon the vapour of the nitrite of amyl, I observed a curious cloudiness in the experimental tube when the beam was sent through it. For a moment this appearance troubled me; for it required a little reflexion to assure me that in my previous publications actions had not been sometimes ascribed to pure cloudless vapour which were really due to such nebulous matter as was then before me. The appearance, however, immediately declared itself to my mind as a product of chemical action then and there exerted on the vapour.

The nitrite vapour was then intentionally subjected to the action of the light. The beam employed was convergent. As the mixture of air and vapour reached the point of greatest concentration of the beam, cloudy matter was there precipitated, which was afterwards whirled by the moving air into the more distant parts of the tube. The cloud thus carried away was incessantly renewed, and after the mixed air and vapour had ceased to enter, precipitation occurred all along the cone of rays in front of the focus.

The lamp was then extinguished, and the mixture of air and nitrite vapour permitted to enter the tube in the dark. When the tube was full, the condensed beam was sent through it. For a moment the light seemed to pass through a vacuum; but after a moment's pause a white cloud fell suddenly upon the conical portion of the beam, causing it to flash forth almost like an illuminated solid.

When the beam, previous to allowing it to enter the vapour, was caused to pass through a red or yellow glass, the action though visible was feeble; it was much more energetic when the beam passed through a blue glass. A convergent beam was sent through a red glass, and the feeble effect was observed. A blue glass was then added, and by the concert of both the light was completely cut off. On withdrawing the red glass, a very beautiful blue cloud came down upon the conical beam. The experiment proved that in this case, as in so many others, the blue rays are the 'chemical rays.'

Solar light, as might be expected, produces all the effects of the electric-light, and in regions more favoured than London may be employed in continuous researches of this nature. When the parallel beams of the sun are duly concentrated, the precipitation which they invoke in passing through nitrite-of-amyl vapour is copious and immediate.

§ 3.

Description of Apparatus.

The simple apparatus employed in these experiments will be at once understood by reference to Figure 31. S S′ is the glass experimental tube, which has varied in length from 1 to 5 feet, and which may be from 2 to 3 inches in diameter. From the end S the pipe *p p′* passes to an air-pump. Connected with the other end we have the flask F, containing the liquid whose vapour is to be examined; then follows a U-tube, T, filled with fragments of clean glass wetted with sulphuric acid; then a second U-tube, T′, containing fragments of marble wetted with caustic potash; and finally a narrow straight tube *t t′*, containing a tolerably tightly-fitting plug of cotton-wool. To save the air-pump gauge from the attack of such vapours as act on mercury, as also to facilitate observation, a separate barometer tube was employed.

Through the cork which stops the flask F, two glass tubes, *a* and *b*, pass air-tight. The tube *a* ends immediately under the cork; the tube *b*, on the contrary, descends to the bottom of the flask and dips into the liquid. The end of the tube *b* is drawn out so as to render very small the orifice through which the air escapes into the liquid.

The experimental tube S S′ being exhausted, a cock at the end S′ is carefully turned on. The air passes slowly through the cotton-wool, the caustic potash, and the sulphuric acid in succession. Thus purified it enters the flask F, and bubbles through the liquid. Charged with vapour it finally passes into the experimental tube, where it is submitted to examination. The electric lamp L, placed at the end of the experimental tube, furnished the necessary beam.

§ 4.

The Floating Matter of the Air.

Prior to the discovery of the foregoing action, and also during the experiments just referred to, the nature of my work compelled me to aim at obtaining experimental tubes

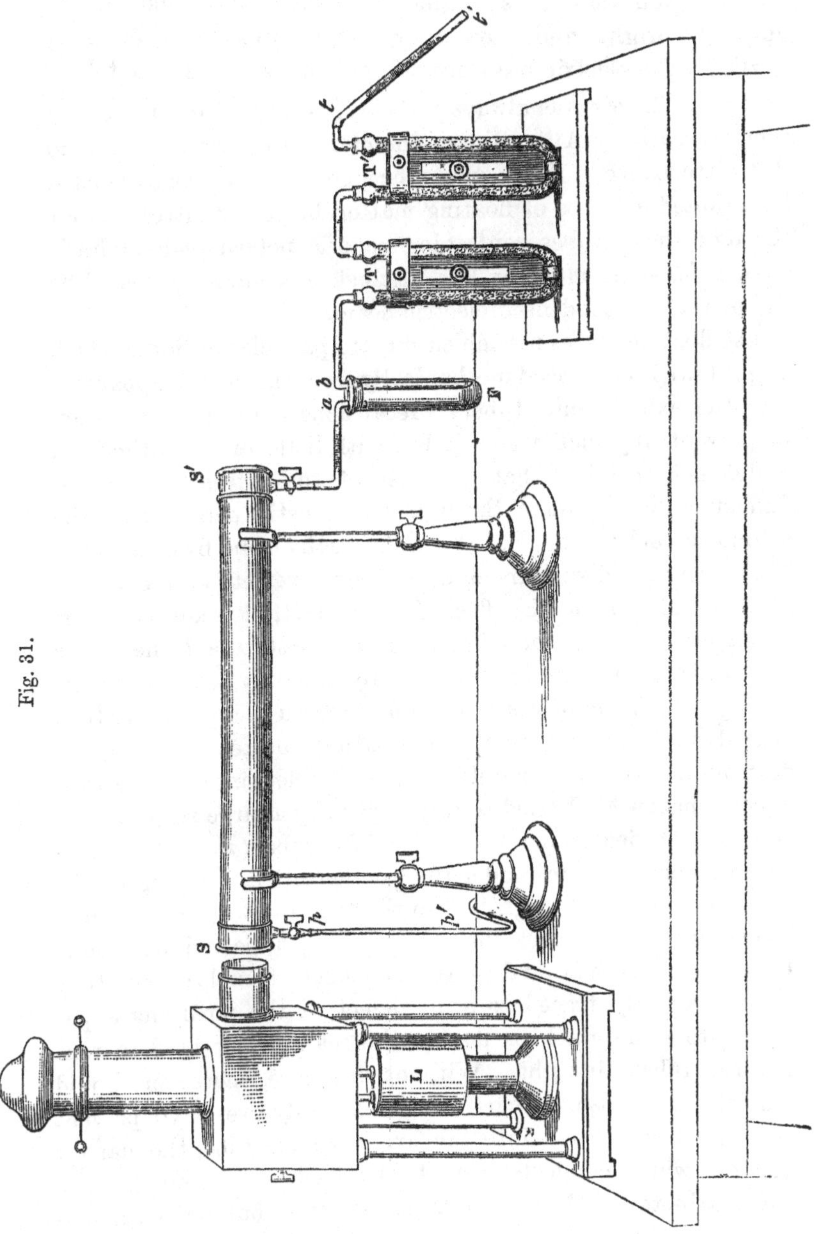

Fig. 31.

absolutely clean upon the surface, and absolutely empty

within. Neither condition is, however, easily attained. For however well the tubes might be washed and polished, and however bright and pure they might appear in ordinary daylight, the electric beam infallibly revealed signs and tokens of dirt. The air was always present, and it was sure to deposit some impurity. All ordinary chemical processes are open to this disturbance. When the experimental tube was exhausted it exhibited no trace of floating matter, but on admitting the air through the U-tubes containing caustic potash and sulphuric acid, a *dust-cone* more or less distinct was always revealed by the powerfully condensed electric beam.

The floating motes resembled minute particles of liquid which might have been carried mechanically from the drying apparatus into the experimental tube. Precautions were therefore taken to prevent any such transfer, but with little or no mitigation. I did not imagine that the dust of the external air could find such free passage through the caustic potash and the sulphuric-acid tubes. But the motes really came from without. They also passed with freedom through a variety of ethers and alcohols placed in the flask F. In fact, it requires long-continued action on the part of an acid first to *wet* the motes and afterwards to destroy them. By carefully passing the air through the flame of a spirit-lamp or through a platinum tube heated to bright redness, the floating matter was sensibly destroyed. It was therefore combustible—in other words, *organic* matter.* I tried to intercept it by a large respirator of cotton-wool tied round the end of the tube $t\,t'$. Close pressure was necessary to render the wool effective. A plug of the wool rammed pretty tightly into the tube $t\,t'$ was finally found competent to hold back the motes. They appeared from time to time afterwards and gave me much trouble; but they were invariably traced to some defect in the purifying-apparatus—to some crack or flaw in the sealing-wax used to render the tubes air-tight. Without due care, moreover, liquid particles may also be carried mechanically over. To prevent the entrance of such into the experimental tube, the narrow conduit which connects it with the flask F is plugged with clean asbestos. Thus through proper care, but not without a

* Mr. Dancer has recently examined microscopically the dust of Manchester, and found it to consist almost wholly of organic particles.

great deal of searching out of disturbances, the experimental tube, even when filled with pure air or vapour, contains nothing competent to scatter the light. The space within it has the aspect of an absolute vacuum.

An experimental tube in this condition I call *optically empty*.

Here follows one of the numerous experiments illustrative of this subject. A platinum tube 9 inches long, 0·4 of an inch wide, and having within it a roll of platinum gauze, was placed in a gas-furnace where it could be intensely heated. One end of this tube was connected with the entry stopcock of the experimental tube S S′, fig. 31; the other end was open to the air of the laboratory. The air was sent first through the platinum tube cold, then through the same tube heated to various degrees of redness, into the experimental tube, where it was subjected to the scrutiny of the concentrated electric beam. Here are the results:—

Quantity of Air	State of Platinum Tube	State of Experimental Tube
15 in. of mercury.	Cold.	Full of floating particles.
15 „	Red-hot.	Optically empty.
15 „	Cold.	Full of floating particles.
15 „	Red-hot.	Optically empty.
15 „	Cold.	Full of particles.
15 „	Dull red.	Optically empty.
15 „	Intensely heated.	Optically empty.
30 „	Intensely heated.	Optically empty.
15 „ (admitted quickly).	„	A perfectly polarized blue cloud.
15 „ (quickly).	Barely visible redness.	Particles.
15 „ (quickly).	Intensely heated.	Blue cloud.
15 „ (slower).	„	A very fine blue cloud.
15 „ (very slow).	„	Optically empty.
15 „	Cold.	Full of particles.
15 „ (quickly).	Red-hot.	Blue cloud.*

The polarization of light by such clouds as the blue ones here mentioned will receive due attention subsequently.

A remarkably fine experiment may be thus made:—Placing a spirit-lamp underneath the cylindrical beam of the electric lamp as it marks its track through the illuminated dust of the atmosphere, torrents of what would be infallibly mistaken for

* In subsequent experiments I found that this 'cloud' arose in great part from the action of the heated air upon the india-rubber joint which connected the platinum tube with the experimental tube.

black smoke rise from the flame into the beam. A Bunsen's flame produces the same effect. But the action of a red-hot poker placed underneath the beam is precisely similar; the action of a hydrogen flame, moreover, where smoke is out of the question, is not to be distinguished from that of the spirit-lamp flame. The apparent smoke rises even when the flame or the poker is placed at a good distance below the beam. The action is really due to the destruction of the floating matter by contact with the heated body. It sends upwards streams of air from which everything competent to scatter the light has been removed. This optically pure air, in passing through the beam, jostles aside the illuminated particles, the space it occupies being *black* in contrast with the adjacent luminosity. The experiment is capable of various instructive modifications, and may of course be executed with sunlight.

It is needless to dwell upon the possible influence of the floating organic matter of the air upon health. Its quantity, when illuminated by a powerful and strongly concentrated beam, sometimes appears enormous. One recoils from the idea of placing the mouth at the intensely illuminated focus and inhaling the swimming dirt revealed there. Nor is the disgust removed by the reflexion that at a distance from the focus, though we do not see the dirt, we are breathing precisely the same air. The difficulty of wetting it, before referred to, may render this suspended matter comparatively harmless to the lungs, but when these are sensitive its mere mechanical irritation must go for something. Perhaps a respirator of cotton-wool might in some cases be found useful.*

§ 5.

Deportment of Nitrite of Amyl.

I now return to the nitrite of amyl. The action of light upon the vapour of this substance is exceedingly prompt and energetic. It may be illustrated by simply blowing the vapour into a concentrated sunbeam. Or the experiment may be made to take the following form:—Connecting the tube *b* of the flask

* Since this paper was forwarded to the Royal Society these experiments have been greatly extended. See Proceedings of the Royal Institution, January 1870; also *Fragments of Science*, 1871–72.

F with the pipe of a bellows, after inflating the latter a sharp tap upon its board sends a puff of vapour through the tube *a* into the air. In a moderately lighted space nothing is seen; but when the puff is projected into a concentrated sunbeam, or into the beam from the electric lamp, on crossing the limiting boundary of light and shade it is instantly precipitated as a *white ring*. The ring has of course the same mechanical cause as the smoke-rings puffed from the mouth of a cannon, but it is *latent* until revealed by actinic precipitation.*

In every one of the numerous experiments made with the nitrite of amyl, the chemical energy appeared to exhaust itself in the frontal portion of the experimental tube. A dense white cloud would fall for a distance of 12 or 15 inches upon the beam, while beyond this distance the tube would appear almost empty. This absence of action might naturally be ascribed to the diffusion of the beam beyond the focus; but when the light was so converged as to bring the focus near the distant end of the tube the effect was the same. When, moreover, a concave mirror received a parallel beam which had traversed the tube, and returned it into the vapour in a high state of luminous concentration, the light was ineffectual. The passage of the beam through a comparatively small depth of the vapour appeared to extract from it those constituents which produced decomposition. That the vapour was present at the distant end of the tube was proved by the fact that both with the sun and with the electric-light the reversal of the tube instantly brought down a heavy cloud. As regards the chemical rays, nitrite of amyl is the *blackest* substance that I have yet encountered. It rapidly extinguishes them, leaving behind a beam of sensibly undiminished photometric intensity, but powerless as a chemical agent as far as the nitrite is concerned.

In these experiments air was employed as the vehicle of the nitrite-of-amyl vapour. By varying the quantity sent into the experimental tube, it was possible to vary in a remarkable manner the character of the resulting decomposition. The most splendid diffraction colours could be thus produced, and the finest texture could be imparted to the clouds. When pure oxygen or pure nitrogen was used, the effect was almost the same

* By a special arrangement it is easy to obtain such rings 2 inches and more in diameter.

as with air. With hydrogen the clouds appeared more delicate and lustrous; and they sometimes fell immediately after their formation in nebulous festoons to the bottom of the tube. This doubtless is to be ascribed to the lightness of the atmosphere in which they floated. In many cases, however, the particles remained suspended, and some of them continued to float even after the tube had been so far exhausted as to produce a tolerably good air-pump vacuum.

An additional effect of considerable beauty and interest was obtained in the following way. Permitting the convergent beam to play for a time upon the mixture of air and nitrite-of-amyl vapour, or, better still, upon a mixture of hydrogen and vapour, a coarse cloud is formed. Suspending the action of the lamp for a minute or so, a new distribution of the vapour appears to occur; for, on re-igniting the lamp, along its convergent beam, and *within* the old cloud, a new cloud is precipitated. The tint of this new cloud is a delicate bluish-white, and its texture is of exquisite fineness. This precipitation of one cloud within another may be obtained a dozen times in succession. Or, permitting a parallel beam to pass for a time through the coarser cloud, on pushing out the lens so as to concentrate the light, the fine cloud comes suddenly down upon the beam about its place of greatest concentration. This effect also may be obtained several times with the same charge of vapour.

No phenomena of this kind have, I believe, been hitherto observed. The necessary conditions for their production are, first, that the light should decompose the vapour, and secondly, that one or more of the products of decomposition should either be a solid, or should possess a boiling-point so high as to ensure its precipitation when set free. For though chemical action might occur, and be even energetic, if the products of decomposition be vaporous and colourless they will remain unseen. In the case just considered, the *nitrate* of amyl is in all probability a product of the decomposition of the *nitrite.* The boiling-point of the latter is estimated at from 91° to 96° C., that of the former being 149° C. The nitrite, therefore, can maintain itself as true vapour in a space where the nitrate, at the moment of its liberation, must fall as a cloud.

§ 6.

Iodide of Allyl and Iodide of Isopropyl.

An exceedingly fine example of actinic action is furnished by the vapour of the iodide of allyl. The effect of light upon this substance was observed on October 7, 1868, but I did not then know the meaning of the 'thin cloud like a kind of smoke' which showed itself in the experimental tube. On satisfying myself regarding the deportment of nitrite of amyl, the iodide of allyl occurred to me, and on it experiments were immediately made.

The decomposition of this vapour was slower than that of the nitrite of amyl. The slowness, moreover, augmented rapidly as the quantity of vapour was diminished. When only a few inches of the mixed air and vapour were in the experimental tube the action was very slow. The clouds were formed both in oxygen and in air. After the action had been continued for some time, the fine purple colour of iodine exhibited itself at the end of the tube most distant from the source of light. When hydrogen was the vehicle, the clouds were particularly lustrous and beautiful. Here and there also, amid the white and coarser sections of a cloud, spaces of delicate blue would reveal themselves, reminding one of the colour of a pure sky. The words 'wonderful,' 'beautiful,' 'lustrous,' and others of a similar nature, occur frequently and naturally in my notes of this period; for in those earlier experiments the cloud-forms obtained were so amazing, and their colours and textures so fine, as to rivet attention upon them alone.

With long-continued action the colour due to the discharge of iodine became very intense. It was strong enough to empurple the beam which passed through the air of the laboratory after its transmission through the experimental tube, and to colour deeply a white screen on which the beam was permitted to fall. In what condition was this iodine? It could be liberated by a beam deprived almost wholly of its calorific rays. The temperature of the experimental tube was indeed so moderate that a quantity of iodine placed within it and permitted to saturate the space with its vapour, produced a barely perceptible flush on a piece of white paper. The far more deeply coloured iodine revealed by the beam in the actinic cloud must, I think, have been for the most part liquid, and not vaporous iodine. I say liquid, because the substance was probably dissolved by the particles of the cloud with which it was so intimately mixed. Di-allyl, for example, is a powerful solvent of iodine, and it was probably one of the products of decomposition.

The iodide of isopropyl also capitally illustrates the action of light upon vapours. It is more slowly acted upon than either the nitrite of amyl or the iodide of allyl; nevertheless, in sufficient quantity, its decomposition is very brisk and energetic. Purified air which had bubbled through the liquid iodide was conducted into the experimental tube. When the pressure was 1 inch of mercury, the light playing upon the vapour for five minutes produced no action; but when it was 10 inches a blue cloud made its appearance in two minutes, and in ten minutes it had almost filled the tube. When the pressure was 20 inches, the action commenced more quickly, and the cloud generated was more dense. The whirling motions of this cloud appeared to be more brisk than that of the others examined. With 30 inches of the mixed air and isopropyl the action began in a quarter of a minute, and in five minutes a dense cloud was formed throughout the tube. The purple of the discharged iodine was also very plain in this cloud.

§ 7.

Deportment of Liquids and of their Vapours towards Rays of High Refrangibility.

In the preliminary notice of these experiments laid before the Royal Society in June 1868,* considerable stress is laid upon the fact that the same rays are absorbed by the nitrite of amyl in the liquid and in the vaporous state. A layer of the liquid not more than one-eighth of an inch in thickness was found competent to withdraw from a powerful beam nearly all the constituents which could effect the decomposition of its vapour.

I endeavoured at the time to apply this fact to the solution of the question whether the absorption of chemical energy was the act of the molecule as a whole, or of its constituent atoms.† I tried to show that on the first of these assumptions it is

* See page 425. † See '*Physical Considerations,*' pp. 427 and 428.

impossible for the selfsame rays to be absorbed by a liquid and its vapour. For absorption depends upon the rate of molecular vibration, and reaches its maximum when this rate synchronises perfectly with the rate of succession of the æthereal waves. Now as the rate of molecular vibration depends upon the elastic forces exerted between the molecules, and as it could hardly be imagined that these forces would remain undisturbed during the passage of a vapour to the liquid condition, the fact of the liquid nitrite of amyl and its vapour absorbing the same rays indicated that the absorption was not molecular. We were thus driven to conclude that it was *atomic*;* and this conclusion was fortified by the consideration already adverted to—that were the absorption the act of the molecule as a whole, no mechanical ground could be assigned for the falling asunder of its atoms. Thus actinic action itself pointed out the seat of the absorption.

A wide, if not entire generality was anticipated for the proposition that the same rays are absorbed by a liquid and its vapour. When this anticipation was first expressed I believed that liquids in general would be found so destructive of the effectual rays as to render transmission through moderate depths of them sufficient to rob a beam of all power to act upon their vapours. This idea, entertained though not expressed, has not been verified, and the deportment of iodide of allyl may be taken as representative of a class of facts which contradict it.

Glass cells were employed varying from one-eighth of an inch to an inch in width. Filled with the transparent iodide, these cells were placed between the electric lamp and the experimental tube charged with the iodide vapour. The rays after traversing an inch of the liquid produced copious decomposition in the tube. A marked distinction was thus proved to exist between the liquid iodide of allyl and the liquid nitrite of amyl. *But the same distinction extends to their vapours.* The ex-

* When I use the word 'atomic' in contrast with 'molecular,' I by no means pledge myself to an absolute limit of divisibility. The molecule may resemble a house, the atoms the hard bricks composing that house. But, while it is both convenient and correct to regard the house as constituted of bricks definitely bounded, it is by no means essential to regard the bricks themselves as absolutely indivisible. The divisibility or non-divisibility of the atoms does not in the least affect the atomic theory as a *working conception.*

ceeding absorbent avidity of the nitrite-of-amyl vapour, and the rapidity with which it deprives a powerful beam of its effective constituents, have been already noticed. It is quite different with the iodide of allyl. A tube 5 feet long was charged with the vapour of this substance, and after it, in the same line, was placed another tube 3 feet long charged with the same vapour. On sending a beam through both tubes in succession, the 5-foot tube, through which the light first passed, was filled immediately with an actinic cloud; but a similar cloud was at the same time falling in the second tube. A transmission through 5 feet did not seem to diminish very materially the power of the beam. A passage through 1 foot of the nitrite of amyl would have been far more destructive.

As these actions are representative and, I believe, most important, some recent confirmatory experiments executed with these two substances may be here summed up.

1. The nitrite-of-amyl vapour absorbs with such avidity the rays competent to decompose it, that a comparatively small depth of the vapour quenches the efficient rays of a powerful beam of solar or electric light.

2. The iodide-of-allyl vapour, on the contrary, permits a beam to traverse it for long distances without very materially diminishing the chemical power of the beam.

3. The liquid nitrite of amyl, in a stratum one quarter of an inch thick, quenches all the rays which could act chemically upon its vapour.

4. The liquid iodide of allyl, on the contrary, in a stratum of four times the thickness just mentioned, does not materially diminish the power of the beam to act upon its vapour.

5. A very marked difference exists between the deportment of the nitrite of amyl alone, and its deportment when mixed with hydrochloric acid. The *chemical penetrability* of the mixture is far greater than that of the pure vapour. The actinic cloud, which with the vapour alone is confined to the anterior portion of the experimental tube, extends in the case of the mixture through the entire tube.

6. A beam, moreover, which has been transmitted through a quarter of an inch of the liquid nitrite is also competent to act chemically upon the mixture, and to produce in it dense actinic clouds.

The action in this last case, though not stopped by the liquid nitrite, is retarded. Employing first the liquid screen, it was interesting to observe the sudden development of a fine-grained luminous cloud, and its violent tumbling about by the decomposing beam, the moment the liquid was withdrawn. The action of a solution of the yellow chromate of potash is substantially the same as that of the liquid nitrite. By the successive introduction and removal of a cell containing either substance successive flashes of actinic energy may be produced a dozen times and more in the same vapour.

The molecular relationship of a liquid and its vapour receives new illustration from these experiments. Whatever alters the action of the one appears to change in a proportionate degree the action of the other.

§ 8.

Influence of a Second Body on the Actinic Process.

Carbonic acid is decomposed by the solar beams in the leaves of plants; but here it is in presence of a substance, chlorophyll, ready, as it were, to take advantage of the loosening of the atoms by the solar rays. The present investigation has furnished numerous cases of a similar mode of action. All the vapours examined may be more or less powerfully affected in their actinic relations by the presence of a second body with which they can interact. The presence, for example, of nitric acid, or of hydrochloric acid, may either greatly intensify or greatly diminish the visible action of the light on many vapours which, alone or when mixed with air, are decomposable; while the presence of the one or the other of the same acids may provoke energetic action in substances which are wholly inactive when left to themselves.

We need not go beyond the nitrite of amyl for an example of this kind. For, prompt and copious as the decomposition of this substance is when mixed with air, the energy and brilliancy of the action are materially augmented by the presence of hydrochloric acid. Let a quantity of the nitrite vapour mixed with air be sent into the experimental tube till the mercury column sinks, say, 8 inches. Then let the flask containing the liquid nitrite be removed and one containing strong hydrochloric acid put in its place. Let purified air be carried through the acid into the experimental tube, until a further depression of 8 inches is obtained. On allowing the convergent beam to play upon this mixture, a cloud of extraordinary density and brilliancy is precipitated. The beam appears to pierce like a shining sword the nebulous mass of its own creation, tossing the precipitated particles in heaps right and left of it. This experiment is very easily made, and nothing could more finely or forcibly illustrate the phenomena here under consideration.

By varying the proportions of the vapour to the acid we vary the effects. For example, the proportion of 1 inch of the nitrite vapour to 15 inches of the hydrochloric acid did not produce so brilliant an effect as the proportion 8 : 8. The same is true of the proportion 15 inches of nitrite vapour to 1 inch of hydrochloric acid. But in this latter case, though the general action was less intense than in the case of 8 : 8, the iridescences due to diffraction were much finer. No doubt for each particular substance a definite proportion exists corresponding to the maximum of actinic action.*

The nitrite of butyl affords another striking example of the influence of a second body. With air, or alone, it was not visibly affected by the light; there was no cloud formed by its exposure. It was also mixed with nitric acid in various proportions, but no visible effect was produced by the beam.

It was then mixed with air which had been permitted to bubble through pure hydrochloric acid, in the following proportions :—

1.	1 inch of air and vapour to 15 inches of air and acid.					
2.	8 inches	„	„	8	„	„
3.	15 inches	„	„	1 inch	„	„

In the first case a dense and brilliant cloud was immediately precipitated. In the second case the precipitation of the fine white cloud was confined to the convergent luminous cone, coarser particles being scattered through the rest of the tube. In the third case the cloud was very coarse and very scanty. The experiment indicates that the best effect is obtained when a small quantity of the vapour is mixed with a considerable quantity of the acid.

* This might form the subject of an interesting inquiry.

Benzol is also a good example of a substance which, when alone, defies the power of the light, but which in the presence of other substances is readily decomposed. During the earlier stages of this inquiry a vast number of experiments were made with benzol and *commercial* hydrochloric acid. The results well illustrate actinic action, but they are not to be accepted as indicative of the action of *pure* hydrochloric acid. Indeed with the pure acid and benzol vapour there is no visible action.

On the 16th of November, 1868, 2 inches of air and benzol vapour were sent into the experimental tube, and afterwards the tube was filled with air which had bubbled through the commercial acid. My notes, written at the time, describe the action of light upon the mixture as producing a cloud of an exquisite sky-blue colour, only more luminous and æthereal than the sky. The figure of the cloud was also very wonderful.

This cloud was permitted to remain for fifteen hours in the experimental tube uninfluenced by light. After this interval it was found still floating, being composed of curiously shaped granular sections joined together by others of more delicate hue and texture. The renewed light set the cloud immediately in motion, the granular parts disappeared, and the whole for a length of 18 inches resumed its primitive delicate hue and texture. At some places it turned to white or whitish-grey, but at others it was a pure firmamental blue. It became very dense as the light continued to act, and finally developed itself into a form of astonishing complexity and beauty.

The experimental tube had then a current of dry air swept through it, and it was afterwards exhausted. Two inches of the benzol vapour were admitted as before, and dry air was added until the tube was full. It required five minutes' action of the light to develop the faintest visible cloud; even after ten minutes' action the cloud was very faint.* The tube was again cleansed and exhausted, 2 inches of the benzol vapour were admitted, followed by air and hydrochloric acid until the tube was full. On starting the light chemical action began almost immediately, and ended by the formation of a cloud throughout the tube. The influence of the *commercial* hydrochloric acid is here demonstrated. The interaction of nitric acid and benzol will be immediately referred to.

* It was certainly due to a residue of the previous charge.

Bisulphide of carbon is also an illustration in point. Alone or mixed with air it resists the action of the light; in the presence of hydrochloric or of nitric acid it is responsive to that action. On the 17th of November, 1868, for example, the pure vapour was admitted into the experimental tube until a depression of 2 inches of the mercury column was observed. A powerful light was permitted to act for twelve minutes upon the vapour, but no action was observed. A quantity of air which had passed through aqueous hydrochloric acid was then admitted into the tube. Six minutes' subsequent action of the light developed a cloud of considerable density. Toluol and other substances might here be mentioned in further illustration of this mode of decomposition. But I pass over hundreds of these earlier experiments which were made chiefly to instruct myself and to secure me from error. Some definite results will be given further on.

§ 9.

Generation of Artificial Skies.

I have now to introduce, though only for partial treatment, a subject which might with advantage be kept isolated, but which is so mixed up with my notes of 1868 as to be inseparable from the descriptions of chemical action which they contain. I refer to the blue colour always exhibited at the birth of clouds obtained from small quantities of the vapours of active substances, and often from large quantities in the case of substances of slow decomposition. The first distinct record of this appearance occurs in my notes for October 10, 1868. On the 9th I had been engaged upon the iodide of allyl with reference to its interaction with hydrochloric acid. Small quantities only of the vapour had been employed; and it was found that when the acid was fresh and strong the action was vigorous, that it declined in energy as successive charges of dry air were sent through the acid, becoming vanishingly feeble on the fifth filling of the experimental tube.

On the morning of the 10th the tube used on the preceding day was washed with distilled water, and swept out by a current of dry air. A mixture of air and hydrochloric acid was then sent into it, no vapour of any kind being employed. When the light first passed through it, and for some time afterwards, the experimental tube appeared perfectly empty. Slowly and gradually, however, upon the condensed beam a cloud was formed which passed in colour *from the deepest violet, through blue, to whiteness.* To this record of my note-book the remark is added, '*connect this blue with the colour of the sky.*'

In fact it was impossible to avoid seeing the relationship of both. Previous to this entry the blue had attracted my attention. It was unfailing in its appearance when the action was slow. The blue colour was in all cases the herald of the denser actinic cloud. I took a pleasure in developing it in connexion with general actinic action, and in determining whether, in all its bearings and phenomena, the blue light was not identical with the light of the sky. This to the most minute detail appears to be the case. *The incipient actinic clouds are to all intents and purposes pieces of artificial sky,* and they furnish an experimental demonstration of the constitution of the real one.

Reserving the fuller discussion of the subject for a subsequent paper, it may be stated in a general way that all the phenomena of polarization observed in the case of skylight are manifested by these blue actinic clouds; and that they exhibit additional phenomena which it would be neither convenient to pursue, nor perhaps possible to detect, in the actual firmament. They enable us, for example, to follow the growth and modification of the phenomena of polarization from their first appearance in the barely visible blue to their final extinction when the cloud has become so coarsely granular as no longer to scatter polarized light.

These changes, as far as it is now necessary to refer to them, may be thus described:—

1. The incipient cloud, as long as it continues blue, discharges polarized light in all directions, but the direction of *maximum* polarization is at right angles to the illuminating beam.

2. As long as the colour of the cloud remains distinctly blue, the light discharged from it normally is *perfectly polarized*; this light may be utterly quenched by a Nicol's prism, the cloud from which it issues being caused to disappear. Any deviation of the line of vision from the normal enables a portion of the light to reach the eye in all positions of the prism.

3. The plane of polarization of the perfectly polarized light is parallel to the direction of the illuminating beam. Hence a plate of tourmaline with its axis parallel to the beam stops the light, and with its axis perpendicular to the beam transmits it.

4. A plate of selenite placed between the Nicol and the cloud shows the colours of polarized light, and as long as the cloud continues blue these colours are most vivid in the direction of the normal.

5. The particles of the incipient cloud are immeasurably small, but they gradually grow in size, and at a certain period of their growth cease to discharge perfectly polarized light. For some time afterwards the light that reaches the eye, when the Nicol is in its position of minimum transmission, is of a magnificent blue colour. It is called in the following pages the *residual blue.*

6. Thus the waves *that first feel the influence of size,* both at the minor and major polarizing limits of the growing particles, are the *smallest waves* of the spectrum. These waves are the first to accept polarization and the first to escape from it.

7. As the actinic cloud grows coarser in texture the direction of maximum polarization changes from the normal, enclosing an angle more or less acute with the axis of the illuminating beam.

8. In passing from section to section of the same cloud the plane of polarization often undergoes a rotation of 90°. In the following pages this is designated as a change from positive to negative polarization, or the reverse.

§ 10.

Changes of Polarization in Atinic Clouds.

The experiments on benzol vapour and hydrochloric acid now to be described are of interest on optical rather than on chemical grounds. They were preceded by other experiments in which the vapour was mixed with nitric acid, and a minute residue of the latter lingering in the experimental tube may have influenced the results. The hydrochloric acid employed, moreover, was the commercial acid, and could not be regarded as pure. Thus though the decomposition of *a* vapour was certain, that it was not the pure vapour of benzol mixed with pure hydrochloric acid gas may be taken for granted. Indeed other experiments executed with the pure acid reduced the action to *nil.*

Dry air charged with the benzol vapour was permitted to enter the tube till a depression of *one inch* of the mercurial column was obtained; half an atmosphere of air charged with hydrochloric acid was then added. The action of light on this mixture was very powerful. The tube was for a moment optically empty, but its transparent contents were immediately shaken into a dense and luminous cloud. The normal polarization was feeble, the oblique strong; the selenite colours in the former case were weak, in the latter brilliant. When the line of vision was transverse, the colours seemed mainly limited to red and green.

The tube was swept with dry air and exhausted. *Half an inch* of air and benzol vapour was admitted, and after it half an atmosphere of air and hydrochloric acid. A fine blue colour soon appeared, and as long as it continued the direction of maximum polarization was along the normal. But a luminous white cloud was rapidly generated, the normal polarization becoming feeble and the oblique strong. The distant end of the cloud, however, continued blue, and in passing from it to the white cloud the plane of polarization changed 90°.

The tube was again exhausted, and *a quarter of an inch* of air and benzol vapour was permitted to enter it, followed by a quarter of an atmosphere of air and hydrochloric acid. The incipient cloud showed an exceedingly fine blue, the polarization along the normal being a maximum. The cloud gradually thickened at the centre, and finally the polarization there disappeared. As before, when the normal polarization became feeble the oblique became strong.

The tube was once more cleansed and *one-tenth of an inch* of air and vapour was admitted, followed by one-tenth of an atmosphere of hydrochloric acid and air. The blue of the incipient cloud was here superb, and it lasted longer than in the last case. The selenite tints produced by the normally polarized light were exceedingly brilliant; but they faded gradually as the cloud passed from blue to whitish-blue. At the centre of the cloud the normal polarization first fell to *nil* and then reappeared, having changed, however, from positive to negative, the two ends remaining as before. The influence of attenuation on the production of the blue colour is here strikingly exemplified.

The tube containing the benzol vapour was again cleansed and exhausted, and the last experiment was repeated; that is to say, one-tenth of an atmosphere of the air and vapour was mixed with one-tenth of an atmosphere of hydrochloric acid. After ten minutes' action the actinic cloud was found divided into five segments, alternately blue and white. Every two adjacent segments of the cloud were oppositely polarized, being divided from each other by a section of no polarization. The rectangle (fig. 32) represents the several divisions of the cloud; the letters B and W denoting the blue and white segments respectively. The transverse lines represent the neutral sections.

Fig. 32.

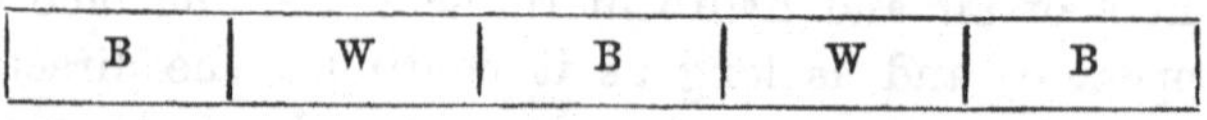

B	W	B	W	B

On the 9th of December, 1868, some experiments were made with the nitrite of butyl which merit a passing notice.

Atmospheric air was permitted to bubble through the nitrite until the experimental tube was quite filled with the mixture. Fifteen minutes' exposure produced a very slight action, an exceedingly scanty and coarse precipitate being formed. When due care is taken the action entirely disappears.

One inch of the mixed air and vapour was now admitted into the experimental tube, and after it half an atmosphere of air which had bubbled through aqueous hydrochloric acid. The instant the beam passed through the experimental tube an intensely white cloud was precipitated.

The tube being cleansed, one-tenth of an inch of the nitrite and air, followed by one-tenth of an atmosphere of air and hydrochloric acid, was sent into it. The blue of the incipient cloud was in this instance perfectly superb. The polarization at right angles to the beam was perfect, and the selenite colours exceedingly vivid. As the cloud thickened the polarization along the normal disappeared, but it became strong obliquely. Two neutral points were observed *by oblique vision* in the case of this cloud. This effect is not uncommon.

The tube was withdrawn from the light for six minutes; on re-examination the cloud was found to have lost its beauty of form; and now the cloud-centre, by normal vision, polarized the light in a plane opposite to that of the two ends.

Twelve bubbles of the air and nitrite vapour were then sent into the exhausted experimental tube, and after them thirty-six bubbles of air and hydrochloric acid; several minutes' exposure produced no action. Three inches of hydrochloric acid were then added, and the same superb blue as that noticed in the last experiment soon made itself manifest. It faded gradually as the cloud became more dense, and finally merged into whiteness.

The mixture of nitrite of amyl and hydrochloric acid was also examined in small quantities; but though the blue was fine, it had not the splendid depth and purity of colour obtained with the nitrite of butyl.

§ 11.

Early Difficulties and Sources of Error. Action of Infinitesimal Quantities of Vapour.

The whole of the autumn of 1868 was devoted to the investigation from which I have taken the foregoing brief extracts. During this period 100 different substances must, I think, have been subjected to examination, and in the case of many of them the experimental tube must have been exhausted and refilled from 50 to 100 times. In some instances, indeed, the largest of these numbers falls considerably short of the truth. For a time I had no notion of the delicacy of the inquiry, nor of the caution required to prevent the action of infinitesimal residues and impurities from being mistaken for the decomposition of substances really inert. The necessity of thoroughly cleansing, or renewing, every tube and every stopcock, on passing from one substance to another, became gradually apparent. Water, alcohol, caustic potash, and acids were successively employed to cleanse the experimental tubes; but the method found most convenient, and that finally adopted, was the thorough lathering and sponging-out of the tubes with soft soap and hot water, and the flooding of them with pure water afterwards. They are then dried with clean towels, and finally polished by passing to and fro within them, by means of a ramrod, a clean silk handkerchief. The stopcocks are cleansed by suitable brushes; fresh cocks, a fresh tube, and a fresh plug of asbestos being employed for each fresh substance.

From the draft of the present memoir, written in February, I take a few notes indicative of the difficulties caused by small impurities. Wishing to set my mind at rest with regard to nitric acid and hydrochloric acid, I operated for a time upon these substances unmixed with any vapour. Fifteen inches of air which had been permitted to bubble through aqueous nitric acid were sent into the experimental tube. The decomposing beam was first sent through a stratum of the liquid acid an inch in thickness. It screened the vapour effectually; no visible decomposition was produced. In this case, at the beginning of the experiment, a few scattered particles were in the tube.

The cell containing the liquid acid was removed, and a minute afterwards a delicate blue colour began to shed itself among the floating particles. It augmented in intensity for five minutes, but during that time it could be entirely quenched by the Nicol, the particles floating in the blue being left intact.

These floating particles (mechanically carried in) extended only about 6 inches down the experimental tube. Beyond them was a streak of fine actinic blue perfectly polarized, and beyond this again a dusky grey cloud, which showed no trace of polarization.

After ten minutes' action the cloud had assumed a fair density, but it suggested doubts whether it was due purely to the nitric acid or to the interaction of the acid and some accidental impurity. The experiment was repeated four times with substantially the same results. In all cases the beam when passed through the liquid acid proved powerless; but always on the removal of this screen, or on displacing it by a cell of water, action was manifested. To all appearance the nitric acid alone generated an actinic cloud.

The experiments, however, did not quite set my mind at rest. The tube was cleansed and the stopcocks heated to redness. When subsequently exposed the nitric acid required a much longer time to develop a cloud. After five minutes' exposure, with no cell interposed, the faintest blue was visible. After ten minutes' exposure the cloud, at first seen with difficulty, was evident for some distance down the tube. By the complete removal of residues and by strict attention to the cerate employed to make the tubes air-tight, the action thus lessened was caused finally to disappear. In each of the experiments with nitric acid recorded in the following pages the acid itself was first tried, and not until its perfect visible inertness had been proved was it permitted to mix with the vapour.

I also wished to set my mind at rest regarding the action of hydrochloric acid. Several experimental tubes were sponged with soap and hot water, washed with alcohol, and finally flooded with hot water. They were then thoroughly dried and mounted. On a first trial most of them showed a feeble actinic action, which on a second trial usually disappeared. In one case the light generated a fine blue cloud which stretched throughout the entire length of an experimental tube 3 feet long. One whitish spot only of the cloud discharged imperfectly polarized light. The cloud could be utterly quenched by the Nicol, with the exception of a small patch of residual blue about 2 inches long, which was left curiously suspended in the general darkness of the tube.

On thoroughly cleansing with dry air the tube containing the cloud, and trying the acid a second time, an exposure of twenty minutes was found to produce no action. This and many other similar experiments demonstrate the inertness of pure hydrochloric acid.* The inert acid of the foregoing experiment was permitted to remain in the experimental tube all night. Next morning, when the beam was permitted to play upon it, a blue streak became visible in less than a minute. In ten minutes the tube was filled with a delicate cloud. This was an almost everyday occurrence at the time here referred to. There must have been something in the tube in the morning which was not there on the preceding night. An infinitesimal residue had crept out of the stopcocks, or the hydrochloric acid had acted on the cerate employed to render the tube air-tight.

And here I would allude in passing to an effect which at a future stage of this inquiry will be found suggestive of the mechanism by which the complex cloud-forms are produced. I touched the top and bottom of the experimental tube for a moment with my two fingers; the cloud, which was of exceeding lightness, immediately showed responsive convexion. It was wonderfully sensitive to the slightest local change of temperature. Once started in this simple way the motions of the

* I had previously reported it active: hence these laborious experiments.

cloud went on, and ended in the development of a splendid cloud-figure.

The influence of a minute residue is also strikingly illustrated by the following fact:—Fifteen inches of mixed hydrochloric acid and air, exposed for fifteen minutes to a powerful beam, showed not the slightest trace of action. A small pellet of bibulous paper, not half the size of a pea, was moistened with the iodide of allyl. I held the pellet between my fingers till it became almost dry, then inserted it into a connecting piece, and sent a little air over it into the experimental tube. On stopping the flow of air a blue cloud began to form immediately, and in five minutes the rich colour had extended quite through the experimental tube. This cloud was 3 feet long and discharged a good body of light, but for some minutes it could be completely quenched by the Nicol. At the end of fifteen minutes a white massive cloud filled the experimental tube. Considering the amount of matter concerned in the production of this nebula, it seemed like the development of a cloud-world out of nothing.

But this is not all. The pellet of bibulous paper was removed, and the experimental tube was cleansed by allowing a current of dry air to sweep through it. *The current passed through the connecting-piece in which the pellet of bibulous paper had rested.* The supply of air was at length cut off and the experimental tube exhausted. Fifteen inches of hydrochloric acid were sent into the tube *through the same connecting-piece.* It is here to be noted, 1st, that the whole quantity of iodide of allyl absorbed by the pellet was exceedingly small; 2ndly, that I had allowed almost the whole of this small quantity to evaporate; 3rdly, that the pellet had been cast away, and the tube in which it had rested had been rendered the conduit of a strong current of pure air. It was such a residue as could linger after all this in the connecting-piece that was carried by the hydrochloric acid into the tube, and there acted on by the light.

A minute after the ignition of the lamp a chemical action declared itself by the formation of a faint cloud. It appeared first at the focus. In a couple of minutes more a faint blue, perfectly polarized along the normal, filled the anterior portion of the tube. The blue also extended from the place of most vigorous action down the tube. An amorphous cylinder of cloud soon filled the first 10 inches of the tube, and pushed gradually down it. It was followed by a complicated cloud-figure, and this again by a vase-shaped nebula, fainter than either. At the end of fifteen minutes a body of light, which, considering the amount of matter involved, was simply astonishing, was discharged from the cloud. In one position of the Nicol this cloud was a salmon-colour, in the other a blue-green. When a plate of tourmaline, with its axis parallel to the beam, was passed along in front of the cloud, at some places it showed a particularly vivid blue-green. When placed perpendicular at these places, the field of the crystal was a yellow-green.

I doubt whether spectrum analysis itself is competent to deal with more minute traces of matter than those revealed by actinic decomposition. If the weight of the cloud formed in this experiment were multiplied by trillions it would probably not amount to a single grain. Bodies placed behind it were seen undimmed through the cloud. The flame of a candle suffered no sensible diminution of its light. *It was easy to read through the cloud a page which the cloud itself illuminated.* In fact the cloud was a comet's tail on a small scale; and it proved to demonstration that matter of almost infinite tenuity is competent to shed forth light of similar quality, and in far greater quantity than that discharged by the tails of comets.*

These facts render the statement intelligible that even when all reasonable precautions appear to have been taken it is not easy to escape every trace of chemical action on first charging the experimental tube even with an inert substance. In my earlier experiments, when distilled water only was employed to cleanse the tube, the first experiment with air alone was sure to develop an actinic cloud of a beautiful fern-leaf pattern. And even now, after the most careful employment of the soft soap and hot water, the first charge of pure nitric, or of pure hydrochloric acid often developes an exceedingly delicate blue actinic cloud. As regards the *optical* question, these irregular clouds exhibit some of the finest effects.

One additional fact to illustrate the disturbances incidental to this work. Pure nitric acid had been proved over and over again to exhibit no visible action; but after its inertness had been

* The action here referred to has been since developed into a provisional hypothesis of cometary phenomena. I shall return to the subject.

demonstrated, a case occurred where it produced rather dense actinic clouds five times in succession. Indeed there seemed to be no end to their possible development. The only thing to which this change from inertness to activity could be ascribed, was a change in the cerate used to render the ends of the tube air-tight. On examination it was found that the infinitesimal effluvium yielded by the new cerate to the nitric acid was the sole cause of the anomaly. Nitric acid, then, produces no actinic cloud; hydrochloric acid produces no actinic cloud; air passed through potash and sulphuric acid produces no actinic cloud, no matter how powerful or how long-continued the action of the light may be.

I had hoped during the present year to be able to go over again a vast amount of ground rendered debateable by the discovery of such irregular actions as those here recorded. An accident in the Alps has unfortunately disqualified me from doing this. But as ardent workers have already entered this new field of inquiry, I think it right not to postpone the publication of this first part of my researches. Not only the descriptions of the deportment, but even the names of the vast majority of the substances with which I have experimented, are omitted. I confine myself to eight or ten closely examined and well-established cases of actinic decomposition, putting aside for reconsideration all such matters as are in the least degree likely to require subsequent correction.

§ 12.

Details of Experiments.

The vapours of the substances mentioned in this section were sent into the tube in the manner described in § 3. They were mixed, in the proportions stated, with air which had been permitted to bubble through aqueous nitric acid, and the effect produced by exposure to the condensed beam of the electric lamp is in each case described.

TOLUOL ($C_7 H_8$) :—*A transparent colourless liquid.*

Contents of Experimental Tube.

I. Air with toluol vapour . . . . 1 inch; then
Air with aqueous nitric acid . . 15 inches.

On igniting the lamp the experimental tube was optically empty.

After thirty seconds the track of the beam through the experimental tube became blue; the blue was about as pure as that of an ordinary cloudless sky in England. After two minutes the colour began to change to a whitish-blue.

The light discharged normally by the blue cloud continued to be perfectly polarized for four minutes after the first appearance of the cloud. A rich residual blue was afterwards observed when the Nicol was in its position of minimum transmission.

At the end of ten minutes the residual colour was no longer blue, but bluish-white. Hence the light which first exhibited perfect polarization, and which first escaped from perfect polarization, was blue.

At the end of fifteen minutes a very beautiful cloud-figure was developed. The denser portions of the cloud were very luminous.

II. Air and toluol vapour . . . 8 inches; then
Air and aqueous nitric acid . . 8 inches.

The experimental tube was optically empty for a moment at starting, but the action was so rapid that in two or three seconds the tube was filled with a heavy cloud. At the beginning the colour of the cloud was blue. The incipient cloud, which whirled round the beam, discharged for two or three seconds perfectly polarized light, but the perfection ceased almost immediately.

The cloud for a time was divided from beginning to end into two longitudinal lobes, separated from each other by an apparently empty space about a quarter of an inch wide. When the cloud was looked at *obliquely* in a vertical plane, one of these lobes was found to polarize the light positively, the other negatively. In passing from the one to the other the selenite tints were reversed.

The quantity of light scattered by this cloud was very considerable; it brightly illuminated the walls and ceiling of the laboratory. As the cloud became denser, the central empty space, which at first divided it into two lobes, gradually disappeared.

Looked at *normally* the polarization of the one-half of this cloud was positive, and that of the other negative. Between the two a neutral section existed. The oblique polarization of the dense cloud was strong.

III. Air and aqueous nitric acid . . 1 inch; then
Air and toluol vapour . . . 15 inches.

The action here was not so prompt as in the last case, nor was the cloud generated so dense. The cloud-particles, moreover, were coarser, and showed iridescent colours. Still the chemical action was distinct and copious.

Looked at normally, a portion of the scattered light was salmon-coloured. The selenite bands appeared to be of this colour, and its complementary greenish tint.

BISULPHIDE OF CARBON (CS_2):—*A transparent colourless liquid.*

Contents of Experimental Tube.

I. Air and bisulphide-of-carbon vapour . 1 inch; then
Air and aqueous nitric acid . . 15 inches.

On starting the experimental tube was optically empty; but in a minute afterwards the track of the beam became blue, which was particularly deep and rich in the middle portion of the tube.

The blue light discharged normally was perfectly polarized, but the least deviation of the line of vision from the normal caused a portion of the light to pass through the Nicol.

The growth of this cloud and the gradual brightening, and subsequent whitening, of the blue were very instructive.

The light discharged normally remained perfectly polarized for seven minutes after the first appearance of the blue colour. A faint but rich residual blue was seen for some time afterwards.

The selenite colours were exceedingly vivid with this cloud. When, moreover, a plate of tourmaline was placed with the crystallographic axes parallel to the beam, it was black; placed at right angles to the beam, a large portion of the light of the cloud was transmitted.

After ten minutes' exposure the cloud itself still showed a distinct trace of blue. The residual blue was then particularly rich and pure. After fifteen minutes the selenite colours were still vivid, though the cloud had then become greyish-white.

II. Air and bisulphide-of-carbon vapour . 8 inches; then
Air and aqueous nitric acid . . 8 inches.

When the lamp was ignited the experimental tube was found optically empty; but the chemical action commenced three-quarters of a minute afterwards, the convergent beam assuming the appearance of a fine blue spear. The action was more energetic than in the last case, though the battery was sensibly sinking in power.

The light discharged normally remained perfectly polarized for two minutes after its first appearance. The selenite colours were rich and vivid, and the tourmaline in its two characteristic positions showed the same striking contrast observed in the last experiment.

In five or six minutes the entire tube was filled with cloud, the residual blue being then perfectly gorgeous.

III. Air and aqueous nitric acid . . 1 inch; then
Air and bisulphide-of-carbon vapour 15 inches.

The tube was optically empty when the lamp was ignited. The chemical action soon commenced, a series of layers of blue cloud stretching through the entire tube. The action was less energetic than in the former cases, this being due in part to the sinking of the battery. The light discharged normally remained perfectly polarized for ten minutes.

Cyanide of Ethyl ($C_2 H_5 Cn$) :—*A transparent colourless liquid.*

Contents of Experimental Tube.

I. Air and cyanide-of-ethyl vapour. . 1 inch; then
Air and aqueous nitric acid . . 15 inches.

The tube was optically empty when the lamp was ignited. In a minute and a half the track of the beam became distinctly blue. The blue light was at first perfectly polarized.

The beam was crossed by a series of disks, which were denser and more whitish than the general mass of the cloud. The extinction of these disks by the Nicol was curious and interesting.

The growth of the particles in this case was so slow that the light emitted normally continued perfectly polarized for thirteen minutes after the first appearance of the cloud. A faint residual blue was afterwards developed.

II. Air and cyanide-of-ethyl vapour	.	8 inches; then
Air and aqueous nitric acid .	.	8 inches.

The experimental tube was optically empty for two seconds after the starting of the lamp; a fine blue colour was then observed upon the upper boundary of the convergent beam. The light emitted normally did not remain perfectly polarized for more than half a minute. In two minutes the tube was filled with cloud, the anterior portion being white, and the posterior portion bluish. The posterior portion could be utterly extinguished by the Nicol long after the anterior portion had begun to show a residual blue. Passing with the Nicol from the densest to the least dense portion of the cloud, the residual colour changed from a bright blue through a gorgeous Alpine skyblue to absolute extinction.

Looked at obliquely in a vertical plane, the two semicylinders into which the cloud was longitudinally divided were found in opposite states of polarization.

This was a truly splendid action. The chemical effect was exceedingly vigorous, and the cloud-form fine.

III. Air and aqueous nitric acid .	.	1 inch; then
Air and cyanide-of-ethyl vapour	.	15 inches.

On starting the light the experimental tube was found optically empty. In a quarter of a minute, however, the track of the beam, which previously had been invisible, was coloured blue. The chemical action appeared to exert itself with almost the same intensity throughout the entire length of the experimental tube.

For a brief interval the whole of the light emitted normally was polarized. Then for a time about three-fourths of the length of the cloud could be quenched by the Nicol, the remainder showing a fine residual blue. This sank from a brilliant azure at the densest portion of the cloud through deep rich blue to entire extinction.

The selenite bands were exceedingly vivid long after this cloud had ceased to be blue. An immense quantity of polarized light was discharged normally, even after the cloud had become white. Placed between the cloud and the eye, a plate of tourmaline with its axis parallel to the beam was practically black, while when placed across the beam a bright green light was copiously transmitted.

In one position of the Nicol this cloud was yellow, in the rectangular position it was blue. Here also the chemical action was very vigorous, and the cloud-form very fine.

BENZOL ($C_6 H_6$) :—*A transparent colourless liquid.*

Contents of Experimental Tube.

I. Air and benzol vapour . . . 1 inch; then
Air and aqueous nitric acid . . 15 inches.

Nitric acid is known to form with benzol nitro-benzol, a liquid possessing a high boiling-point. But though the mixed vapours were allowed to remain together for ten minutes before starting the lamp, when the beam passed through the experimental tube it was optically empty.

Chemical action commenced a quarter of a minute after the ignition of the lamp; a very delicate blue light was then discharged from the beam, the centre of which was particularly bright and transparent. The light emitted normally remained perfectly polarized for one minute.

I looked through the Nicol towards the cloud. For a minute it was absolutely extinguished. Continuing to look in the same direction the residual colour appeared, and passed from a rich deep violet to a hard whitish-blue. It was exceedingly interesting to watch the growth and change of the residual colour. At a certain period of its existence it rivalled the richest blue of the spectrum.

In two or three minutes the anterior portion of the tube was filled by a thick cloud generated by the beam. The cloud rapidly diminished in density as the more distant end of the tube was approached. It was composed of two longitudinal lobes, which, looked at obliquely in a vertical plane, discharged light polarized in planes at right angles to each other.

When the cloud was looked at normally, the line of vision being horizontal, on one side of the centre the polarization was positive, on the other side negative. Moved to and fro across the neutral section, the sudden expansion and contraction of the selenite bands was very curious.

After twenty minutes' action the neutral section was abolished, and the normal polarization (now feeble) became the same throughout the entire length of the cloud.

II. Air and benzol vapour . . . 8 inches; then
Air and aqueous nitric acid . . 8 inches.

On sending the light through it, the tube was not optically empty, but crowded with particles. Through them the beam appeared to force its way like a spear, bringing down upon itself a finer cloud, which soon swathed and masked the coarser spherules.

This experiment was many times repeated, but it was found impossible to bring the benzol and nitric acid together in the quantities here employed without the formation of a crowd (cloud would hardly be the word) of coarse particles. Chemical action had manifestly set in *without the intervention of the light.*

I then varied the quantity of benzol vapour and nitric acid. When 2 inches of each were admitted into the experimental tube, no particles were seen when the lamp was ignited. A quarter of a minute after the starting of the lamp the track of the beam became blue. This light remained perfectly polarized for a minute. In three minutes a dense cloud had filled the tube. In the two rectangular positions of the Nicol the cloud exhibited a salmon-colour and a hard bluish-greenish-white.

When the quantities of the two vapours were 4 inches each, there were no particles in the tube when the lamp was ignited. No doubt the substances were ready to attack each other, and in less than a quarter of a minute the beam precipitated the attack. The action was exceedingly vigorous. For a moment, and only for a moment, the polarization was perfect. In less than a minute the rapid thickening of the cloud and the quick growth of its particles abolished almost all traces of polarization.

When the quantities were 5 inches to 5, particles were found in the experimental tube on starting; and the same occurred with all greater quantities. When, for example, the quantities were 6 inches to 6, 10 inches to 10, or 15 inches to 15, there were invariably particles. In some of the experiments it seemed as if the chemical attractions were satisfied before the light began, the subsequent action being very feeble. In other instances this did not seem to be the case; for though the particles existed, the spaces between them became immediately filled by a fine dense cloud when the beam passed among them. In some instances the precipitation was exceedingly sudden and copious. Mr. Cottrell, who has assisted me with zealous intelligence in these experiments, thus describes one result:—'Some coarse particles were in the tube on commencing, and these, when the light was started, remained perfectly tranquil for a moment; but after an instant's pause the beam appeared to pierce like a ploughshare the cloud it had formed, throwing right and left of it heaps of precipitated particles. This cloud filled the tube almost instantaneously.'

To give the benzol and nitric acid more time to act upon each other, on Tuesday evening, the 16th of February, 2 inches of each were admitted into the experimental tube, and allowed to remain there through the night. Sixteen hours subsequently the beam was permitted to act upon the mixture. The tube which contained it was to all appearance absolutely empty; no particles whatever had formed during the night. In a quarter of a minute after starting the lamp chemical action began, and in five minutes the beam had filled the tube with a dense cloud.

The deportment of benzol may be thus summed up:—

Benzol.	Nitric acid.	
2 inches.	2 inches.	No particles; strong actinic action.
4 „	4 „	No particles; very strong actinic action.
5 „	5 „	Particles; dense actinic cloud precipitated among them.
6 „	6 „	„ sometimes „ „
10 „	10 „	„ sometimes „ „
15 „	15 „	„ sometimes „ „
1 „	15 „	No particles; strong actinic action.
15 „	1 „	Particles.

Iodide of Allyl ($C_3 H_5 I$):—*A transparent yellowish liquid.*

Contents of Experimental Tube.

I.	Air and iodide-of-allyl vapour . .	1 inch; then
	Air and nitric acid	15 inches.

The beam traversed the tube for an instant as if the space within it were a vacuum, but in the fraction of a second a brilliant shower of particles fell upon the beam. The cloud became coarse immediately. The action occurred in the anterior part of the tube, the most distant part being apparently free from action. This is quite different from the deportment of iodide of allyl and hydrochloric acid. On reversing the tube another cloud, of finer texture than the first, was precipitated. The cloud assumed beautiful and curious forms. The inner portions of its two longitudinal lobes were shaped like screws; they moreover rotated like screws, moving as if they were pushed mechanically into the mass of cloud in front of them. The whole effect was very fine, and the action extremely vigorous. As might be expected from the density of the cloud, the normal polarization was almost *nil*.

II. Air and iodide-of-allyl vapour . . 8 inches; then
Air and nitric acid . . . . . 8 inches.

The tube was optically empty at first, but the action, though not so brilliant as in the last case, was very prompt and energetic. A very coarse cloud was rapidly formed throughout the entire tube, upon the bottom of which the particles appeared to fall in showers.

The cloud having apparently ceased to thicken, the action of the lamp was suspended. On its accidental re-ignition a fine cloud, dense and luminous, was suddenly precipitated among the coarser particles. On again suspending the lamp the finer cloud vanished, but the coarser particles remained. On re-ignition the fine white cloud was precipitated as before, entirely masking the coarser one by its superior density and closeness of texture. This action was repeated several times in succession.

Allowing the *parallel* beam from the lamp to act for a time upon the cloud, on changing it to a convergent one the superior intensity of the light immediately caused a fine, dense, and luminous precipitation. By rendering the beam alternately parallel and convergent, this action could be reproduced several times in succession.

III. Air and nitric acid . . . . . 1 inch; then
Air and iodide-of-allyl vapour . 15 inches.

Immediately after igniting the lamp the action commenced, and spread through the entire tube in less than two minutes. The falling of the particles in vertical showers occurred here also.

After a time the lamp was extinguished, and the tube was permitted to remain quiescent for an hour. On re-igniting the lamp the tube appeared to be quite empty. The cloud that had previously filled it had entirely disappeared. Half a minute's action of the beam brought down upon it copious precipitation, a revival of the action occurring afterwards throughout the entire tube.

IODIDE OF ISOPROPYL $CH(CH_3)_2I$.

Contents of Experimental Tube.

I. Air and iodide-of-isopropyl vapour . 1 inch; then
Air and nitric acid . . . . 15 inches.

After a moment of apparent emptiness a very splendid action set in. A cloud of exceeding brightness suddenly filled the space occupied by the convergent beam. The light scattered by this anterior cloud was very powerful. At the distant end of the tube the action was feeble. I reversed the tube, but the precipitation here was by no means so prompt and copious as at the other end, into which the vapour had been evidently swept by the air and nitric acid.

The lamp was suspended for about five minutes; on re-igniting it a coarse cloud was found within the tube; but instantly through this coarseness a finer cloud of exquisite colour, luminousness, and texture was shed. A violent whirling motion was set up at the same time. The longitudinal lobes in this case were very curiously found.

II. Air and iodide-of-isopropyl vapour . . 8 inches; then
Air and nitric acid . . . . . 8 inches.

Tube optically empty, but in the fraction of a minute a shower of very coarse particles had fallen upon the beam. They augmented up to a certain point and then appeared to diminish. The reversal of the tube caused fresh precipitation. The rendering of the beam more convergent also caused augmented precipitation, but nothing so fine as in the last experiment. The action, indeed, was altogether inferior to the last in beauty and energy.

The action of the lamp being suspended for a few minutes; on re-igniting it the tube appeared empty, but in a moment a cloud much finer than that at first obtained was precipitated on the beam. Curious masses of particles gushed at irregular intervals upon the beam. On reversing the tube the action was decidedly finer than at first.

Thus, extinguishing the lamp after it has been acting for a time, the vapour during the period of suspension undergoes a change which enables it to fall as a finer and more visibly copious cloud than at the beginning of the action.

III.	Air and nitric acid	1 inch; then
	Air and iodide-of-isopropyl vapour .	15 inches.

The action commenced immediately, and in less than a minute the beam had filled the tube with an unbroken cloud. The beam was rendered parallel, and the action permitted to continue for eight minutes. The end nearest the light became rapidly empty, while in the distant half of the tube the particles fell in heavy showers. The whole tube subsequently became almost empty; the disappearance of the dense cloud first generated was very striking. It would appear as if after the first sudden precipitation evaporation had set in and restored the particles to the gaseous condition.

Nitrite of Amyl ($C_5H_{11}ONO$):—*A transparent yellowish liquid.*

Contents of Experimental Tube.

I.	Air and nitrite-of-amyl vapour . .	1 inch; then
	Air and nitric acid	15 inches.

The tube was optically empty at starting; the action began in half a minute, the cloud particles formed being very coarse. In four minutes the anterior two-thirds of the tube were filled with a very coarse cloud, the remaining third with a finer one. The whole rotated round a longitudinal axis, and the finer portion was rolled into a curious spiral form, and was tinged throughout with iridescent colours. The normal polarization was almost *nil*, except in the finer part of the cloud, which was slightly blue.

II.	Air and nitrite-of-amyl vapour. .	8 inches; then
	Air and nitric acid	8 inches.

The tube was optically empty for an instant only, a dense precipitation occurring immediately upon the concentrated beam. The distant part of the tube, however, was but scantily filled, showing the sifting action of the nitrite vapour. On reversing the tube copious precipitation occurred. After ten minutes' exposure the particles tended to settle at the bottom of the tube.

III.	Air and nitric acid	1 inch; then
	Air and nitrite-of-amyl vapour .	15 inches.

The tube was optically empty only for an instant; as in the last experiment, a dense cloud was immediately precipitated on the cone of rays. Here also the distant end of the tube was protected by the vapour in front.

In all these cases the action was distinctly less energetic than when the nitrite vapour, mixed with air alone, was exposed to the light; and very much less energetic than when hydrochloric acid was mixed with the vapour.

Nitrite of Butyl (C_4H_9ONO) :—*A transparent yellowish liquid.*

This substance gives no sensible action with nitric acid; but with hydrochloric, as already mentioned, the action is vigorous and brilliant. Here are a few of the results:—

Contents of Experimental Tube.

I.	Air and nitrite-of-butyl vapour . .	1 inch; then
	Air and hydrochloric acid . . .	15 inches.

The action began a quarter of a minute after starting, a very white and brilliant cloud forming upon the concentrated beam and quickly spreading throughout the tube.

II.	Air and nitrite-of-butyl vapour .	8 inches; then
	Air and hydrochloric acid . .	8 inches.

The action began in about half a minute, a cloud of comparatively fine particles being precipitated in the cone of rays, while the distant part of the tube was filled with coarse particles. The cloud was coarser, and the action less energetic than in the last experiment.

III.	Air and hydrochloric acid . .	1 inch; then
	Air and nitrite-of-butyl vapour .	15 inches.

After four minutes' action a number of coarse particles had formed in the tube together with a faint scroll of cloud. The action was very feeble. For vigorous action with the nitrite of butyl the proportion of the acid to the vapour must be large.

The hydrochloric acid here employed was that ordinarily used by chemists in quantitative analysis. The same series of experiments was executed with *commercial* hydrochloric acid, and the action found distinctly more energetic than when the pure acid was employed.

HYDRIDE OF CAPROYL ($C_6 H_{11} O, H$):—*A transparent colourless liquid.*

Contents of Experimental Tube.

Air and hydride of caproyl	8 inches.
Air and nitric acid	8 inches.

The tube was optically empty at starting. In three quarters of a minute a blue cloud had formed. It remained perfectly polarized for three minutes; then became gradually white, discharging imperfectly polarized light. At the end of ten minutes a dense white cloud filled the tube.

§ 13.

Action of Rays of Low Refrangibility.

For the sake of bringing out the influence of the vibrating period, I thought it worth while, to contrast the action of powerful foci of dark rays with the feeble foci produced by the convergence of the more refrangible rays of the spectrum. A solution of iodine in bisulphide of carbon was employed to hold back the luminous part of the electric beam. A cell containing ammonia-sulphate of copper was employed to hold back the rays of low refrangibility and allow those of high refrangibility transmission. The destructive action of the ammonia-sulphate on the calorific rays is well known. Its depth in the present case was such as to quench completely the red, orange, and yellow of the spectrum, but it allowed transmission to the violet and blue, and a small portion of the green. The vapours employed were mixed with the various acids mentioned in the respective cases.

Nitrite of amyl	8 inches.
Pure hydrochloric acid	8 inches.

The convergent beam of the lamp was sent through the cell containing the solution of iodine, and was permitted to act upon the mixed acid and vapour for ten minutes. The ammonia-sulphate cell was then introduced and the opaque solution removed. For an instant afterwards the tube was optically empty. Then a dense cloud was precipitated, which advanced like a moving ploughshare towards the most distant end of the tube. Within half a minute after the withdrawal of the opaque solution the tube was filled with cloud, which augmented in density for five minutes, when the experiment ceased. A repetition of the experiment yielded the same result.

Iodide of allyl	8 inches.
Nitric acid	8 inches.

Looked at for an instant after the vapour and acid had entered, with the white light of the electric lamp, the experimental tube was seen to be optically empty. The opaque solution was immediately introduced, and the vapour was subjected to the action of the dark rays for ten minutes.

The opaque solution was then removed for an instant, and the tube was seen to be optically empty. The strong calorific rays had produced no action.

The cell containing the blue liquid was then introduced; in less than half a minute the action became visible, and augmented rapidly. In three minutes a cloud stretched quite through the tube from end to end. The scattering of the blue light by the coarse particles of this cloud produced a very pretty effect.

Benzol	4 inches.
Nitric acid	4 inches.

Looked at for an instant after the admission of the vapour and acid the tube was optically empty. The opaque solution was introduced, and the invisible rays permitted to act for ten minutes. The solution was then removed, and the tube was examined for a moment with white light. It was optically empty. The blue liquid being interposed, visible action commenced 2½ minutes afterwards,* and in ten minutes a cloud was formed throughout the tube. A repetition of this experi-

* No doubt it had previously commenced, but it was invisible in the feeble light.

ment confirmed the inaction of the calorific rays, and showed the action of the blue rays to be visible a minute after the introduction of the ammonia-sulphate cell.

Toluol	8 inches.
Nitric acid	8 inches.

Looked at for an instant after the admission of the vapour and acid, the tube was found optically empty. Ten minutes' action of the calorific rays produced no effect. The blue liquid was then interposed, and in two minutes a cloud was visible upon the feeble blue beam. At the end of ten minutes this cloud stretched throughout the tube.

Iodide of β propyl	8 inches.
Nitric acid	8 inches.

The tube was optically empty at the commencement. At the end of ten minutes' exposure to the calorific rays the tube was also empty. The blue cell was introduced, but in two minutes after its introduction, no cloud appearing, the cell was removed for an instant. The action had begun, though the coarse particles of the actinic cloud were too scanty to be seen by the weak blue light. The experiment was repeated. As before, ten minutes' action of the calorific rays proved quite ineffectual. In one minute after the introduction of the blue liquid, no cloud being visible in the tube, the cell was removed. A crowd of particles were then seen upon the cone of light. The cell was again introduced, and after three minutes again withdrawn. The particles had increased considerably. Seven minutes' action rendered them sufficiently numerous to be visible in the blue light. After ten minutes the coarse cloud was very plainly seen. The action was continued with white light after the removal of the blue liquid; it was scarcely more energetic than that produced by the blue rays.

Nitrite of butyl	1 inch.
Hydrochloric acid	15 inches.

Examined for a moment by white light the tube was optically empty. After ten minutes' exposure to the dark rays the tube was again examined by the white beam: it was still optically empty. The blue liquid was then introduced, and in a quarter of a minute a long streak of cloud had formed. In $2\frac{1}{2}$ minutes a dense cloud was produced which filled the entire tube. An exceedingly delicate blue light, and at some parts a deep violet, was scattered by this cloud. After five minutes' exposure to the blue rays an intensely white cloud had formed, which completely filled the tube. The action here was very fine.

Bisulphide of carbon	8 inches.
Nitric acid	8 inches.

The tube was optically empty when the opaque solution was introduced; but after ten minutes' exposure to the calorific rays a faint blue tinge was observed, when the opaque solution was removed.* The experiment was abandoned, and the mixture of vapour and acid was again introduced. At the beginning the tube was optically empty; after ten minutes' exposure to the calorific rays it was also empty. In two minutes after the introduction of the blue cell, a cloud became visible: it quickly increased, and after four minutes extended throughout the tube. After ten minutes' action a dense whitish-blue cloud filled the entire tube. The experiment was repeated twice with the bisulphide, with substantially the same result.

These experiments are quite conclusive as to the inability of the calorific rays to produce actinic clouds: they are the product of the more refrangible rays of the spectrum.

* It is sometimes difficult to get the bisulphide into the tube without this blue tinge. It is certainly due to some impurity. With care it can be caused to disappear wholly.

XI.

AQUEOUS VAPOUR: DISCUSSION RESUMED.

ANALYSIS OF PROFESSOR MAGNUS'S PAPER ON GASEOUS CONDUCTION AND ABSORPTION.

1. *Gaseous Conductivity.*

THE evidence of the action of aqueous vapour on radiant heat adduced in 1864 and the previous years, was in my estimation so conclusive that I resolved to leave the future treatment of the question in the hands of practical meteorologists. But this evidence did not produce the same effect on the mind of Professor Magnus, and there is reason to believe that his subsequent papers raised in other minds a strong conviction in favour of his views. Inasmuch as I have never been able to share this conviction, it behoves me to show the reason for my dissent from it, and for my continued reliance on the truth of my own results. I therefore propose to resume the discussion, and to subject the experiments of Professor Magnus to a more thorough examination than that which I have hitherto bestowed upon them. It is a source of deep regret to me that he is not amongst us to answer with his own pen the observations and arguments now to be adduced.

In the 'Historic Notice' of Memoir I. the circumstances under which Professor Magnus entered upon the investigation of gaseous conduction and absorption are given, and in the body and analysis of Memoir II. certain differences between his results and mine are referred to. It now devolves on me to give a sketch of the origin and character of these differences. It will greatly facilitate comprehension if I reproduce here the drawings of Professor Magnus's first two instruments, one of which, growing out of the other, yielded the results which are at variance with mine.

One of these instruments was devised to determine the conduction of heat by gases, the other to determine the radiation of heat through gases.

The essential parts of the former are the glass vessel A B, Fig. 1 of the annexed Plate, on which is fused the flask C, the top of the one being the bottom of the other. The flask C is filled with water, which is kept boiling by steam. Its bottom is the source of heat employed in the experiments.

The receiver A B is immersed in a vessel, P Q, filled with water preserved at a constant temperature of 15° C. Into it, and at a distance of 35 millimètres (an inch and an half) from the source of heat, is introduced the thermometer *f g*. Above *f g* is a screen *o o*, intended to defend the bulb from the direct radiation from the source of heat. Other thermometers are seen carefully suspended here and there so as to ensure the constancy of the surrounding temperature.

Professor Magnus started with the idea that the screen *o o*, which was usually of cork, but sometimes of metal, would completely cut away the radiation from the source of heat above it, and at the same time impart no sensible heat to the

To face p. 378.

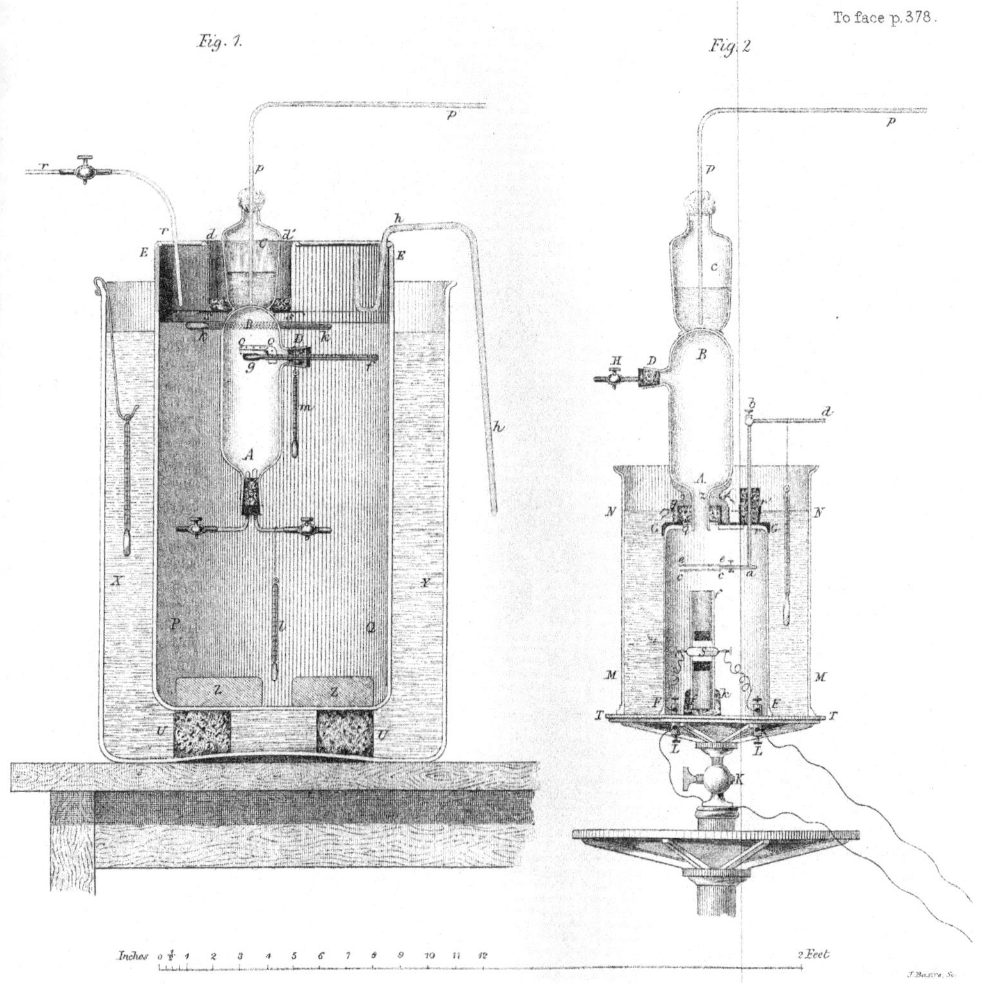

The material originally positioned here is too large for reproduction in this reissue. A PDF can be downloaded from the web address given on page iv of this book, by clicking on 'Resources Available'.

thermometer underneath. He found, however, subsequently that by long-continued action (länger dauernder Einwirkung) the screen became warm, and that a metal screen was materially less heated than a cork one. With the cork screen, however, to use his own words, by far the greater number of the experiments were made.

The vessel A B being exhausted by an air-pump, and the screen in its place, the thermometer *f g* was permitted to assume its maximum temperature. This was found to be 26·7° C, the temperature of the surrounding medium being 15° C. The difference 11·7° C., which expresses the rise of temperature in a vacuum, is set down by Professor Magnus as 100, and with reference to this standard all other temperatures are expressed.

The vessel A B was then successively filled with atmospheric air and other gases, the thermometer, in each case, being allowed to reach its maximum temperature. From 20 to 40 minutes were necessary for this. The observed temperatures, reduced to the standard before mentioned, are here given :—

Vacuum . . .	100	Nitrous oxide. .	75·2
Atmospheric air .	82	Marsh gas . .	76·9
Oxygen . . .	82	Ammonia . .	69·2
Hydrogen . .	111·1	Cyanogen . .	65·2
Carbonic acid. .	70	Sulphurous acid .	66·6
Carbonic oxide .	81·2		

In every case but one the vacuum-temperature was lowered by the introduction of the gas. With hydrogen it was augmented, and this Professor Magnus considers a conclusive proof that hydrogen conducts heat like the metals. He obtains substantially the same result when his vessel is loosely filled with cotton-wool or eider down.

The first question that here arises is: What is this vacuum temperature? How does the heat which raises its temperature 11·7° C. reach the thermometer? Does the vacuum *conduct* it, and are we to consider the hydrogen a better conductor, and all the other gases worse conductors, than a vacuum? Simple as they at first sight appear, it is very difficult to seize upon the exact meaning attached by Professor Magnus to his results. The instrument is protected from the direct radiation of the primary source, what then is the *proximate* source of heat which warms the thermometer? It is in a vessel of 'very thin' glass, surrounded everywhere, save at its upper radiating surface, and the parts closely adjacent, by cold water. There cannot, I think, be a doubt that the thermometer derives its heat from the screen, which is almost in contact with its bulb; and that the screen, in its turn, is heated by the source of 100° C., less than an inch and a half above it.

Professor Magnus excludes all thought of *convection* as having anything to do with the lowering of the temperature by the foregoing gases. He also entertains no doubt that every one of them possesses the power of conduction; but he concludes that, except in the case of hydrogen, this power is masked and overcome by their opacity to radiant heat. 'From this,' he says, 'it follows that these gases oppose a hindrance to the transmission of radiant heat, and that they are athermanous to such a degree that their athermancy exerts a greater influence than their capacity to conduct heat.'*

* I quote from the translation in the *Philosophical Magazine.* Professor Magnus's own words are, 'Daraus folgt, dass diese Gase der Verbreitung der strahlenden Wärme ein Hinderniss entgegensetzen, und dass sie in solchem Masse ätherman sind, dass ihre Aethermansie einen grössern Einfluss übt als ihre Fähigkeit die Wärme zu leiten.'

The first great difference between Professor Magnus's conclusions and mine reveals itself here. In a tube over four feet in actual length, and, because it is polished within, much more than four feet in virtual length, I found, in those early days, the absorption of radiant heat by atmospheric air and oxygen to be 0·33 per cent.; whereas, according to Professor Magnus, *a stratum of air or oxygen an inch and a half in thickness* (the distance of his thermometer from his source of heat) *is able, through its athermancy, to reduce the heat reaching the thermometer from* 100 *to* 82, *or to cut off* 18 *per cent. of the total radiation.*

2. *Gaseous Diathermancy.*

The foregoing conclusion from his experiments on conduction led Professor Magnus at once to the subject of radiation. The same source of heat was employed. But here the the vessel A B is mounted on another, F G, Fig. 2, resting upon the plate, T T, of an air-pump. S is a thermo-pile with its two cylindrical tubes attached to it, the lower one being fixed in a piece of cork resting on the air-pump, and the upper one opening towards the source of heat above B. Over the thermo-pile is the screen *ee*, *cc*, which can be moved aside, or introduced between the pile and the source of heat, at pleasure. The vessel F G is immersed in another, M N, filled with water at the constant temperature of 15° C.

The mode of experiment was this: A B and F G being exhausted by the air-pump, the screen *ee*, *cc* was moved aside, and the rays of heat passing through the tubulure *z* impinged upon the pile. The consequent deflection was noted. The gases were then admitted in succession through the cock H, and the deflection produced by the radiant heat passing through each of them was observed.

After the removal of the screen, about *two minutes* elapsed before the needles of the galvanometer took up a fixed position.

Calling the radiation through the exhausted vessel 100, the radiations through the respective gases are given in the following table:—

Gas		Gas	
Vacuum	100	Nitrous oxide	74·06
Atmospheric air	88·88	Marsh gas	72·21
Oxygen	88·88	Cyanogen	72·21
Hydrogen	85·79	Olefiant gas	46·29
Carbonic acid	80·23	Ammonia	38·88
Carbonic oxide	79·01		

In this table there is an absorption of 11·12 per cent. ascribed to atmospheric air, and an absorption of 14·21 per cent. to hydrogen.

Let us compare this result with that already deduced from the experiments on conduction. A B is 160 millimètres high, and F G is 175 millimètres; this makes the sum of both 335 millimètres. But judging from the drawing, the upper face of the pile was about 60 millimètres higher than the bottom of F G; hence the distance through which the heat had to pass to reach the pile was about 275 millimètres. In the former experiments the distance between the thermometer and source of heat was 35 millimètres. The question therefore arises:—If the observed diminution of temperature in both these cases be, as alleged, an effect of athermancy, *how is it that in the one case a layer of* 35 *millimètres effects an absorption of* 18 *per cent.; and that in the other case a layer of* 275 *millimètres effects an absorption of only* 11·12 *per cent. of the total radiation?*

3. *Proof of Convection.*

Professor Magnus, as already stated, excludes convection from the causes in operation; he considers that he has protected himself against this action by heating his gases at the top. In Memoir X. of this series I have had occasion to refer to the production of attenuated clouds by the chemical action of light. Such clouds reveal in a most instructive manner the motion of the air or other gas in which they are suspended. They were contained in glass experimental tubes in which the powerful beam of light which produced them also illuminated them. Placing the warm finger for a moment on the upper surface of the tube, the glass of which might be three, or four, or five millimètres thick, the promptness of response on the part of the cloud was extraordinary. It would lift itself immediately under the point touched, and turn over right and left forming two beautiful cloud spirals, the whole cloud dividing itself at the same time into two halves, separated from each other by a transverse black septum. In view of this and a thousand previous experiments, I should regard as certain that in the vessel A B, heated as it was in Professor Magnus's experiments, every gas that he examined would establish an ascending current along the centre of A B, which would turn over like a mimic trade wind and fall along the sides.

It is easy, however, to transfer the question from the domain of inference to that of ocular demonstration. When a little incense smoke, or, still better, a small quantity of the fumes of chloride of ammonium, is introduced into the vessel A B, crossed by a powerful beam which has been sifted by passing through a solution of alum; the currents *are seen* exactly as, on *à priori* grounds, we should expect them to appear. Here also we have spirals sometimes formed by the eddies of convection. The smoke in this experiment must not be uniformly diffused, but in streaks or striæ.

Thus observed, the deportment of hydrogen is particularly instructive. In this gas the chloride of ammonium rests at the bottom of the vessel until the heat is applied above. Then it shoots up in narrow tongues of smoke, which, when they reach the top, turn over and fall rocket-like downwards. These isolated streamers continued to be formed for a time, but the final action is a steady upward stream through the middle of the cylinder, and a descent along its sides. With a small glass shade, of the height and diameter of A B, and a flask with a concave bottom containing boiling water, this experiment can be made with facility. Filling the shade by displacement with hydrogen, and blowing into it a little of the chloride of ammonium fumes, it may be allowed to stand upon a table. The fumes remain quiescent at the bottom of the shade until the hot flask is applied at the top. The currents then begin in the manner described.

I think it certain that to the currents here revealed is to be ascribed the great discrepancy existing between Professor Magnus and myself as regards the action of atmospheric air, oxygen, and hydrogen upon radiant heat.

The solution it offers seems in every way complete. Take, for example, the experiments on conduction. The warming of the thermometer $f g$ will depend upon the rapidity with which the convection currents, falling right and left from the hot source, return to the bulb. If during their downward and upward journey sufficient time elapse to enable them to yield up their warmth to the walls of the vessel, both the thermometer and the source of heat will be chilled by their contact. If the convection be rapid, as it is in the case of hydrogen, the gas will reach the bulb before it has wholly parted with its heat. This, I submit, and not the conductivity of hydrogen, is the cause of the augmentation of temperature observed in the case of this gas.

Take again the experiments on radiation. Between A B and F G we have the narrow tubulure *z*, which checks the propagation of the convection into F G. Indeed the smoke in F G may be observed sensibly quiescent, while currents are active in A B. In this case the hot hydrogen does not reach the face of the pile at all. But the very power of abstracting and transporting heat which enabled it to warm a thermometer close to the source of heat operates in a precisely opposite fashion when the thermoscopic instrument is at a distance from the source of heat. Here the gas takes heat from the source without communicating it to the pile. But the consequent diminution of the galvanometric deflection has, I submit, nothing to do with absorption.

The very ingenious argument founded on the supposed action of hydrogen upon radiant heat, and urged by Professor Magnus against me,* must, I think, also be abandoned. The contention is that, inasmuch as hydrogen is more athermanous than atmospheric air, the greater heating of the thermometer in the first experiments could only be due to the greater conductivity of hydrogen.†

4. *Experiments in Glass Tubes with Glass Ends.*

Thus far we have dealt with the experiments in which the source of heat was a glass surface raised to the temperature of boiling water. Professor Magnus subsequently changed his apparatus, and executed a second series of experiments. The recipient of his gases was a glass tube 1 mètre long and 35 millimètres in diameter. Its ends were closed by plates of glass 4 millimètres thick. At one end of the tube was placed a strong gas-flame with double draught; at the other the thermo-pile. In one set of experiments the tube was left naked within, the heat being reflected from its interior surface towards the pile; in another set the tube was lined with black paper, which in great part abolished the interior reflexion. The arrangement is, in fact, similar to that of Dr. Franz already referred to, and it was accepted by Professor Magnus without misgiving.

The following numbers express the quantities of heat which reached the pile when the two tubes were filled successively with various gases; the radiation through the exhausted tubes being set down as 100:—

	Blackened Tube	Unblackened Tube
Vacuum	100	100
Atmospheric air	97·56	85·25
Oxygen	97·56	85·25
Hydrogen	96·43	83·77
Carbonic acid	91·81	78·08
Carbonic oxide	91·85	72·05
Nitrous oxide	87·85	75·50
Marsh gas	95·87	76·61
Olefiant gas	64·10	59·96
Ammonia	58·12	55

* *Phil. Mag.* vol. xxvi. p. 28.

† Excluding as he did convection from his thoughts, the reasoning of Professor Magnus is strictly logical. 'Die Wärme,' he says, 'welche von der untern Fläche des Gefässes C ausgeht, verbreitet sich in demselben allein durch Strahlung, oder durch Strahlung und Leitung.' If this were conceded, his position would be unassailable. When the eider down and cotton wool were employed, every filament of them heated by the source must have started a convection current.

In general the results yielded by the two tubes are widely different from each other. Thus in the blackened tube atmospheric air and oxygen absorbed only 2·44 per cent. of the incident heat, while in the unblackened tube they absorbed 14·75 per cent. Hydrogen in the blackened tube absorbs 3·75 per cent., while in the unblackened one it absorbed 16·27 per cent. These great discrepancies are ascribed by Professor Magnus to a change in the quality of the heat by the act of reflexion.

In the memoirs here placed before the reader, these results are ascribed to a very different cause; they are in *some* cases wholly, and in *all* cases partly, the result of convection. With such a source of heat and in such a tube the action of properly purified atmospheric air on radiant heat is absolutely incapable of measurement by the method described, or even by far more powerful methods. This is demonstrated by experiments with glass tubes introduced expressly for the purpose of testing both the facts and explanations of Professor Magnus. The measurements in the case of the other gases are of course all affected by the circumstances here referred to. Take one example, that of ammonia. According to the foregoing table, its absorption in the unblackened tube is about 3 times that of air; the multiple ought to be raised, at the very least, from 3 to 1000. Similar remarks would apply to the other gases.

The course of the discussion, as far as it is pursued in the preceding Memoirs, is indicated in the 'analyses;' but from time to time experiments and reasonings not thus touched upon, but which merit a passing reference, were introduced on both sides. Of this character is the paper by Professor Magnus referred to in the following observations.

Observations on Professor Magnus's Paper 'On the Influence of the Absorption of Heat on the Formation of Dew.'

1. *Explanatory Remarks.*

In Poggendorff's 'Annalen' for 1866, vol. cxxviii,* Professor Magnus published a paper 'On the Influence of the Absorption of Heat on the Formation of Dew.' After speaking of the difficulty of determining the radiation of heat by gaseous bodies, he goes on to say: 'I have, however, made a few determinations of the radiation of dry and moist air, and some other gases and vapours. Up to the present time (February 1866), the capacity of these bodies to transmit heat has alone been determined.' For the sake of avoiding misapprehensions which he would not countenance, it may be remarked that in my first Memoir (January 1861), I not only deal with *transmission*, but publish direct experiments on *radiation*, and show the order of radiation to be the same as that of absorption. It is to these experiments that Professor Magnus refers in such flattering terms in the letter quoted at page 61. In my second Memoir, moreover (January 1862), the dynamic radiation and absorption of both gases

* *Phil. Mag.* vol. xxxii. p. 111.

and vapours are very amply illustrated, the method of heating being so gentle, and of such absolute constancy, that the most unstable vapours could bear it without change, and yield accurate results. These are shown to harmonise perfectly with others obtained by totally different methods. In my third Memoir also (June 1863) elaborate experiments of the same character are recorded, and among other forms of the question, vapours and gases dynamically heated in one chamber of the experimental tube are caused to radiate through gases and vapours enclosed in a second chamber.

After speaking of former methods and results, Professor Magnus expresses himself thus:—'Yet there is a complete discordance between the results which I had obtained for aqueous vapour by this method, and those which Professor Tyndall has obtained by the use of rock-salt plates, and as this physicist, although the influence of rock-salt plates is easily ascertained, continually reverts to the statement that heat is absorbed by aqueous vapour with several thousand times greater energy than air,* . . . I have considered it a duty incumbent on me to compare, if possible, in another manner, the absorption of heat by aqueous vapour with that by air.'

It is needful for me to explain the words 'several thousand times' here employed. The statement they refer to was made by me in a lecture given at the Royal Institution on January 23, 1863, and they were used to render the comparative action of aqueous vapour clear and emphatic to an intellectual, but not wholly scientific audience.† The statement was a perfectly definite one. One hundred parts of our atmosphere are described as being composed of 99½ parts of oxygen and nitrogen; and it is stated that the remaining 0·5 per cent. exerts, not several thousand times, but 80 times the action of the air in which it is diffused. I then compare the *atom* of oxygen and nitrogen with the *molecule* of aqueous vapour, and it is the absorptive energy of the molecule that is affirmed, and I believe rightly affirmed, to be several thousand times that of the atom. This explanation will probably dissipate much of the wonder and the doubt which would rightly attach themselves to the vague and unqualified statement 'that heat is absorbed with several thousand times greater energy by aqueous vapour than by air.'

Finally, I may be allowed to say one word regarding the slight *timbre* of reproach which many will detect in the words: 'and this physicist, although the influence of rock-salt plates is easily ascertained, continually reverts to the statement,' &c. This, I fear, will be understood to mean that I had persisted in a defective mode of experiment when the proofs of its defects were easily attainable. I do not think, however, that Professor Magnus would willingly write a sentence that would bear this interpretation. The hygroscopic qualities of rock-salt are conceded; care is necessary in the use of it, and especial care when air artificially moistened is *blown* over it. But as I use it the necessary caution is by no means inordinate. I have, for example, permitted my assistant, Mr. Cottrell, to experiment for a week at a time with dry and moist air, and to detach every evening the rock-salt plates from the experimental tube while the latter was filled with moist air. In these experiments no more trace of moisture, or its effects, was found upon the plates of salt than could be discerned upon the

* 'Da dieser Physiker, obgleich der Einfluss der Steinsalz-Platten leicht zu constatiren ist, immer wieder darauf zurückkommt,' &c.

† Under the heading, 'Radiation through the Earth's Atmosphere,' this lecture is printed farther on.

most carefully dried and polished surface of flint-glass or rock-crystal. It will also be borne in mind that in the autumn of 1862 Professor Magnus himself examined my plates of salt without being able to detect a symptom of that 'influence' which in the above passage he declares to be 'easily ascertained.'

2. *Discussion of Paper.*

In this inquiry Professor Magnus seeks to determine the absorptive power of aqueous vapour by experiments on its radiative power. Compared with air, Professor Frankland had found this to be high. Professor Magnus now takes up the subject, pursuing a method of his own. 'The gases and vapours were passed through a brass tube 15 millimètres in internal diameter, placed horizontally, and heated by gas-flames. One end was bent upwards so that the heated air ascended vertically, while at a distance of 400 millimètres from the vertical current was placed the thermo-electric pile.'

When dry air was forced through this tube, the deflection of a delicate reflecting galvanometer was 3 divisions of the scale; when air which had previously passed through water of 15° C. was forced through the tube, the deflection rose to 5 divisions. When the water was from 60° to 80° C. in temperature, the deflection was 20 divisions; while when the water was caused to boil violently, the deflection was 100 divisions. But in this last case a *mist* was always visible in the air; the large deflection, therefore, could not be considered as caused by radiation from *vapour.*

But in the case of the 20-division deflection no mist was visible: still Professor Magnus 'is tempted to conclude that this deflection also depended on the formation of a mist at the boundary of the ascending current.'

But the most important—and it is really an important—consideration urged by Professor Magnus has reference to the formation of dew. Having described various experiments made with the heated tube, he brings forward his main argument with much force and clearness. 'I do not,' he says, 'think that these experiments were needed, for if aqueous vapour were really so good an absorber of heat as Professor Tyndall maintains, dew could never be formed, for the aqueous vapour which is indispensable for dew would at the same time form a covering on the surface of the earth and prevent its radiation. But just where the atmosphere is particularly rich in water, as in the tropics, dew is principally deposited. . . . It would then,' he adds, 'be inexplicable that clouds could prevent dew.'

I may say that I had long previously looked at this difficulty, and in the face of it had affirmed the absorptive power of aqueous vapour. For, I reasoned, where the vapour is abundant, as, for instance, in the valley of the Nile, an inconsiderable lowering of temperature will suffice to bring it down; while if the vapour were not abundant, the requisite opening to radiation would be offered and the requisite lowering of temperature produced. I have already referred to the influence of wet straw in preventing the formation of ice in India. But a far more striking demonstration of the influence of aqueous vapour on the maintenance of temperature is furnished by the well-reasoned observations of Strachey at Madras, which are recorded in the 'Philosophical Magazine' for July 1866. From them I will choose a particularly striking sample, in which terrestrial radiation was determined at night during a period when 'the sky remained remarkably clear, while great variations in the quantity of vapour took place.' The first column of the annexed table gives the vapour-pressure, and the second the fall of the thermometer from 6.40 P.M. to 5.40 A.M.

Pressure of vapour	Fall of thermometer
0·888 inch	6·0°
0·849 "	7·1°
0·805 "	8·3°
0·749 "	8·5°
0·708 "	10·3°
0·659 "	12·6°
0·605 "	12·1°
0·554 "	13·1°
0·435 "	16·5°

Now it appears to me that where the supply of vapour is abundant and the air nearly saturated, the lowering even of 6° observed in the first observation might bring down copious dew. But the table nevertheless forbids the thought that aqueous vapour does not exert a great influence on the terrestrial radiation, for the halving of its quantity more than doubles the fall of the thermometer. It is to be observed also that in many places where dew falls heavily, the stratum of vapour, though heavy, is *shallow.* I have often seen the orange trees of Mentone teeming with dew near the level of the sea, while very little was to be seen on the hills 100 feet above the sea level.

General Strachey also showed the influence of aqueous vapour on solar radiation. In March 1850, the following augmentations of temperature between 5.40 A.M. and 1.40 P.M. were observed, the corresponding vapour-pressure being appended:—

Pressure of vapour	inch 0·824	0·737	0·670	0·576	0·511	0·394
Rise of temperature from 5.40 A.M. to 1.40 P.M.	Fahr. 12·4°	15·1°	19·3°	22·2°	24·3°	27·0°

Showing the freer transmission of the solar rays as the vapour diminishes.

I think there is little doubt that the gaps observed by Sir J. Herschel in the ultra-red region of the spectrum, which have been so successfully investigated of late by M. Lamansky,* a pupil of Helmholtz, arise from the attack of aqueous atmospheric vapour. Indeed M. Lamansky noticed them to become deeper on days when the humidity was great.

In passing, I may remark that the radiation from aqueous vapour which is in part the subject of Professor Magnus's paper, may be simply illustrated by the hydrogen flame; its emission, however, being less than one-half that of a *luminous* gas-flame of equal size. Why this feeble emission? Simply, I believe, because of the extreme attenuation of a flame of hydrogen. The flame is not necessary to exhibit the radiation. I placed an argand burner consuming hydrogen in front of a thermo-electric pile, and defending the pile from all radiation from the flame itself, allowed the ascending column of aqueous vapour to radiate against it. The needle was sent up to its stops at 90°. At a distance of ten or twelve feet from the column of hot vapour the action was very sensible. In causing the vapour to ascend into a powerful

* *Monatsberichte* for December 1871.

beam of light, it could be observed, not as a white cloud or mist, which would be infallibly revealed by this method of illumination, but by a space *as black as night*, from which the floating matter of the air had been ejected. Neither at the edges nor in the middle was the slightest trace of precipitation. The radiation is due to the perfectly invisible mixture of air and vapour. I think it quite likely, however, that in Professor Frankland's experiment mist may have been precipitated, or what is still more probable, water in fine particles may have been carried mechanically along with the jet of air.

Appended to the translation of Professor Magnus's paper, which was published in the 'Philosophical Magazine' for August 1866, are the following remarks written by me at the time :—

First Remarks on the Paper of Professor Magnus.

I SHOULD refrain for the present from making any remark upon the paper of my friend Professor Magnus, did I not fear that my silence might be misconstructed by meteorologists, and that they might be withheld, through a doubt as to their value, from prosecuting observations which, I think, are sure to expand the boundaries of their science.

For an abstract of the experiments and reasonings by which each successive objection which has been urged against my experiments on the action of aqueous vapour on radiant heat has been met, I would refer to the second edition of my work 'On Heat,' pp. 381–421. With the desire there manifested to get at the bottom of the difference between us, I approach the latest objections of Professor Magnus, regretting only that, being on the point of quitting London, I can do no more than jot down a few of the more obvious reflexions suggested by his own description of his experiments.

Professor Magnus now infers the absence of absorption from the absence of radiation. He employs as source of heat a stream of air, which is first urged through water at the ordinary temperature, and afterwards caused to pass through a hot brass tube 15 millimètres in diameter. On its emergence from the tube it radiates against a thermo-electric pile placed at a distance of 400 millimètres. When dry air was urged through the tube, the deflection was exceedingly small; when air moistened as above was employed, the deflection was but slightly augmented.

Now, in the first place, the amount of vapour taken up by air in its passage through cold water is so small, and the stream of such air employed by Professor Magnus is so thin, that the heat radiated from the vapour must be excessively minute. Supposing the vapour compressed only to the density of ordinary atmospheric air, the average thickness of the radiating layer would be less than $\frac{1}{300}$th of an inch. Even assuming the rays from this source of heat to reach the pile without impediment, its action would be inconsiderable, if not insensible.

But the rays were not permitted to reach the pile without impediment. I assume that Professor Magnus did not deem it necessary to *pry* the air intervening between his source of heat and his pile; otherwise he would have mentioned a precaution of such importance. Here, then, we have the vapour of a cylindrical column of air, 15 millimètres thick at its widest part, radiating through a layer 400 millimètres thick, of a substance intensely opaque to its radiation. Considering, then, the feebleness of its origin and the difficulties in its way, it is not surprising that the radiation from the source of heat chosen by Professor Magnus failed to produce any very sensible impression upon his galvanometer.*

It must be borne in mind that, to obtain copious radiation from a substance so attenuated as aqueous vapour, *a considerable length of it must be employed.* An example will illustrate this. When enclosed in a tube 3 feet long, the radiation of sulphuric-ether vapour, at 0·5 of an inch of pressure, exceeds that of olefiant gas at the pressure of the atmosphere. In a tube 3 inches long, on the contrary, the radiation from the gas is more than treble that from the vapour.† That carbonic acid gas excels aqueous vapour in the experiments of Professor Magnus, does not therefore surprise me.

I have at present no means of judging of the validity of the assumption of Professor Magnus that the air urged through water at a temperature of 60° or 80° C., produced its effect by precipitation. There is none visible. By a similar assumption, he explains the experiment of Professor Frankland, in which aqueous vapour was discharged along the axis of a cylinder of hot air and hot carbonic acid, being thus protected by its gaseous envelope.

With regard to the formation of dew, the amount deposited depends on the quantity of vapour present in the air; and where that quantity is great, a small lowering of temperature may cause copious precipitation. Supposing 50, or even 70, per cent. of the terrestrial radiation to be absorbed by the aqueous vapour of the air, the uncompensated loss of the remaining 30 would still produce dew, and produce it copiously where the vapour is abundant. Attenuated as aqueous vapour is, it takes a good length of it to effect large absorption. I have already risked the opinion that at least 10 per cent. of the earth's radiation is intercepted within 10 feet of the earth's surface; but there is nothing in this opinion incompatible with the observed formation of dew. A surface circumstanced like that of the earth, and capable of sending unabsorbed 80 or 90 per cent. of its emission to a distance of 10 feet from itself, must of necessity become chilled, and must, if the vapour be abundant, produce precipitation.

I would now leave to others the further development of this question, feeling assured that, once fairly recognised by field meteorologists, the evidence in favour of the action of aqueous vapour on solar and terrestrial radiation will soon be overwhelming. An exceedingly important instalment of this evidence was furnished by Lieut.-Colonel Strachey in the 'Philosophical Magazine' for June 1862. It is especially gratifying to me to find my views substantiated by so excellent an observer and so philosophical a reasoner.

Let me say, in conclusion, that nothing less than a conviction based on years of varied labour and concentrated attention could induce me to dissent, as I am forced to do, from so excellent a worker as Professor Magnus. Hitherto, however, our differences have only led to the shedding of light upon the subject; and as long as this is the result, such differences are not to be deprecated.

Royal Institution, July 2, 1866.

* We have no means of judging the humidity of the radiating air as compared with that of the air through which it radiated. If the water employed to saturate the air were very cold, the latter might be the greater of the two.

† Section 14, Memoir II.

Professor Wild's Experiments.*

I intended the foregoing remarks on the paper of Professor Magnus to be my last contribution to this discussion, believing that in the hands of practical meteorologists the alleged power of aqueous vapour would soon receive either confirmation or refutation. But already a new experimenter had appeared in the person of Professor Wild, of the University of Berne. He describes with consummate clearness the state of the question; and especially points out the radical difference between Professor Magnus and myself in the experiments on dry and humid air with open tubes. Professor Magnus, it will be remembered, obtained results diametrically opposed to mine, observing heat where I had observed cold, and cold where I had observed heat, and notwithstanding the verification of my results by Professor Frankland, there was no indication on the part of the Berlin philosopher that he had either corrected his experiments or changed his opinion. It was to this point of difference between us that Professor Wild addressed himself.

From three distinct series of experiments, executed with the utmost skill and care, Professor Wild drew a conclusion which is best expressed in his own words:—'In conclusion, I may say that in all my experiments conducted according to Tyndall's method, which included more than a hundred distinct observations, I have *never* obtained deflections of the galvanometer needle in contradiction to the statements of Professor Tyndall; that, further, my measurements give approximately the same ratio of the absorption by moist air to that by coal-gas; and that, lastly, I consider that certain objections which might have been raised against the conclusiveness of Professor Tyndall's method have been removed by means of appropriate changes in his method of experimenting.

'This complete verification of Tyndall's results rendered it more desirable to investigate the absorption of dry and moist air by the method of Professor Magnus.'

He does so, and corroborates all that I had stated regarding dynamic heating and chilling. He also corroborates my statements regarding convection. But his observations and conclusions are best stated in his own words: 'The observations with the apparatus again showed strong deflections of the magnet-mirror of the galvanometer when air was leaving or entering the tube, the former case indicating a chilling, and the latter a warming of the upper face of the thermo-electric pile. These deflections were much more intense than those produced by the total radiation of the upper source of heat of 100° C. against the pile.' This is exactly what I had found them to be. Professor Wild could not obtain a stationary temperature, and the irregularities were sufficient to mask any difference existing between dry and moist air. He makes this noteworthy observation: 'Coal-gas, one of the most powerful absorbers, produced a deflection of only 40 divisions of the scale, whereas the difference of action between air at the pressure of the atmosphere and the same air at 16 millimètres pressure, amounted to from 100 to 200 divisions of the scale.' He rightly concluded that this enormous difference could not be a difference of *absorption*, but one of *convection*.† When the air which chilled it was removed, the source rose to its pristine temperature.

* *Pogg. Ann.* vol. cxxix. p. 57; *Phil. Mag.* vol. xxxii. p. 118.

† Professor Magnus replies to this that M. Wild's tube, though heated at the top, was of metal, and conducted the heat from the source a certain distance downwards;

He winds up thus: 'The insufficient sensitiveness on the one hand, and the convection-currents on the other, caused me at length to abandon the experiments by Magnus's method. And although this method of investigating absorption may, in the hands of so experienced and expert an experimenter as Professor Magnus, be an appropriate one, I feel bound from my own experience to give a decided preference to Tyndall's method, not only on account of the greater facility with which it furnishes qualitative [quantitative] results, but also because of its greater delicacy. It is principally in consequence of this greater delicacy that, notwithstanding the negative results furnished by Magnus's method, I maintain that the greater power of moist air, as compared with dry, has been fully established by the experiments made according to Tyndall's method; and I am of opinion that meteorologists may without hesitation accept this new fact in their endeavours to explain phenomena which had hitherto remained more or less enigmatical.'

Professor Magnus's Last Paper.*

Stimulated by the investigations of Professor Wild, Professor Magnus again took up the subject in 1867. He arranged his apparatus after the pattern of the Swiss philosopher, and verified with it the experiments of Professor Wild, which were substantially those that he had seen during his visit to London in 1862.

But he soon observed an effect new to him, and furnishing what he considered a complete explanation of the results obtained by Professor Wild and me.

The thermo-pile, with its two conical reflectors, was placed between two open metal tubes; at the ends of the tubes most distant from the pile were placed two cubes of boiling water, which radiated through the tubes against the opposite faces of the pile. The arrangement, in fact, was the method of compensation. By means of a movable screen the radiation from the two sources of heat could be caused to neutralize each other, and by means of a bellows moist air, or dry, could be forced into the two tubes. Professor Magnus found that after the needle had been brought to zero, and when it might be expected that the blowing *of the same kind of air* into both tubes ought to produce no change in the position of the needle, a deflection occurred. When the tubes were examined the effect was found to depend on the state of the interior surface.

Acting upon a first hint of this kind, he lightly coated one of his tubes with

whereas his tube was of glass and very thin. I have remarked in another place that Professor Magnus's glass vessel must at all events have been thick enough to bear the pressure of the atmosphere, for it had been frequently exhausted. There could only have been *a difference of degree* between M. Wild and him. In his account of the apparatus employed in his experiments on conduction, Professor Magnus expressly says, 'The portions of the surface adjacent to the vessel of boiling water are heated by conduction.' It is the selfsame apparatus that he employs in his experiments on radiation, with this difference, that the heat must extend further in the latter case, because here his apparatus, instead of being surrounded by *water*, is surrounded by *air*. See Plate facing page 378.

* *Pogg. Ann.* vol. cxxx. p. 207; *Phil. Mag.* vol. xxxiii. p 413.

lampblack, and left the other polished. When moist air was blown into both, the deflection of the needle proved the loss of heat in the polished tube to be greater than in the blackened one. In fact, while the one withdrew 1·4 per cent. the other withdrew 3·75 per cent. of the total radiation.

Coating the tube with a still stronger layer of lampblack, he found the chilling by moist air not only *nil*, but negative; that is to say, a *heating* effect was observed. This was also the case when the tube was lined with cotton velvet. When a smooth pasteboard lining was inserted into the experimental tube neither heating nor chilling was observed, while a second pasteboard tube coated with coarse paper produced a heating effect. It was known that porous bodies, such as lampblack and bibulous paper, were heated by the condensation of aqueous vapour in their pores, and to an effect of this class Professor Magnus rightly ascribed the warming of the interior of his tube by the moist air.

But in a previous inquiry he had added another fact, which was not so well known; namely, that metals, as well as non-metallic bodies, were also warmed when moist air was blown against them, and chilled by a current of dry air of the same temperature. The first action he concludes is due to the formation of water and the liberation of latent heat; the second action he ascribes to the evaporation of the water thus formed.

Placing a linear thermo-pile in contact with his experimental tube, Professor Magnus found that heat was produced when moist air was forced through it, and cold when it was swept by dry air. Hence the following explanation of the results which I had ascribed to aqueous vapour:—By the condensing action of the interior surface of the experimental tube, that surface covers itself with a *liquid* layer of the very highest degree of opacity to radiant heat; and it is the *water* acting on the rays impinging against the surface of the tube that produces the effect ascribed by me to the *vapour* of water. Professor Magnus does not consider this liquid layer to be continuous, but broken into parcels, which not only *absorb*, but *scatter* the incident heat.

It is to be borne in mind that the layer of liquid is altogether invisible, and indeed eludes the eye when the most powerful means are employed to detect it. Still it covers all substances—resinous, vitreous, or metallic—at all times and in the driest climates. It exists, for example, on the glass of the electrician after he has taken the utmost pains to warm and dry it with a view to its insulation. This is not stated in so many words by Professor Magnus, but this, and much more than this, is involved in his explanation.

Professor Magnus adds to those just referred to, another remarkable observation. Air at a temperature of 16° or 17° C. was saturated with moisture, forced through a hot tube which raised its temperature to 38° C., and then caused to impinge upon a thermo-pile possessing the same temperature of 38°. When the air was moist it warmed the face of the pile, when dry it chilled it. This occurred not only when the face of the pile was coated with lampblack, but also after the coating had been removed, and the metal ends of the elements exposed to the current of air. Professor Magnus infers from this that aqueous moisture is condensed upon a surface 22° C. higher in temperature than the dew point of the vapour.*

This property of condensing vapours on the surface of bodies is named by

* To render this experiment secure, no moist air must impinge against the space immediately surrounding the pile, but it is not said that this space was protected.

Professor Magnus vapour-hesion. He extended his experiments to alcohol, found in its case the vapour-hesion to be far stronger than in the case of aqueous vapour, and the amount of heat withdrawn by the liquid layer overspreading the interior surface of the experimental tube much greater.

But in the tube a length of aqueous vapour is always associated with the assumed layer of water; hence it would be exceedingly desirable to break this association and to determine the action of the water stratum, pure and simple. By means of concave and plane mirrors, Professor Magnus reflected the rays from a source of heat to his pile. Dry and moist air were urged in succession against the mirrors, but unless he visibly *wetted* them no change whatever was observed in the amount of the reflected heat.

This is accounted for by saying that the loss of heat in a single passage to and fro across the layer, vanishes in comparison with the heat withdrawn by the multiple reflexions within the experimental tube. So frequent are these reflexions in the case of some of the heat-rays that Professor Magnus pictures them as describing a spiral on the surface of the tube. He does not, however, consider that the greater distance traversed by the heat-rays in the polished tubes can have contributed in any sensible degree to the greater loss of heat observed in them. The quenching of the heat is the act of the liquid layer, and not of augmented distance.

While denying any sensible action to aqueous vapour, Professor Magnus concedes it to alcohol, which he finds competent to absorb heat even when it is discharged in the open air. Many years previously this experiment had been made, not only with alcohol, but with every vapour that I had subjected to examination. It was indeed one of the many checks introduced to protect me from error.

REMARKS ON PROFESSOR MAGNUS'S LAST PAPER.

I SHOULD like to make one remark on the historic opening of this paper, which runs thus:—'Mr. Tyndall found that aqueous vapour absorbs heat to so extraordinary an extent that, when a single atom of oxygen and nitrogen is compared with a single molecule of aqueous vapour, the latter absorbs 16,000 times more heat than the former. He subsequently says that humid air in a tube 4 feet long absorbs from 4·2 to 6 per cent. of the radiation.' It is to be borne in mind that the two statements placed here in juxtaposition are in entire harmony with each other. The word 'extraordinary' has a *relative* significance, and might perhaps be more fitly applied to the transparency of the air, which is truly astonishing, than to the opacity of the vapour. In the place from which Professor Magnus quotes,* it is observed that, for every molecule of aqueous vapour, there are about 200 atoms of oxygen and nitrogen; and as experiment showed the total vapour to exert eighty times the effect of the total air, it followed that the single molecule exerted 16,000 times the effect of the single atom.

* The article entitled 'Radiation through the Earth's Atmosphere,' p. 422.

I now pass on to the experiments described in the paper and the inferences drawn from them.

Let us in the first place inquire whether the fact of heating observed by Professor Magnus implies such condensation as must of necessity play the important part assigned to it by Professor Magnus.

The question may be answered experimentally. A thin plate of highly polished silver was laid against the face of a thermo-pile. Directed towards the plate was a glass tube connected with a holder of hydrogen. The gas was forced through two drying tubes filled with fragments of glass moistened by sulphuric acid, and caused to impinge in a gentle current against the silver. The needle of the galvanometer was at the same time observed, and at the beginning of the experiment pointed to a high figure, indicating that the face of the pile in contact with the plate was colder than the opposite one.

The moment the dry gas touched the plate of silver the needle moved to a still higher figure, declaring, therefore, the further chilling of the pile. But an opposite action immediately set in. The needle stopped, returned, fell rapidly to zero, and went up to 90° on the other side of it. It could, in fact, be made to swing quite round the dial by the generated heat.

A highly polished plate of brass was then substituted for the silver. On causing a jet of dry hydrogen to impinge against it, precisely the same effects were observed. The needle, as before, could be swung quite round its dial by the heat generated through the contact of gas and metal.

Dry air, or dry carbonic acid, produced no heating of this kind, but, on the contrary, the chill already observed by Professor Magnus.

When the jet of hydrogen continued to play upon the plate of brass for a considerable time, the needle slowly sank towards zero, and went up to its original position on the side of cold. The attraction which produced the first heating being satisfied, either the attracted film remained there, the heat produced by it being dispersed, or it was incessantly displaced, the chill of its evaporation balancing the heat of condensation. Probably, however, the film, once seized upon by the surface, remained attached, the succeeding gushes of the gas gliding over it without coming into any real contact with the metal.

What then is the effect of this powerful condensation upon the transmission of radiant heat through a polished brass experimental tube? None whatever. Oxygen is inactive, hydrogen is inactive, the mixture of oxygen and hydrogen is inactive. It will be remembered that Professor Magnus *infers* both the existence of the liquid film, and the absorption by the film, from the fact of heating. The foregoing experiments prove that the inference is not necessarily correct, that we may have heating, probably more powerful than that observed by Professor Magnus, without any sensible interference with the transmission.

Let us now pass on to vapours. Not only does aqueous vapour produce the observed heating effect, but alcohol, according to Professor Magnus, produces it in a much greater degree. This he first infers from the energy of its absorption, and he afterwards confirms the inference by direct experiment. In a former paper he had indeed shown that 'results perfectly similar to those obtained with vapour of water were also obtained with the vapour of alcohol, of ether, and other vapours.' *

Now, in the first memoir of this series, which was written twelve years ago,

* *Phil. Mag.* vol. xxvii. p. 249.

it is stated that the fear lest a change of the reflecting surface of my experimental tube might have some share in the production of the observed effects, caused me to compare the deportment of all the vapours in a polished tube with their action in a blackened one. *The substantial identity observed in the two cases renders the conclusion impossible that condensation on the surface of the polished tube could have had any material influence on the results.*

But let us give the argument its entire logical value. It may be urged that though the influence of the interior surface of the tube is proved practically *nil* by these experiments, the surfaces of the rock-salt remained, and the liquid films may have collected on *them.* A conclusive reply to this criticism is given by the numerous experiments recorded in Memoir V. For were the measured absorptions due to films on the plates of salt, and not to the vapour between them, the mere augmentation of the distance between the plates would produce no augmentation of the absorption. But that the reverse is the truth is amply established in Memoir V. The case of sulphuric ether will suffice. It was measured in layers varying from 0·05 of an inch to 2 inches in thickness, the absorption rising with the distance traversed by the rays from 2 per cent. to 35 per cent. of the total radiation.

It is thus, I think, established that the particular kind of surface condensation invoked by Professor Magnus, and which consists of a film, not only invisible to ordinary sight by ordinary light, but which I can affirm to be invisible even when extraordinary means are taken to reveal it, has no disturbing influence on the reflexion from the interior polished surface of a brass experimental tube.

LATEST EXPERIMENTS.

THE tube with which the polished one was compared, was, as stated above, coated with lampblack, but a still more conclusive series of experiments remains to be mentioned. An experimental tube 38 inches long and 6 inches in diameter had two ends attached to it, each perforated by an aperture 2·6 inches in diameter. These apertures were closed with plates of rock-salt. The source of heat was a platinum spiral, well defended from air-currents, and heated to redness by an electric current. In front of the spiral was a rock-salt lens, which sent a slightly convergent beam through the tube. Behind the most distant plate was formed a sharply defined image of the spiral, its size being such *that it was wholly embraced by the plate of salt.* Here then was a beam of heat passing through an experimental tube *without coming into contact either with the surface of the tube itself, or with any coating or lining of that surface.* With this apparatus all my old experiments on vapours have been frequently repeated. There is no substantial difference between the results thus obtained, and those obtained with an experimental tube, where nineteen-twentieths of the heat which reached the pile, was reflected heat. If, therefore, a film be deposited on the interior surface, and if its action be at all sensible, that action must be precisely proportional to the action of the pure vapour, and can therefore introduce no disturbance into comparative measurements.

I may add that experiments made with the same tube, source of heat, and lens entirely confirm the results announced in Memoir VI., where it is shown that the order of absorption of liquids and their vapours is the same.

Experiments have been also made with this wide experimental tube on dry and moist air. I had hazarded the estimate that 10 per cent. of the earth's radiation is absorbed within ten feet of the surface, and I wished to test this surmise with a source of heat whose rays should closely resemble in quality the earth's emission. This, at all events in England, is mainly derived from moist surfaces: the radiation from *water* thus forming a principal part of the earth's emitted heat. From experiments recorded in Memoir VI. it may be safely inferred that the radiation of a hydrogen flame is similar in quality to that of water. This, therefore, was chosen as a source of heat in applying the test referred to.

Dry and moist air in succession were permitted to enter the experimental tube, the sides of which as before, were entirely withdrawn from the radiation. A long series of concurrent experiments yielded the mean result that the aqueous vapour of a column of humid air, a yard in length, intercepted 8 per cent. of the total radiation from a hydrogen flame. The tube was, as usual, stopped with plates of rock-salt, which on being detached showed not the slightest trace of moisture.

The air here entered the experimental tube through a stopcock placed at its centre. Removing the plates of salt, and gently forcing dry air and moist successively into the tube, the difference between them amounted to 5 per cent. of the total radiation.

But a noteworthy circumstance is now to be mentioned. To an observer looking at the needle as the moist air entered the *closed* experimental tube, the action of the aqueous vapour would in some cases appear to be absolutely *nil.* At starting and at concluding the needle would point to zero, or nearly so. But when this was the case it was always found that the entrance of *dry air* caused a deflection of seven or eight degrees *on the side of heat.* This was entirely due to the dynamic heating of the interior surface of the tube by the collision of the air; and this effect, *plus* the slightly additional effect of surface condensation, had to be overcome by the moist air. In short, when the needle remained at zero with the moist air, the deflection produced by the dynamic heating of the dry air became the measure of the absorption.

This fact is capable of application. With a lining of smooth pasteboard and moist air Professor Magnus found his needle to remain motionless. Does this prove the absence of absorption? No more than the case just cited. May it not be that the warming of the pasteboard tube exactly made good the absorption? Professor Magnus begins with a lightly coated tube and finds the absorption fall from 3·75 to 1·4. With smooth pasteboard tube he finds the absorption *nil,* while with a tube more thickly coated than the first he crosses the zero of neutrality, and finds the pile *heated* instead of chilled. Is it to be assumed that the heating thus manifested in the third tube was entirely absent from the first? If not the figure 1·4 does not express the absorption, which ought to be 1·4 *plus* the radiation from the interior surface. With the thickly coated tube the increase through surface radiation amounted to 1 per cent; and when no external source of heat was employed the increase was still greater. How much greater is not stated; and we are left without the means of analysing the composite effect produced by the true source of heat and the warmed interior surface.

CONCLUDING REMARKS AND SUMMARY.

I HAVE thus endeavoured to unravel a very tangled skein, and I think these Memoirs testify that I have not shrunk from the necessary labour. It may, moreover, be taken for granted that the work here recorded constitutes but a small fraction of that really done. Indeed scientific literature is so voluminous, and human life so short, that I thought it right to confine myself to the bare essentials of experimental evidence.

Professor Magnus and myself approached this question from different points of view. He had not experimented with vapours: I had. He had not varied the density of the more powerful gases: I had. The consequence was, that I approached aqueous vapour thinking that it would be found to exert a sensible action, while he approached it with the certainty that it would exert no action.* Both expectations were fulfilled; he found the action of aqueous vapour *nil*, while I found it to be 15 times that of the air in which it was diffused. By purifying the air and thus lowering its action, the absorption of aqueous vapour became subsequently 30, 40, 50, even 90 times that of the pure air with which it was mixed.

Professor Magnus first explained the discrepancy between us by ascribing my results to the condensation of moisture in the liquid form on my plates of rock-salt. By special experiments he proved it possible to render rock-salt dripping wet in a sufficiently humid atmosphere. It was a matter of everyday experience with me that such a result was possible; but I replied by showing that my plates of salt were as dry during my experiments as plates of glass or rock-crystal. After experimenting with moist air I submitted the plates of salt to Professor Magnus himself, and he could detect no trace of moisture upon them.

The objection to the rock-salt plates was still further met by abolishing them altogether, and showing that precisely the same results could be obtained with a tube open at both ends. These results were shown to Professor Magnus, but he did not note whether the deflections which I brought under his notice were due to heat or cold;† and on repeating the experiments he obtained deflections similar to mine in magnitude, but opposite in direction. His results were proved to be due to the condensation of aqueous vapour on the face of his pile when humid air was urged against it, and to the evaporation of the condensed moisture when dry air was urged against the pile.

I was perfectly well acquainted with the source of error pointed out by Professor Magnus, and, warned by this knowledge, I avoided the error. But I did not content myself with reaffirming the correctness of my results. The apparatus was handed over to Dr. Frankland, who, having entire control of it, repeated my experiments with critical care. He verified them all. They were subsequently proved true by Professor Wild, and still later by Professor Magnus himself.

* 'Obgleich mit Bestimmtheit vorauszusehen war, dass die geringe Menge von Wasserdampf, welche die Luft bei gewöhnlicher Temperatur aufzunehmen vermag, von keinem Einfluss auf die Durchstrahlung sein könne.'—*Pogg. Ann.* vol. cxii. p. 539.

† *Phil. Mag.* vol. xxvi. p. 23.

In conversation he had mentioned to me that the mechanical impurities of London air might be the cause of the action which I had ascribed to aqueous vapour. I replied by operating on air from Epsom Downs, the Isle of Wight, and other places. London air, moreover, was purified until it became neutral, and afterwards, without coming into contact with any source of impurity whatever, was led over fragments of clean glass moistened with distilled water. Thus charged with vapour, its action was proved even greater than that of pure country air.

It was intimated to me as early as 1862 that condensation on the interior surface of the experimental tube might, by diminishing the reflexion, have produced my results. This source of error was examined by Professor Magnus in 1863, and abandoned.* I replied to the objection, when first mentioned, by showing that the absorption was accurately proportional to the quantity of humid air present in the experimental tube, a result almost physically impossible to be produced by condensation.

The *quality* of the evidence was then varied. Comparing the absorption of liquids with that of their vapours, both were found with follow the same order. In no case whatever, where the volatility of the liquid is sufficient to render its vapour manageable, has an exception to this law been discovered. The position of a liquid, therefore, as an absorber of radiant heat, fixes practically that of its vapour; and as water has been proved by Melloni to be of all liquids the most energetic absorber of radiant heat, it may be safely concluded that the vapour of water has a corresponding power.

So constant is the relation between vapourous and liquid absorption that in the various and singular shiftings of diathermic position revealed in Memoir VI. the vapour followed with undeviating precision the fluctuations of the liquid from which it was derived.

Nor is this constancy of relation confined to the thermal rays of low refrangibility. In the experiments on rays of high refrangibility recorded in Memoir VII. it is shown that the 'penetrability' of liquids to the chemical rays has its exact parallel in the penetrability of their vapours to the same rays.

Such facts prove, I think to demonstration, both radiation and absorption to be, in the main, the acts of the constituent atoms of the molecules of compound bodies. For were they to any great extent the acts of the molecule taken as a whole, the change from the gaseous to the liquid state, altering so profoundly the relations between molecule and molecule must introduce changes in the order of absorption. If this be true, then the liberation of the molecules of water from the liquid condition, cannot destroy the radiant and absorbent power of the constituent atoms; though the unparalleled attenuation of aqueous vapour must enormously diminish the action. (At page 428 this subject is further considered).

To this attenuation, in the first place, to the thickness of the strata employed, and to the fact that the experiments were conducted in air already charged to a great extent with aqueous vapour, the negative results obtained by Professor Magnus in his radiation experiments are, in my opinion, to be ascribed.†

* *Phil. Mag.* vol. xxvi. p. 23.

† In these experiments Professor Magnus carried his air through water. Even then the quantity of vapour taken up is very small. The volatility of a liquid, at all events the ease with which it yields its vapour to dry air, is by no means indicated by the boiling point of the liquid. The quantity of aqueous vapour taken up by dry air is a minimum. Other liquids

Further evidence regarding the absorption of heat by aqueous vapour, based upon the principle that the emission from any vapour is absorbed with peculiar avidity by that vapour, is adduced. It is proved in Memoir VII. that, despite enormous differences of temperature between the radiant and the absorbent, a gas or vapour is intensely opaque to the radiation from itself. Thus carbonic acid, one of the feeblest absorbers of heat from ordinary sources, transcends all other gases in its action on the heat emitted by a carbonic-oxide flame. Hence the fair conclusion that aqueous vapour will be found specially opaque to the emission from a hydrogen flame. This inference of reason is completely confirmed by experiment. When a hydrogen flame is the source of heat, the absorption is treble that exerted on the emission from a platinum wire, with a temperature not higher than it assumes when plunged into a flame of hydrogen; while the actual immersion of such a wire so changes the quality of the heat as to reduce the absorption to one-half that exercised upon the emission from the flame above.

To Professor Magnus's latest experiments, in which he seeks to show that my results were due to the formation of a broken layer of water on the interior surface of my experimental tube, I reply by employing an experimental tube, with a diameter two and a half times that of the plates of salt used to close it, sending through these plates a beam *which touches neither the surface of the tube, nor any lining used to clothe that surface,* and proving by this crucial test that, not only aqueous vapour, but all the other vapours, behave exactly as they had been found to behave in a tube which made *reflected heat* nineteen-twentieths of the total radiation.

Finally, the phenomena of meteorology all speak in favour of the action ascribed to aqueous vapour. The great thermometric range in the interior of Australia; the night chill of Sahara; the temperature of the table-land of Asia; the great fluctuations observed by Hooker in the Himalayas, and a multitude of similar effects observed at other places, are perfectly accounted for by the alleged action. To this evidence Professor Magnus offers the single reply, that the checking of the terrestrial drain of heat is due to 'nebulous' instead of vaporous matter. When Leslie affirms that his *æthrioscope* on days *equally clear* indicates widely different degrees of terrestrial radiation, Professor Magnus replies that the days are *not* equally clear. The same reply is offered to Strachey's observations. The meteorologist vainly appeals to the visible purity of the sky; when the firmament is of the deepest blue there is still unseen cloudy matter in the air, and to it is due the stoppage of terrestrial radiation.

It is needless to say that, if we except the experiments of Professor Magnus, which have just been subjected to analysis, there is not the slightest warrant

with boiling points far higher than that of water yield far greater quantities of vapour to dry air passing through them. They are more speedily exhausted, and they evaporate more quickly when exposed in the common air. A great number of experiments illustrating this point have been executed. Glass fragments, placed in two U-tubes, and moistened with equal weights of iodide of allyl (boiling point 101° C.) and water, had the same quantity of dry air carried over them. The loss by evaporation of the iodide was 97 per cent., while the loss of water was only 19 per cent. of the total weight. Toluol (boiling point 110·3° C.) and water treated in the same manner showed a loss of 88 per cent. of the former to a loss of only 14 per cent. of the latter. The nitrite of amyl with a boiling point of 97° C. showed a loss of 90 per cent. as contrasted with a loss of 12 per cent. by water. Even the *nitrate* of amyl with a boiling point of 149° C. is taken up in far greater quantities by dry air than water is.

for this assumption. My experiments lead me to conclude that the cloudy matter capable of producing the observed effects would infallibly be detected. As stated in Memoir VI., the dry smoke of London is found to exert but a small comparative action on radiant heat. Such smoke if raised aloft in the atmosphere would assuredly sully the purity of the sky. And if, for every particle of carbon, a particle of water were substituted, the effect upon the firmamental blue could not, in my opinion, escape observation. This, however, is a matter which can be completely set at rest by meteorologists; for if, during days of palpable firmamental impurity and great dryness, the radiation should be found greater than on humid days with an unsullied azure overhead, the assumption will be left without any basis.

There is, I believe, proof that in India and Australia the radiation is often greater when the sky is dimmed with dust, than when it is visibly free from floating matter. It is not the *dry* winds which bring this dust that check the radiation, but the *humid* winds which carry no dust but render the sky clear.

Bearing upon this subject is an article on Ice-making in the Tropics, recently published in 'Nature' by Mr. Wise. I hope the author will allow me to show my interest in his communication by reproducing the greater part of it here. The care taken to secure local dryness is very manifest. The ground is first dried by the sun, and the straw which covers the ground and on which the dishes rest, is also carefully kept dry. A glance at the map of India will explain the efficacy of the N.N.W. wind, and the hindrance offered by the southerly and easterly winds to the formation of ice. In the former case the air not only crosses the hills mentioned by Mr. Wise, but in all probability has rolled over the Himalayas from Thibet; in the latter case it comes from the adjacent ocean. The thermometric difference between 48°, the temperature of the air a yard above the straw-beds, and 27°, the temperature of the beds themselves, is greater than I had supposed it to be. It is also interesting to know that the wind which resists the formation of ice is often colder than the one which favours it.

Ice-making in the Tropics.

By J. A. Wise.

The following method is employed by the natives of Bengal for making ice at the town of Hooghly near Calcutta, in fields freely exposed to the sky, and formed of a black loam soil upon a substratum of sand.

The natives commence their preparations by marking out a rectangular piece of ground 120 feet long by 20 broad, in an easterly and westerly direction, from which the soil is removed to the depth of two feet. This excavation is smoothed, and *is allowed to remain exposed to the sun to dry,* when rice straw in small sheaves is laid in an oblique direction in the hollow, with loose straw upon the top, to the depth of a foot and a half, leaving its surface half a foot below that of the ground. Numerous beds of this kind are formed, with narrow pathways between them, in which large earthen water-jars are sunk in the ground for the convenience of having water near, to fill the shallow unglazed earthen vessels in which it is to be frozen. These dishes are 9 inches in diameter at the top, diminishing to $4\frac{3}{10}$ inches at the bottom, $1\frac{3}{10}$ deep, and $\frac{3}{10}$ of an inch in thickness; and are so porous as to become moist throughout when water is put into them.

During the day the loose straw in the beds above the sheaves is occasionally turned up, *so that the whole may be kept dry*, and the water-jars between the beds are filled with soft pure water from the neighbouring pools. Towards evening the shallow earthen dishes are arranged in rows upon the straw, and by means of small earthen pots, tied to the extremities of long bamboo rods, each is filled about a third with water. The quantity, however, varies according to the expectation of ice—which is known by the clearness of the sky, and the steadiness with which the wind blows from the N.N.W. When favourable, about eight ounces of water is put into each dish, and when less is expected, from two to four ounces is the usual quantity; but, in all cases, more water is put into the dishes nearest the western end of the beds, as the sun first falls on that part, and the ice is thus more easily removed, from its solution being quicker.

There are about 4,590 plates in each of the beds last made, and if we allow five ounces for each dish, which presents a surface of about 4 inches square, there will be an aggregate of 239 gallons, and a surface of 1,530 square feet of water in each bed.

In the cold season, when the temperature of the air at the ice-fields is under 50° F., and there are gentle airs from the northern and western direction, ice forms in the course of the night in each of the shallow dishes. Persons are stationed to observe when a small film appears upon the water in the dishes, when the contents of several are mixed together, and thrown over the other dishes. This operation increases the congealing process; as a state of calmness has been discovered by the natives to diminish the quantity of ice produced. When the sky is quite clear, with gentle steady airs from the N.N.W., which proceed from the hills of considerable elevation near Bheerboom, about 100 miles from Hooghly, the freezing commences before or about midnight, and continues to advance until morning, when the thickest ice is formed. I have seen it seven-tenths of an inch in thickness, and in a few very favourable nights the whole of the water is frozen, when it is called by the natives solid ice. When it commences to congeal between two and three o'clock in the morning, thinner ice is expected, called paper-ice; and when about four or five o'clock in the morning, the thinnest is obtained, called flower-ice.

Upwards of two hundred and fifty persons, of all ages, are actively employed in securing the ice for some hours every morning that ice is procured, and this forms one of the most animated scenes to be witnessed in Bengal. In a favourable night upwards of 10 cwt. of ice will be obtained from one bed, and from twenty beds upwards of 10 tons.

When the wind attains a southerly or easterly direction, no ice is formed, *from its not being sufficiently dry*; not even though the temperature of the air be lower than when it is made with the wind more from a northern or western point. The N.N.W. is the most favourable direction of wind for making ice, and this diminishes in power as it approaches the due north, or west. In the latter case more latitude is allowed than from the N.N.W. to the north. So great is the influence of the direction of the wind on the ice, that when it changes in the course of a night from the N.N.W. to a less favourable direction, the change not only prevents the formation of more ice, but dissolves what may have been formed. On such occasions a mist is seen hovering over the ice-beds, from the moisture over them, and the quantity condensed by the cold wind. A mist in like manner forms over deep tanks during favourable nights for making ice.

Another important circumstance in the production of ice is the amount of wind. When it approaches a breeze no ice is formed. This is explained by such rapid currents of air removing the cold air, before any accumulation of ice has taken place in the ice-beds. It is for these reasons that the thickest ice is expected when during the day a breeze has blown from the N.W., which thoroughly dries the ground.

The ice-dishes present a large moist external surface to the dry northerly evening air, which cools the water in them, so that, when at 61°, it will in a few minutes fall to 56°, or even lower. But the moisture which exudes through the dish is quickly frozen, when the evaporation from the external surface no longer continues radiative; a more powerful agent then produces the ice in the dishes.

The quantity of dry straw in the ice-beds forms a large mass of a bad conductor of heat, which penetrates but a short way into it during the day; and as soon as the sun descends below the horizon, this large and powerfully-radiating surface is brought into action, and affects the water in the thin porous vessels, themselves powerful radiators The cold thus produced is further increased by the damp night air descending to the earth's surface, and by the removal of the heating cause, which deposits a portion of its moisture upon the now powerfully radiating and therefore cold surface of the straw, the water, and the large moist surface of the dishes. When better radiators [?] of heat were substituted, as glazed white, or metallic dishes, the cold was greater, and the ice was thicker, and the dishes were heavier in the morning than the common dishes. Any accumulation of heat on their surface from the deposit of moisture is prevented by the cold dry north-west airs which slowly pass over the dishes. The winds quickly dries the ground, and declines towards night to moderate airs. The influence of these causes is so powerful that I have seen the mercury in the thermometer placed upon the straw between the dishes descend to 27°, when three feet above the ice-pits it was 48°.

So powerful is the cooling effect of radiation on clear nights in tropical climates, that in very favourable mornings, during the cold season, drops of dew may sometimes be found congealed in Bengal upon the thatched roofs of houses, and upon the exposed leaves of plants. In the evening the cooling process advances more rapidly than could be supposed by one who has not experienced it himself, and proves the justness of his feelings, by the aid of the thermometer. In the open plain on which the ice is made, I have seen the temperature of the air, four feet above the ground, fall from 70·5° to 57°, in the time the sun took to descend the last two degrees before his setting.

[*Here follow a few shorter papers which are either referred to in the foregoing Memoirs, or contain something which the Memoirs do not contain.*]

XII.

RECENT RESEARCHES ON RADIANT HEAT.

XII.

RECENT RESEARCHES ON RADIANT HEAT.*

THE last number of Poggendorff's 'Annalen' contains a short paper by Professor Magnus, 'On the Passage of Radiant Heat through Moist Air,' a translation of which appears in the present number of the 'Philosophical Magazine.' This paper has excited considerable interest and some discussion among the scientific men of London, and it is on many accounts desirable that I should not delay attempting to offer an explanation of the differences which exist between my eminent friend and myself. A brief sketch of the history of the subject is also considered desirable; and this, as far as the extremely limited time at my disposal will admit of, I shall also endeavour to supply.

On the first perusal of Melloni's admirable work 'La Thermochrose,' which came into my hands soon after its publication, the thought of investigating the action of gases on radiant heat occurred to me. Melloni, it will be remembered, failed to obtain any evidence of the absorption of radiant heat by a column of atmospheric air 18 or 20 feet long. My attention was further fixed upon this subject by the discussion carried on in 1851 between Professors Stokes and Challis, regarding Laplace's correction for the theoretic velocity of sound. Professor Challis, it will be remembered, contended that Laplace had no right to his correction, because the heat evolved in condensation would be instantly wasted by radiation in a mass of air of indefinite extension. In the first lecture of my first course at the Royal Institution in 1853, the compression of air in a rock-salt syringe was proposed to decide the question; and in a paper presented quite recently to the Royal Society, this

* *Philosophical Magazine*, for April, 1862. This paper is referred to at the end of the 'Historic Remarks' on Memoir I.

point has been solved in a manner which I hope Professor Challis himself will deem conclusive, the mode of solution resembling in some respects the device of 1853. In 1854 the action of gases and vapours on radiant heat was a frequent subject of conversation between my scientific friends and myself; and some of these still remember my remarks at the time, the hopes entertained regarding the subject, and the devices by which it was proposed to meet its difficulties. I was, however, prevented by other engagements from attacking the subject at this time; and not till the early spring of 1859 were my ideas brought to practical definition. Then, however, I devised and applied the apparatus which, with some modifications and improvements, has been used ever since.

This apparatus immediately opened to me a large and rich field of experimental inquiry; and the greatest pleasure this discovery gave me, and which was often expressed to Mr. Faraday at the time, was, that it placed me in possession of a subject in the prosecution of which I could not possibly interfere with the claims of any previous investigator. The first notice of my researches is published in the 'Proceedings of the Royal Society' for May 26, 1859. On June 10 following, they were made the subject of a Friday evening discourse at the Royal Institution. The late lamented Prince Consort was present on this occasion, and with characteristic kindness interested himself afterwards to obtain plates of rock-salt for me. I then executed many of my experiments in presence of a large audience; and an account of the discourse is published in the 'Proceedings of the Royal Institution' of the date referred to. I also communicated an account of the investigation to my friend Professor de la Rive, and he published a translation of my letter in the *Bibliothèque Universelle.* The investigation was also described in *Cosmos,* in the *Nuovo Cimento,* and in other journals. When I reached Paris in 1859, the subject had attracted a greater degree of attention than I could have hoped to see bestowed upon it. In short, the publicity of my mode of experiment and results was quite general.

I will here ask permission to cite a number of these results obtained during the month of July 1859, after the main difficulties of my apparatus had been surmounted. [They are certainly closer to the truth than those published in 1861 by

Professor Magnus.] The method employed was substantially the same as that described in my last memoir.* The heat passed from the radiating surface through a vacuum into the experimental tube; the principle of compensation was also employed; the length of the tube used to receive the gas was 12 inches; and from the galvanometric deflection consequent on the admittance of the gas or vapour its absorption was deduced.

I. *Gases.*

Name of gas	Deflection
Atmospheric air	6°.
Oxygen	8°; 8°; 7°; 7°.
Nitrogen, 20th July	6°; 5°.
Again, 25th July	7°; 7°.
Hydrogen	10°; 10°.
Carbonic oxide	34°; 34°; 34°.
Carbonic acid	37·5°; 35°; 37·5°; 37°.
Nitrous oxide	57·5°; 57·5°; 57·5°.
Olefiant gas, 1 inch pressure	43°; 43°.
,, ,, 5 inches ,,	62·5°; 62·5°.
,, ,, 30 inches ,,	74°.
Coal-gas, 1 inch pressure	28°.
,, 5 inches ,,	54°; 53°.
,, 30 inches ,,	74°; 74°.
Total heat	79·8°.

The figures separated from each other by semicolons indicate the results of different experiments; and their close agreement shows the accuracy which, even in this early stage of the inquiry, the experiments had attained. The above deflections represent the following absorptions, at a common pressure of 30 inches of mercury:—

II. *Gases.*

Name of gas	Absorption per 100	Name of gas	Relative Absorption
Atmospheric air	6	Carbonic acid	37
Nitrogen	6	Nitrous oxide	110
Oxygen	7	Olefiant gas	345
Hydrogen	10	Coal-gas	345
Carbonic oxide	34		

The vapours of the following substances were also examined in the same month, at a common pressure, and the annexed results were obtained.

* *Philosophical Transactions*, February 1861; *Philosophical Magazine*, Sept. 1861.

III. *Vapours.*

Name of vapour	Deflection
Bisulphide of carbon	16°; 17°.
Bichloride of carbon	33°; 33°.
Iodide of methyl	37·5°; 37·5°.
Chloroform	40°; 41°; 40°.
Benzol	43°; 43°; 44°.
Amylene	55°; 55°.
Wood-spirit	55°; 55°.
Methylic alcohol (from Dr. W.)	55·5°; 55°.
" " (from Dr. H.)	63·5°; 64° (impure).
Ethylic ether	63·5°; 63°.
Absolute alcohol	64·5°; 64·5°.
Ethyl-amylic ether	65°; 65°.
Sulphuric ether	67°; 67°.
Propionate of ethyl	68°: 68°.
Acetate of ethyl	70°; 70°.
Double brass screen	79·8°

These deflections correspond to the following absorptions, omitting decimals:—

IV. *Vapours.*

Name of vapour	Absorption per 100	Name of vapour	Relative Absorption
Bisulphide of carbon	17	Ethylic ether	200
Bichloride of carbon	33	Absolute alcohol	210
Iodide of methyl	38	Ethyl-amylic ether	216
Chloroform	44	Sulphuric ether	237
Benzol	50	Propionate of ethyl	252
Amylene	84	Acetate of ethyl	282
Pure methylic alcohol	84		

These results, *which followed many thousand undescribed experiments*, were all obtained before the end of July 1859; and I should certainly have published them and many others *in extenso* at the time, had I not felt that the wide circulation the general description of the inquiry had obtained relieved me from this necessity. I wished to impart the last finish to my apparatus, and to pursue the subject with that deliberation and thoroughness which its difficulty and importance demanded. Not until the close of 1860 was the full account of the investigation drawn up; and the memoir in which it was embodied bears the receipt of the Royal Society for January 10, 1861. It afterwards formed the Bakerian Lecture for the year.

For months I was harassed by the discordant results obtained with gases generated in different ways. The nitrogen obtained from the passage of air over heated copper turnings gave me at first many times the effect of the air itself; that obtained from the combustion of phosphorus in air differed from both; while the nitrogen obtained from the nitrate of potassa could not be made to agree with its fellows. In like manner, the oxygen obtained from the chlorate of potash and peroxide of manganese differed from electrolytic oxygen; the hydrogen obtained from sulphuric acid and zinc differed from electrolytic hydrogen; the carbonic oxide obtained from chalk and carbon differed from that generated from the ferrocyanide of potassium, while carbonic acid from different sources showed similar anomalies. It will be borne in mind *that at this time nothing whatever was known of the vast action which a small amount of certain impurities can exert, and that my own experiments were the first to exhibit this action.*

Further, my drying apparatus first consisted of sixteen feet of glass tubing filled with chloride of calcium, and a large U-tube filled with fragments of pumice-stone moistened with sulphuric acid. Sometimes the chloride of calcium was used alone, sometimes the sulphuric acid, and sometimes both were used together. Every morning it was necessary to allow the air to pass through the drying apparatus, and fill the experimental tube several times before the results became constant; and even after they had become tolerably constant with the chloride of calcium, the introduction of the sulphuric acid caused a considerable variation of the absorption. This might naturally be ascribed to the more perfect desiccation of the air by the acid, but this does not account for the effects obtained. For when both were used, the magnitude of the absorption was found to depend on the circumstance whether the air entered the sulphuric-acid tube or the chloride-of-calcium tube *first*. I will here give an example of this irregularity:—

	Absorption
Air passed through Ca Cl alone	7
When SO^3 was added	4
Through new Ca Cl tube	7
New SO^3 tube added	4
Through another Ca Cl tube alone	7
A fresh tube of SO^3 added	5
Reversed current of air, and sent it through SO^3 first	10

The fluctuations above referred to are here distinctly exhibited; and the last experiment shows that, without changing the tubes in any way, but merely by reversing the direction in which the current of air passed through them, the absorption was doubled. Difficulties almost innumerable of this kind had to be overcome. I finally abandoned the chloride of calcium and the pumice-stone altogether, and made use of fragments of pure marble for my caustic potash, and of pure glass for my sulphuric acid. But with these also a long time elapsed before I was master of the anomalies which from time to time made their appearance. The dust of a cork; a fragment of sealing-wax, so minute as almost to escape the eyesight; the moisture of the fingers touching the neck of the U-tube, in which the sulphuric acid was contained—these, and many other apparently trivial causes, were sufficient entirely to vitiate the results in delicate cases, giving me on many occasions effects which I knew to be large multiples of the truth. Thus, while perfectly safe as regards the stronger gases whose energy of action masked small errors, prolonged experiment was needed to connect these with the feebler ones, and to refer them to air as a standard. In short, I thought it due both to the public and myself to abstain from giving more than a clear general account of the inquiry until every anomaly that had arisen had been mastered. I cannot regret having exercised this patience, more especially when one of the ablest and most conscientious experimenters of modern times is found falling into error on some of the points which most perplexed me.

A few weeks subsequent to the receipt of my paper by the Royal Society, that is to say, on February 7, 1861, an account of experiments on the transmission of radiant heat through gases was communicated by Professor Magnus to the Academy of Sciences in Berlin. In this inquiry the absorption of heat by vapours was left untouched, nor did it embrace the reciprocity of radiation and absorption which my investigation revealed. But as regards absorption by gases, Professor Magnus and myself had operated on the same substances; and, considering the totally different methods employed, the correspondence between our results must be regarded as remarkable.

Previous to occupying himself with the transmission of heat through gases, Professor Magnus had made an investigation on the *conduction* of heat by gases, and he was led naturally by this inquiry to take up the question of gaseous diathermancy. My knowledge of his great skill and extreme caution as an experimenter entirely ratifies a statement which he has repeated more than once in his published memoir, namely, that his results on the diathermancy of gases were already obtained at the time he communicated his results on conduction to the Academy of which he is a member, that is to say, in the month of July 1860. In fact, the very experiments intended to determine their conduction really revealed the absorption of the gases. I am quite persuaded that the results of Professor Magnus are independent of mine, and that, had I published nothing on the subject, his own inquiries would have led him to the discoveries which he has announced. That my researches preceded his by more than a year, is simply to be ascribed to the fact of my attention having been directed to the radiation of heat through gases long before even his researches on conduction had commenced. It is needless to dwell upon the value of such a general corroboration as that which subsists between Professor Magnus and myself. However private interests may fare, science is assuredly a gainer when independent courses of experiment lead, as in the present instance, to the same important results.

But while furnishing, by an independent method, a highly valuable general corroboration of my results, there are some special points on which Professor Magnus differs from me; and one of these (the action of aqueous vapour on radiant heat) he has made the subject of special examination. My first experiment gave the action of the vapour of the London air on a November day to be 15 times that of the air itself. Only a few weeks subsequently Professor Magnus announced, and cited very clear experiments in support of his statement, that the amount of aqueous vapour capable of being taken up by air at a temperature of 15 C. has no influence whatever upon the absorption. This announcement caused me to repeat my experiments with more than usual care; and I found the absorption of the vapour not 15 times, but 40 times that of the air. This result was mentioned incidentally in my letter to Sir John Herschel; and Professor Magnus, induced by this mention to take up the question again, corroborates his former results, and finds, by repeated experiments, that the aqueous vapour of the atmosphere has no influence whatever upon radiant heat, 'and that the rays of the sun, so long as the air is clear, reach the earth in the same manner whether the atmosphere is saturated with vapour or not.'

The more I experiment, the farther I seem to retreat from the position of my friend; for in a paper quite recently presented to the Royal Society, the action of the air of the laboratory of the Royal Institution is set down not at 15, nor at 40, but often at 60 times that of perfectly dry air. In fact, the greater my experience and the stricter the precautions I take to exclude impurities, the more does atmospheric air, in its action on radiant heat, approach the character of a vacuum, and consequently the greater, by comparison, becomes the action of the aqueous vapour of the air.

In the paper which has suggested this communication, Professor Magnus assigns as a possible source of error on my part, that the aqueous vapour may have been precipitated in a liquid form upon my plates of rock-salt. He cites experiments of his own to show the hygroscopic nature of this substance; and refers to Melloni's experiments in proof of the highly opaque character of a solution of rock-salt for the obscure rays of heat. In a series of experiments made with the express intention of wetting the plates of salt by precipitation, Professor Magnus exalts the absorption to four times that of air; but though the plates were visibly wet, no nearer approach than this could be made to my result, which makes the absorption of aqueous vapour forty, fifty, and even sixty times that of air. It was only on the inner surface of the salt, which came into contact with the saturated air, that the moisture was precipitated in the experiments of Professor Magnus; the outer surface, which was in contact with the common air of his laboratory, remained dry; and even the wetted surface, when exposed for a time to the same air, became dry also. Now *it is with this common outer air, and not with air artificially saturated with moisture, that I find the absorption of aqueous vapour to be fifty or sixty times that of the air in which it is diffused.*

I think it would be hardly possible for a person of any experimental aptitude whatever to work for three years with plates of rock-salt, which must be kept polished and bright, without becoming aware of all the circumstances referred to by Professor Magnus. But the truth is that I was well acquainted with the peculiarities of rock-salt many years before this investigation commenced.* A slight consideration of the conditions of the case will, I think, show how improbable it is that a precipitation, such as that surmised, could take place in my experiments. First, then, the common air of the laboratory, according to Professor Magnus, does not produce the effect which he considers may be active in my case; this, as already stated, is the air employed in all kinds of weather, dry as well as moist. Secondly, this air is introduced into a tube through which is passing a flux of heat from the radiating source. Thirdly, the air on entering the tube is heated by the stoppage of its own motion, and thereby rendered more capable of maintaining its vapour in a transparent state. The exterior surface of my terminal plate of salt was, moreover, always open to inspection, and it was never found wet; much less could the inner surface be wetted when the laboratory air was used, because the temperature within the tube was higher than that without.

But I have not relied on the inspection of the outer surface alone of the rock-salt plates. The apparatus has been taken asunder more than fifty times, on occasions when I had most reason to expect precipitation, but no trace of moisture has been found on my plates.

This, however, did not entirely satisfy me, and I therefore made an arrangement of the following kind:—An india-rubber bag was filled with air and subjected to gentle pressure. By a suitable arrangement of cocks and T-pieces, this air could be forced either through a succession of tubes containing fragments of marble moistened with caustic potash and fragments of glass moistened with sulphuric acid; or through a similar series in which fragments of glass were moistened with distilled water. A current of either dry air or damp air could be thus obtained at pleasure; and my object then was to introduce either the dry air or the wet air, under precisely the same conditions, into an

* The action of moisture upon rock-salt was unhappily made strikingly evident to me some months ago; for through a chink in the roof of the laboratory some water entered which destroyed two of my plates, and left me more or less a cripple ever since.

open tube. To effect this, matters were so arranged that either current could be discharged into the same narrow glass tube. This glass tube was left in undisturbed connexion with one end of my experimental tube, while the other end was connected with the air-pump. *The plates of salt were entirely abandoned,* the experimental tube was separated from the 'front chamber' described in my memoir, and a distance of a foot intervened between the radiating surface and the adjacent open end of the tube. In front of the other open end of the experimental tube was my thermo-electric pile, the 'compensating cube' being applied in the usual way. By pressing the bag and gently working the pump, dry air could, to a great extent, be displaced by moist, and moist air by dry. And in this way, *without any plates of rock-salt whatever,* all the results obtained with them were verified. Similar experiments have been executed in the case of all other vapours examined: with them, as well as with aqueous vapour, my plates of rock-salt are perfectly to be relied on.

Whence, then, the difference between Professor Magnus and myself? I am quite persuaded that no greater care could be bestowed upon scientific work than Professor Magnus bestows upon his; and it is the perfectly accurate nature of his experiments which renders the explanation of the differences between us an easy task.

Let me, however, first ask attention to what may be called a case of *internal evidence.* The mere inspection of the drawing of my apparatus (*Frontispiece*) will show that there was a good deal of thought and labour expended in its construction. To one part of it especially I would direct attention. In front of the experimental tube is a chamber which is always kept exhausted, the radiant heat thus passing through a vacuum into the experimental tube. To obtain that chamber gave me great trouble: it was necessary to unite its anterior wall with silver solder to its sides; and this, moreover, had to be done for every special source of heat employed. The chamber had also to pass through a copper vessel, being soldered water-tight at its place of entrance and of exit. This vessel had to be connected by a tube 20 feet long with the water-pipes of the Institution, so as to get a supply; and to carry off the water, I had the stone floor of the laboratory perforated, and one of our drains connected by a second tube with the vessel.

As already known, the water-vessel was intended to prevent the heat of the source from reaching the first plate of rock-salt. To introduce this plate air-tight between the front chamber and the experimental tube was also a difficult matter, which required special means to meet it. Now, let me ask, *what could have induced me to go to all this trouble*? The obtaining of suitable plates of rock-salt has been one of my greatest difficulties; why then did I expend time in seeking for *a pair* of them? Why did I not content myself with a single plate to stop the remote end of my tube, and allow the latter to form a continuous whole from the radiating surface to the remote end? Nay, *why did I not abandon both plates, and simply cement my pile air-tight into the remote end of my tube*? All these devices passed through my mind, and formed subjects of experiment at an early stage of this inquiry. These experiments taught me that by bringing the gases *into direct contact with my source of heat*, or into direct contact with the face of my pile, I entirely vitiated my results. And this arrangement, which in my case would have been perfectly fatal as far as accuracy is concerned, is that which Professor Magnus has adopted, and is, I believe, the sole source of the differences which have shown themselves between his results and mine.

His chief apparatus may be thus described:—A glass vessel fits like a receiver with its ground-edge on the plate of an air-pump. To the top of this receiver a second glass vessel is fused, and partially filled with water. Into this water steam is conducted, which causes the water to boil, a temperature of 100° C. being thus imparted to the bottom of the vessel, which is at the same time the top of the receiver. On the plate of the air-pump a thermo-electric plate is fixed with its face turned upwards, so as to receive the radiation from the heated top of the receiver. The face of the pile can be screened off at pleasure from the radiation from above. From the pile, wires proceed through the plate of the air-pump to the galvanometer. The receiver is first exhausted and the screen removed; the consequent deflection gives the amount of heat radiated against the pile through a vacuum. Air, or some other gas, is then admitted, and the reduction of the deflection is regarded as due to the absorption of the gas.

Air at the common laboratory temperature is here admitted into direct contact with the radiating source of heat possessing a temperature of 100° C.; chilling of that source is the immediate consequence. And no matter how long the gas may remain there, the hot surface can never attain its pristine temperature. Professor Magnus, it will be observed, experiments in the ordinary way, making use of one face only of his pile. I entirely failed to obtain any absorption by air or any of the elementary gases by this mode of experiment, while he obtains for oxygen and air an absorption of 11 per cent., and for hydrogen an absorption of 14 per cent. My apparatus enables me to measure an absorption of 0·1 per cent.; and surely with it an action so gross as the above could never have escaped me. Nor could it have escaped Melloni, who operated upon a column of air fifteen times the length of that used by Professor Magnus, and still found no absorption. With a column of air more than double the length of his I obtain for oxygen only $\frac{1}{110}$th of the absorption ascribed to it by Professor Magnus, and only $\frac{1}{140}$th of what he finds for hydrogen.

The greater action of hydrogen is quite in accordance with the known chilling-power of that gas. While ascribing their results to a different cause, some experiments of my own, which are briefly described in the paper recently presented to the Royal Society, completely corroborate those of Professor Magnus. In these experiments the gases were allowed to come into direct contact with the radiating source of heat, and here the action of hydrogen bore to that of oxygen the precise ratio found by Professor Magnus. The tube used in these experiments was 8 inches long; and had I been tempted to ascribe the results to absorption, a tube of 8 inches would have yielded fifty times the effect observed in a tube of 33 inches, in which the gases were withdrawn from contact with the source of heat.

The negative results of Professor Magnus, as regards aqueous vapour, are now sufficiently intelligible. The action which he observed in the case of air being due to direct chilling by contact—a process in which *the mass* of the chilling agent is the most important consideration—the action of the minute quantity of aqueous vapour present in the air becomes a vanishing quantity. He makes air more than a hundred times what it ought to be, and the action of the vapour practically disappears.

It is curious and instructive to observe the contrast of opinion between Professor Magnus and myself. He concludes that, even if his experiments did not actually prove it, it must be evident that the small amount of aqueous vapour in the air cannot sensibly affect the absorption; and I apply the same consideration of smallness of quantity to account for the neutrality of the aqueous vapour, when mixed with air, as a chilling agent by contact. With regard to absorption, however, the quantity of vapour usually afloat in the atmosphere is large in comparison with some of the quantities habitually employed in my experiments.

Further, an inspection of these experiments showed me long ago that those substances which, in the liquid condition, are highly absorbent of radiant heat, are also highly absorbent in the vaporous condition. Now, water is proved by Melloni to be the most opaque liquid that he had examined; and it would be perfectly anomalous, on *à priori* grounds, if the vapour of this liquid proved so utterly neutral as the experiments of Professor Magnus make it.

But the exposure of the naked face of the pile to the gas has also been spoken of. My experience of this arrangement is not without instruction.

A square aperture was cut into a tin tube, and the face of a pile, cemented air-tight all round, introduced into the aperture. The tube was closed at the ends and connected with an air-pump. The tube being exhausted and the needle of the galvanometer connected with the pile at zero, on allowing air to enter, the amount of heat generated dynamically and communicated to the pile, was sufficient to dash my needles against their stops at 90°. I do not entertain a doubt of being able to cause my needles to swing through an arc of 500° by the heat thus generated. When, on the contrary, the tube was *full* at the commencement, and the needle at zero, two or three strokes of the pump sufficed to send the needle up against the stops, the deflection now being due to the chilling of the thermo-pile. In fact this very deportment of a gaseous body on entering an exhausted receiver, and on being pumped out of a full one, has enabled me to solve the paradoxical problem of determining the radiation and absorption of a gas or vapour without any source of heat external to the gaseous body itself. *The pile of Professor Magnus was exposed to a similar action to that here described, though he never, to my knowledge, refers to it.** It would be quite impossible for me to carry out my experiments with an instrument thus circumstanced; for after the pile had been either heated or chilled dynamically, it required in some cases hours for the needle to return to zero. I may add that these experiments on dynamic heating and chilling have been made with my needles loaded with pieces of paper, so as to render their motion visible to the most distant members of the large audience of the Royal Institution.

In addition to the experiments made with the apparatus above described, Professor Magnus has made two other series with a glass tube one mètre in length, stopped at its ends by plates of glass. His source of heat in this case was a powerful Argand lamp, the rays of which were collected by a parabolic mirror placed behind it. In one series the tube was covered within by a coating of blackened paper, while in the other this coating was removed, the radiation through the tube being augmented by the reflexion from its sides. With the blackened tube, Professor Magnus corroborates the results already obtained for air by Dr. Franz, who makes the absorption of a column of nearly the same length as that employed by Professor Magnus 3 per cent. of the incident heat.

The difference between this result and that obtained with the other apparatus, which gave an absorption of 11 per cent., might naturally be ascribed to the different kinds of heat employed in the two cases, were it not that in the series of experiments made with his *unblackened* tube, he finds the absorption of oxygen and of air to be 14·75 per cent.; and of hydrogen to be 16·23 per cent. of the incident heat. This great difference between the blackened and unblackened tube, Professor Magnus ascribes to a change of quality which the heat has undergone by reflexion at the interior surface of the tube, and which has rendered the heat more capable of absorption. I have tried to obtain this result with a glass tube of nearly the same length

* See also Professor Wild's Experiments, p. 389.

as that used by Professor Magnus, but have failed to do so. *The absorption of oxygen and air in his tube is* 140 *times, and the absorption of hydrogen is* 160 *times what they show themselves to be in mine.*

Whence these differences? They are plainly to be referred to a source similar to that which caused the former ones. Indeed, I do not know a more instructive example of a single defect running through a long series of conscientious experiments, and so completely accounting for the observed anomalies. Professor Magnus stops his tube with plates of glass 4 millimètres in thickness. Now Melloni has show that 61 per cent. of the rays of a Locatelli lamp are absorbed by a plate of glass 2·6 millimètres in thickness. It is therefore almost certain that 70 per cent. of the entire heat emitted by the lamp of Professor Magnus were lodged in his first glass plate. A much less quantity of the direct heat would be absorbed by his second plate; but here the amount absorbed would be most effective as a secondary source of heat, on account of the proximity of this plate to the thermo-electric pile.

With the blackened tube, then, we had three sources of heat acting directly or indirectly upon the pile: the lamp, the first plate of glass, and the second plate. In reality, however, the sources of heat reduce themselves to *two*. For, glass being opaque to the radiation from glass, the heat emitted by the first plate was expended in exalting the temperature of the second, close to which the pile was placed. On admitting air at the ordinary temperature into this tube, an effect similar in kind to that which takes place in the other instrument must occur: the heated glass plates are chilled, and they are chilled more by the hydrogen than by the air, thus giving us the exact results recorded by Professor Magnus.

The same considerations applied to the unblackened tube explain perfectly the singular result obtained with it. On theoretic grounds it is extremely difficult, if not impossible, to conceive of such a change of quality in the heat as that above referred to. But there appears to be no reason to call in its aid. Professor Magnus himself finds that the quantity of heat transmitted through his unblackened tube is 26 times that which passes through his blackened one where the oblique radiation is cut off. In the case therefore of the naked tube, the flux of heat sent down by the heated glass plate adjacent to the lamp, to its fellow at the other end, and likewise the heat sent directly from the lamp to the same plate, are greatly superior to what they are in the case of the blackened tube. The plate adjacent to the pile becomes therefore more highly heated; and as its chilling is approximately proportionate to the difference of temperature between it and the cold air, the withdrawal of heat will be greatest when the tube is unblackened within. While leaving myself open to correction, I would offer this as the explanation of the extraordinary result which Professor Magnus has obtained. It is, I submit, not a case of absorption, but of direct chilling by the cold air.

It is hardly necessary to say that similar remarks to those made with reference to the blackened tube of Professor Magnus apply to the experiments of Dr. Franz. Dr. Franz never touched the absorption by air at all; his effects are entirely due to chilling by contact. This accounts for his finding the same effect in a tube 45 centimètres long as in a tube of 90 centimètres, for his ranking carbonic acid as low as air, while it is 90 times more powerful, and for making bromine-vapour a greater absorbent than nitrous acid, whereas the absorption by the compound gas is vastly in excess. The heat rendered latent by the evaporation of his bromine, augmented the effect, which in reality he was measuring. In fact, all the differences between the German philosophers and myself appear to be strictly accounted for by reference to a source of error which the application of plates of rock-salt enabled me from the outset to avoid.

XIII.

ON RADIATION THROUGH THE EARTH'S ATMOSPHERE.*

NOBODY ever obtained the idea of a line from Euclid's definition. The idea is obtained from a real physical line drawn by a pen or pencil, and therefore possessing width, the notion of width being afterwards dropped by a process of abstraction. So also with regard to physical phenomena: we conceive the invisible by means of proper images derived from the visible, and purify our conceptions afterwards. Definiteness of conception, even though at some expense to delicacy, is of the greatest utility in dealing with physical phenomena. Indeed it may be questioned whether a mind trained in physical research can at all enjoy peace without having made clear to itself some possible way of imaging those operations which lie beyond the boundaries of sense, and in which sensible phenomena originate.

It is well known that our atmosphere is mainly composed of the two elements oxygen and nitrogen. These elementary atoms may be figured as small spheres scattered thickly in the space which immediately surrounds the earth. They constitute about 99½ per cent. of the atmosphere. Mixed with these atoms we have others of a totally different character; we have the molecules, or atomic groups, of carbonic acid, of ammonia, and of aqueous vapour. In these substances diverse atoms have coalesced to form little systems of atoms. The molecules of aqueous vapour, for example, consist each of two atoms of hydrogen united to one of oxygen; and they mingle as little triads among the monads of oxygen and nitrogen, which constitute the great mass of the atmosphere.

A medium embraces our atoms; within our atmosphere exists a second and a finer atmosphere, in which the atoms of oxygen and nitrogen hang like suspended grains. This finer atmosphere unites not only atom with atom, but star with star; and the light of all suns, and of all stars, is in reality a kind of motion propagated through this interstellar medium. This image must be clearly seized, and then we have to advance a step. We must not only figure our atoms suspended in this medium, but we must figure them vibrating in it. In this motion of the atoms consists what we call their heat. 'What is heat in us,' as Locke has perfectly expressed it, 'is in the body heated nothing but motion.' We must figure this motion communicated to the medium in which the atoms swing, and sent through it with inconceivable velocity. Motion in this form, unconnected with ordinary matter, but speeding through the interstellar medium, receives the name of Radiant Heat; and if competent to excite the nerves of vision, we call it Light.

Aqueous vapour is an invisible gas. If vapour be permitted to issue horizon-

* A public lecture, referred to at pp. 384 and 392; *Proceedings of the Royal Institution*, vol. iv. p. 4.

tally with considerable force from a tube connected with a small boiler, the track of the cloud produced by the precipitation of the vapour is seen. What is seen, however, is not vapour, but vapour condensed to water. Beyond the visible end of the jet the cloud resolves itself again into true vapour. A lamp placed under the jet cuts the cloud sharply off, and when the flame is placed near the efflux orifice the cloud entirely disappears. The heat of the lamp completely prevents precipitation. This same vapour may be condensed and congealed on the surface of a vessel containing a freezing mixture, from which it may be scraped in quantities sufficient to form a small snowball. When a luminous beam is sent through a large receiver placed on an air-pump, a single stroke of the pump causes the precipitation of the aqueous vapour to a cloud within. This, illuminated by the beam, produces upon a screen behind a richly-coloured halo, due to diffraction by the little cloud.

The waves of heat pass from our earth through our atmosphere towards space. These waves meet in their passage the atoms of oxygen and nitrogen, and the molecules of aqueous vapour. Thinly scattered as these latter are, we might naturally think meanly of them as barriers to the waves of heat. We might imagine that the wide spaces between the vapour molecules would be an open door for the passage of the undulations; and that if those waves were at all intercepted, it would be by the substances which form 99½ per cent. of the whole atmosphere. It had, however, been found that this small modicum of aqueous vapour intercepts fifteen times the quantity of heat stopped by the whole of the air in which it was diffused. It was afterwards found that the dry air then experimented with was not perfectly pure, and that the purer the air became the more it approached the character of a vacuum, and the greater, by comparison, became the action of the aqueous vapour. The vapour was found to act with 30, 40, 50, 60, 70 times the energy of the air in which it was diffused; and no doubt was entertained that the aqueous vapour of the air which filled the Royal Institution theatre, during the delivery of this discourse, quenched 90 or 100 times the quantity of radiant heat absorbed by the main body of the air of the room.

Looking at the single atoms, for every 200 of oxygen and nitrogen there is about 1 molecule of aqueous vapour. This 1, then, is 80 times more powerful than the 200; and hence, comparing a single atom of oxygen or nitrogen with a single molecule of aqueous vapour, we may infer that the action of the latter is 16,000 times that of the former. This is a very astonishing result, and it naturally excited opposition, based on the philosophic reluctance to accept a fact of such import before testing it to the uttermost. From such opposition a discovery, if it be worth the name, emerges with its fibre strengthened; as the human character gathers force from the healthy antagonisms of active life. It was urged that the result was on the face of it improbable; that there were, moreover, many ways of accounting for it, without ascribing so enormous a comparative action to aqueous vapour. For example, the cylinder which contained the air in which these experiments were made, was stopped at its ends by plates of rock-salt, on account of their transparency to radiant heat. Now rock-salt is hygroscopic; it attracts the moisture of the atmosphere. Thus, a layer of brine readily forms on the surface of a plate of rock-salt; and it is well known that brine is very impervious to the rays of heat. Breathing for a moment on a polished plate of rock-salt, the brilliant colours of thin plates (soap-bubble colours) flash forth, these being caused by the film of moisture which overspreads the salt. Such a film, it was contended, is formed when undried air is sent into the cylinder; it was, therefore, the absorption of a layer of brine that was measured, instead of the absorption of aqueous vapour.

This objection was met in two ways:—First, by showing that the plates of salt when subjected to the strictest examination show no trace of a film of moisture. Secondly, by abolishing the plates of salt altogether, and obtaining the same results in a cylinder open at both ends.

It was next surmised that the effect was due to the impurity of the laboratory air, and the suspended carbon particles were pointed to as the cause of the opacity to radiant heat. This objection was met by bringing air from Hyde Park, Hampstead Heath, Primrose Hill, Epsom Downs, a field near Newport in the Isle of Wight, St. Catharine's Down, and the sea-beach near Black Gang Chine. The aqueous vapour of the air from these localities intercepted at least 70 times the amount of radiant heat absorbed by the air in which the vapour was diffused. Experiments made with dry smoky air proved that the atmosphere of West London, even when an east wind pours over it the smoke of the city, exerts only a fraction of the destructive powers exercised by the transparent and impalpable aqueous vapour diffused in the air.

The cylinder which contained the air through which the calorific rays passed being polished within, the rays striking the interior surface were reflected from it to the thermo-electric pile. The following objection was raised:—You permit moist air to enter your cylinder; a portion of this moisture is condensed as a liquid film upon the interior surface of your tube; its reflective power is thereby diminished; less heat therefore reaches the pile, and you incorrectly ascribe to the absorption of aqueous vapour an effect which is really due to diminished reflexion of the interior surface of your tube.

But why should the aqueous vapour so condense? The tube within is warmer than the air without, and against its inner surface the rays of heat are impinging. There can be no tendency to condensation under such circumstances.* Further, let 5 inches of undried air be sent into the tube—that is, one-sixth of the amount which it can contain. These 5 inches produce their proportionate absorption. The driest day, on the driest portion of the earth's surface, would make no approach to the dryness of our cylinder when it contains only 5 inches of air. Make it 10, 15, 20, 25, 30 inches: you obtain an absorption exactly proportional to the quantity of vapour present. It is next to a physical impossibility that this could be the case if the effect were due to condensation. But lest a doubt should linger in the mind, not only were the plates of rock-salt abolished, but the cylinder itself was dispensed with. Humid air was displaced by dry, and dry air by humid in the free atmosphere; the absorption of the aqueous vapour was here manifest, as in all the other cases.

No doubt, therefore, can exist of the extraordinary opacity of this substance to the rays of obscure heat; and particularly such rays as are emitted by the earth after it has been warmed by the sun. It is perfectly certain that more than 10 per cent. of the terrestrial radiation from the soil of England is stopped within 10 feet of the surface of the soil. This one fact is sufficient to show the immense influence which this newly-discovered property of aqueous vapour must exert on the phenomena of meteorology.

This aqueous vapour is a blanket more necessary to the vegetable life of

* This was saying too much. Professor Magnus has proved the existence of a kind of condensation under the conditions named.

England than clothing is to man. Remove for a single summer-night the aqueous vapour from the air which overspreads this country, and you would assuredly destroy every plant capable of being destroyed by a freezing temperature. The warmth of our fields and gardens would pour itself unrequited into space, and the sun would rise upon an island held fast in the iron grip of frost. The aqueous vapour constitutes a local dam, by which the temperature at the earth's surface is deepened: the dam, however, finally overflows, and we give to space all that we receive from the sun.

The sun raises the vapours of the equatorial ocean; they rise, but for a time a vapour screen spreads above and around them. But the higher they rise, the more they come into the presence of pure space, and when, by their levity, they have penetrated the vapour screen, which lies close to the earth's surface, what must occur?

It has been said that, compared molecule with atom, the absorption of a molecule of aqueous vapour is 16,000 times that of air. Now the power to absorb and the power to radiate are perfectly reciprocal and proportional. The atom of aqueous vapour will therefore radiate with 16,000 times the energy of an atom of air. Imagine then this powerful radiant in the presence of space, and with no screen above it to check its radiation. Into space it pours its heat, chills itself, condenses, and the tropical torrents are the consequence. The expansion of the air, no doubt, also refrigerates it; but in accounting for those deluges, the chilling of the vapour by its own radiation must play a most important part. The rain quits the ocean as vapour; it returns to it as water. How are the vast stores of heat set free by the change from the vaporous to the liquid condition disposed of? Doubtless in great part they are wasted by radiation into space. Similar remarks apply to the cumuli of our latitudes. The warmed air, charged with vapour, rises in columns, so as to penetrate the vapour screen which hugs the earth; in the presence of space, the head of each pillar wastes its heat by radiation, condenses to a cumulus, which constitutes the visible capital of an invisible column of saturated air.

Numberless other meteorological phenomena receive their solution, by reference to the radiant and absorbent properties of aqueous vapour. It is the absence of this screen, and the consequent copious waste of heat, that causes mountains to be so much chilled when the sun is withdrawn. Its absence in Central Asia renders the winter there almost unendurable; in Sahara the dryness of the air is sometimes such that, though during the day 'the soil is fire and the wind is flame,' the chill at night is painful to bear. In Australia, also, the thermometric range is enormous, on account of the absence of this qualifying agent. A clear day, and a dry day, moreover, are very different things. The atmosphere may possess great visual clearness, while it is charged with aqueous vapour, and on such occasions great chilling cannot occur by terrestrial radiation. Sir John Leslie and others have been perplexed by the varying indications of their instruments on days equally bright—but all these anomalies are completely accounted for by reference to this newly-discovered property of transparent aqueous vapour. Its presence would check the earth's loss; its absence, without sensibly altering the transparency of the air, would open wide a door for the escape of the earth's heat into infinitude.

XIV.

ON A NEW SERIES OF CHEMICAL REACTIONS PRODUCED BY LIGHT.

I WISH to draw the attention of chemists to a form or method of experiment which, though obvious, is unknown, and which, I doubt not, will in their hands become a new experimental power. It consists in subjecting the vapours of volatile liquids to the action of concentrated sunlight, or to the concentrated beam of the electric-light.

Action of the Electric-light.

A glass tube 2·8 feet long, and of 2·5 inches internal diameter, was supported horizontally. At one end of it was placed an electric lamp, the height and position of both being so arranged that the axis of the glass tube and that of the parallel beam issuing from the lamp were coincident. The tube in the first experiments was closed by plates of rock-salt, and subsequently by plates of glass.

As on former occasions, for the sake of distinction, I will call this tube *the experimental tube.*

The experimental tube was connected with an air-pump, and also with a series of drying and other tubes used for the purification of the air.

A number of test-tubes (perhaps fifty in all) were converted into Woulfe's flasks. Each of them was stopped with a cork, through which passed two glass tubes: one of these tubes (*a*) ended immediately below the cork, while the other (*b*) descended to the bottom of the flask, being drawn out at its lower end to an orifice about 0·03 of an inch in diameter. It was found necessary to coat the cork carefully with cement.

The little flask thus formed was partially filled with the liquid whose vapour was to be examined; it was then introduced into the path of a purified current of air.

The experimental tube being exhausted, and the cock which admitted the purified air being cautiously turned on, the air entered the flask through the tube *b*, and escaped by the small orifice at the lower end of *b* into the liquid. Through this it bubbled, loading itself with vapour, after which the mixed air and vapour, passing from the flask by the tube *a*, entered the experimental tube, where they were subjected to the action of light.

The power of the electric beam to reveal the existence of anything within the experimental tube, or the impurities of the tube itself, is extraordinary. When the experiment is made in a darkened room, a tube which in ordinary daylight appears absolutely clean is often shown by the present mode of examination to be exceedingly filthy.

The following are some of the results obtained with this arrangement:—

Nitrite of amyl (boiling-point 91° to 96° C.).—The vapour of this liquid was in the first instance permitted to enter the experimental tube while the beam from the electric lamp was passing through it. Curious clouds were observed to form near the place of entry, which were afterwards whirled through the tube.

The tube being again exhausted, the mixed air and vapour were allowed to enter it in the dark. The slightly convergent beam of the electric-light was then sent through the tube from end to end. For a moment the tube was *optically empty*, nothing whatever was seen within it; but before a second had elapsed a shower of liquid spherules was precipitated on the beam, thus generating a cloud within the tube. This cloud became denser as the light continued to act, showing at some places a vivid iridescence.

The beam of the electric lamp was now converged so as to form within the tube, between its end and the focus, a cone of rays about eight inches long. The tube was cleansed and again filled in darkness. When the light was sent through it, the precipitation upon the beam was so rapid and intense that the cone, which a moment before was invisible, flashed suddenly forth like a solid luminous spear.

The effect was the same when the air and vapour were allowed to enter the tube in diffuse daylight. The cloud, however, which shone with such extraordinary radiance under the electric beam, was invisible in the ordinary light of the laboratory.

The quantity of mixed air and vapour within the experimental tube could of course be regulated at pleasure. The rapidity of the action diminished with the attenuation of the vapour. When, for example, the mercurial column associated with the experimental tube was depressed only five inches, the action was not nearly so rapid as when the tube was full. In such cases, however, it was exceedingly interesting to observe, after some seconds of waiting, a thin streamer of delicate bluish-white cloud slowly forming along the axis of the tube, and finally swelling so as to fill it.

When dry oxygen was employed to carry in the vapour, the effect was the same as that obtained with air.

When dry hydrogen was used as a vehicle, the effect was also the same.

The effect, therefore, is not due to any interaction between the vapour of the nitrite and its vehicle.

This was further demonstrated by the deportment of the vapour itself. When it was permitted to enter the experimental tube unmixed with air or any other gas, the effect was substantially the same. Hence the seat of the observed action is the vapour.

With reference to the air and the glass of the experimental tube, the beam employed in these experiments was *perfectly cold.* It had been sifted by passing it through a solution of alum, and through the thick double-convex lens of the lamp. When the unsifted beam of the lamp was employed, the effect was still the same; the obscure calorific rays did not appear to interfere with the result.

I have taken no means to determine strictly the character of the action here described, my object being simply to point out to chemists a method of experiment which reveals a new and beautiful series of reactions; to them I leave the examination of the products of decomposition. The molecule of the nitrite of amyl is shaken asunder by certain specific waves of the electric beam, forming nitric oxide and other products, of which the *nitrate* of amyl is probably one. The brown fumes of nitrous acid were seen to mingle with the cloud within the experimental tube.

The nitrate of amyl, being less volatile than the nitrite, would not be able to maintain itself in the condition of vapour, but would be precipitated in liquid spherules along the track of the beam.

In the anterior portions of the tube a sifting action of the vapour occurs which diminishes the chemical action in the posterior portions. In some experiments the precipitated cloud only extended halfway down the tube. When, under these circumstances, the lamp was shifted so as to send the beam through the other end of the tube, precipitation occurred there also.

Action of Sunlight.

The solar light also effects the decomposition of the nitrite-of-amyl vapour. On the 10th of October I partially darkened a small room in the Royal Institution, into which the sun shone, permitting the light to enter through an open portion of the window-shutter. In the track of the beam was placed a large plano-convex lens, which formed a fine convergent cone in the dust of the room behind it. The experimental tube was filled in the laboratory, covered with a black cloth, and carried into the partially darkened room. On thrusting one end of the tube into the cone of rays behind the lens, precipitation within the cone was copious and immediate. The vapour at the distant end of the tube was in part shielded by that in front, and was also more feebly acted on through the divergence of the rays. On reversing the tube, a second and similar cone was precipitated.

Physical Considerations.

I sought to determine the particular portion of the white beam which produced the foregoing effects. When, previous to entering the experimental tube, the beam was caused to pass through a red glass, the effect was greatly weakened, but not extinguished. This was also the case with various samples of yellow glass. A blue glass being introduced, before the removal of the yellow or the red, on taking the latter away augmented precipitation occurred along the track of the blue beam. Hence, in this case, the more refrangible rays are the most chemically active.

The colour of the liquid nitrite of amyl indicates that this must be the case; it is a feeble but distinct yellow: in other words, the yellow portion of the beam is most freely transmitted. It is not, however, the transmitted portion of any beam which produces chemical action, but the absorbed portion. Blue, as the complementary colour to yellow, is here absorbed, and hence the more energetic action of the blue rays. This reasoning, however, assumes that the same rays are absorbed by the liquid and its vapour.

A solution of the yellow chromate of potash, the colour of which may be made almost, if not altogether, identical with that of the liquid nitrite of amyl, was found far more effective in stopping the chemical rays than either the red or the yellow glass. But of all substances the nitrite itself is most potent in arresting the rays which act upon its vapour. A layer one-eighth of an inch in thickness, which scarcely perceptibly affected the luminous intensity, sufficed to absorb the entire chemical energy of the concentrated beam of the electric-light.

The close relation subsisting between a liquid and its vapour, as regards their action upon radiant heat, has been already amply demonstrated.* As regards the nitrite of amyl, this relation is more specific than in the cases hitherto adduced; for here the special constituent of the beam which provokes the decomposition of the vapour is shown to be arrested by the liquid.

A question of extreme importance in molecular physics here arises:—What is the real mechanism of this absorption, and where is its seat? †

I figure, as others do, a molecule as a group of atoms, held together by their mutual forces, but still capable of motion among themselves. The vapour of the nitrite of amyl is to be regarded as an assemblage of such molecules. The question now before us is this:—In the act of absorption, is it the molecules that are effective, or is it their constituent atoms? Is the *vis viva* of the intercepted waves transferred to the molecule as a whole, or to its constituent parts?

The molecule, as a whole, can only vibrate in virtue of the forces exerted between it and its neighbour molecules. The intensity of these forces, and consequently the rate of vibration, would, in this case, be a function of the distance between the molecules. Now the identical absorption of the liquid and of the vaporous nitrite of amyl indicates an identical vibrating period on the part of liquid and vapour, and this, to my mind, amounts to an experimental demonstration that the absorption occurs in the main *within* the molecule. For it can hardly be supposed, if the absorption were the act of the molecule as a whole, that it could continue to affect waves of the same period after the substance had passed from the vaporous to the liquid state.

In point of fact the decomposition of the nitrite of amyl is itself to some extent an illustration of this internal molecular absorption; for were the absorption the act of the molecule as a whole, the *relative* motions of its constituent atoms would remain unchanged, and there would be no mechanical cause for their separation. It is probably the synchronism of the vibrations of one portion of the molecule with the incident waves which enables the amplitude of those vibrations to augment until the chain which binds the parts of the molecule together is snapped asunder.

The *liquid* nitrite of amyl is probably also decomposed by light; but the reaction, if it exists, is incomparably less rapid and distinct than that of the vapour. Nitrite of amyl has been subjected to the concentrated solar rays until it boiled, and it has been permitted to continue boiling for a considerable time, without any distinctly apparent change occurring in the liquid.

I anticipate wide, if not entire, generality for the fact that a liquid and its vapour absorb the same rays. A cell of liquid chlorine now preparing for me will, I imagine, deprive light more effectually of its power of causing chlorine and hydrogen to combine than any other filter of the luminous rays. The rays which give chlorine its colour have nothing to do with this combination, those

* *Philosophical Transactions*, 1864.

† My attention was very forcibly directed to this subject some years ago by a conversation with my excellent friend Professor Clausius.

that are absorbed by the chlorine being the really effective rays. A highly sensitive bulb containing chlorine and hydrogen in the exact proportions necessary for the formation of hydrochloric acid was placed at one end of an experimental tube, the beam of the electric lamp being sent through it from the other. The bulb did not explode when the tube was filled with chlorine, while the explosion was violent and immediate when the tube was filled with air. I anticipate for the liquid chlorine an action similar to but still more energetic than that exhibited by the gas. If this should prove to be the case, it will favour the view that chlorine itself is *molecular* and not *monatomic.*

Production of the Blue of the Sky by the Decomposition of Nitrite of Amyl.

When the quantity of nitrite-of-amyl vapour is considerable, and the light intense, the chemical action is exceedingly rapid, the particles precipitated being so large as to *whiten* the luminous beam. Not so, however, when a well-mixed and highly attenuated vapour fills the experimental tube. The effect now to be described was obtained in the greatest perfection when the vapour of the nitrite of amyl was derived from a residue of the moisture of its liquid, which had been accidentally introduced into the passage through which the dry air flowed into the experimental tube.

In this case the electric beam traversed the tube for several seconds before any action was visible. Decomposition then visibly commenced, and advanced slowly. The particles first precipitated were too small to be distinguished by an eye-glass; and, when the light was very strong, the cloud appeared of a milky blue. When, on the contrary, the intensity was moderate, the blue was pure and deep. In Brücke's important experiments on the blue of the sky and the morning and evening red, pure mastic is dissolved in alcohol, and then dropped into water well stirred. When the proportion of mastic to alcohol is correct, the resin is precipitated so finely as to elude the highest microscopic power. By reflected light, such a medium appears bluish, by transmitted light yellowish, which latter colour, by augmenting the quantity of the precipitate, can be caused to pass into orange or red.

But the development of colour in the attenuated nitrite-of-amyl vapour, though admitting of the same explanation, is doubtless more similar to what takes place in our atmosphere. The blue, moreover, is purer and more sky-like than that obtained from Brücke's turbid medium. There could scarcely be a more impressive illustration of Newton's mode of regarding the generation of the colour of the firmament than that here exhibited; for never, even in the skies of the Alps, have I seen a richer or a purer blue than that attainable by a suitable disposition of the light falling upon the precipitated vapour. May not the aqueous vapour of our atmosphere act in a similar manner? and may we not fairly refer to liquid particles of infinitesimal size the hues observed by Principal Forbes over the safety-valve of a locomotive, and so skilfully connected by him with the colours of the sky?

In exhausting the tube containing the mixed air and nitrite-of-amyl vapour, it was difficult to avoid explosions under the pistons of the air-pump, similar to those described as occurring with the vapours of bisulphide of carbon and other substances.* Though the quantity of vapour present in these cases must have been infinitesimal, its explosion was sufficient to destroy the valves of the pump.

* Page 28.

Iodide of Allyl (boiling-point 101° C.).—Among the liquids hitherto subjected to the concentrated electric-light, iodide of allyl, in point of rapidity and intensity of action, comes next to the nitrite of amyl. With the iodide of allyl I have employed both oxygen and hydrogen, as well as air, as a vehicle, and found the effect in all cases substantially the same. The cloud column here was exquisitely beautiful, but its forms were different from those of the nitrite of amyl. The whole column revolved round the axis of the decomposing beam; it was nipped at certain places like an hour-glass, and round the two bells of the glass delicate cloud-filaments twisted themselves in spirals. It also folded itself into convolutions resembling those of shells. In certain conditions of the atmosphere in the Alps I have observed clouds of a special pearly lustre; when hydrogen was made the vehicle of the iodide-of-allyl vapour a similar lustre was most exquisitely shown. With a suitable disposition of the light, the purple hue of iodine vapour came out very strongly in the tube.

The remark already made as to the bearing of the decomposition of nitrite of amyl by light on the question of molecular absorption applies here also; for were the absorption the work of the molecule as a whole, the iodine would not be dislodged from the allyl with which it is combined. The non-synchronism of iodine with the waves of obscure heat is illustrated by its marvellous transparency to such heat. May not its synchronism with the waves of light in the present instance be the cause of its divorce from the allyl? Further experiments on this point are in preparation.

Iodide of Isopropyl.—The action of light upon the vapour of this liquid is at first more languid than upon iodide of allyl; indeed many beautiful reactions may be overlooked in consequence of this languor at the commencement. After some minutes' exposure, however, clouds begin to form, which grow in density and in beauty as the light continues to act. In every experiment hitherto made with this substance the column of cloud which filled the experimental tube was divided into two distinct parts near the middle of the tube. In one experiment a globe of cloud formed at the centre, from which, right and left, issued an axis which united the globe with the two adjacent cylinders. Both globe and cylinders were animated by a common motion of rotation. As the action continued, paroxysms of motion were manifested; the various parts of the cloud would rush through each other with sudden violence. During these motions beautiful and grotesque cloud-forms were developed. At some places the nebulous mass would become ribbed, so as to resemble the graining of wood; a longitudinal motion would at times generate in it a series of curved transverse bands, the retarding influence of the sides of the tube causing an appearance resembling, on a small scale, the dirt-bands of the Mer de Glace. In the anterior portion of the tube those sudden commotions were most intense; here buds of cloud would sprout forth, and grow in a few seconds into perfect flower-like forms.

A gorgeous mauve colour was developed in the last twelve inches of the tube; the vapour of iodine was present, and it may have been the sky-blue produced by the precipitated particles which, mingling with the purple of the iodine, produced this splendid mauve. As in all other cases here adduced, the effects were proved to be due to the light; they never occurred in darkness.

XV.

ON THE BLUE COLOUR OF THE SKY, THE POLARIZATION OF SKY-LIGHT, AND ON THE POLARIZATION OF LIGHT BY CLOUDY MATTER GENERALLY.*

SINCE the communication of my brief abstract 'On a New Series of Chemical Reactions produced by Light,' the experiments upon this subject have been continued, and the number of the substances thus acted on considerably augmented. New relations have also been established between *mixed vapours* when subjected to the action of light.

I now beg to draw attention to two questions glanced at incidentally in the abstract referred to—the blue colour of the sky, and the polarization of sky-light. Reserving the historic treatment of the subject for a more fitting occasion, I would merely mention now that these questions constitute, in the opinion of our most eminent authorities, the two great standing enigmas of meteorology. Indeed it was the interest manifested in them by Sir John Herschel, in a letter of singular speculative power, that caused me to enter upon the consideration of these questions so soon.

The apparatus with which I work consists, as already stated, of a glass tube, about a yard in length, and from 2½ to 3 inches internal diameter. The vapour to be examined is introduced into this tube in the manner described in my last abstract, and upon it the condensed beam of the electric lamp is permitted to act until the neutrality or the activity of the substance has been declared.

It has hitherto been my aim to render the chemical action of light upon vapours *visible*. For this purpose substances have been chosen, *one* at least of whose products of decomposition under light shall have a boiling-point so high that as soon as the substance is formed it shall be *precipitated*. By graduating the quantity of the vapour, this precipitation may be rendered of any degree of fineness, forming particles distinguishable by the naked eye, or particles which are probably far beyond the reach of our highest microscopic powers.

I have no reason to doubt that particles may be thus obtained whose diameters constitute but a very small fraction of the length of a wave of violet light.

In all cases when the vapours of the liquids employed are sufficiently attenuated, no matter what the liquid may be, the visible action commences with the formation of a *blue cloud*. I would guard myself at the outset against all misconception as to the use of this term. The blue cloud here referred to is totally invisible in ordinary daylight. To be seen, it requires to be surrounded by darkness, *it only* being illuminated by a powerful beam of light. This blue

* From the *Proceedings of the Royal Society*, No. 108, 1869.

cloud differs in many important particulars from the finest ordinary clouds, and might justly have assigned to it an intermediate position between these clouds and true cloudless vapour.

With this explanation, the term 'cloud,' or 'incipient cloud,' as I propose to employ it, cannot be misunderstood.

I had been endeavouring to decompose carbonic acid gas by light. A faint bluish cloud, due it may be, or it may not be, to the residue of some vapour previously employed, was formed in the experimental tube. On looking across this cloud through a Nicol's prism, the line of vision being horizontal, it was found that when the short diagonal of the prism was vertical, the quantity of light reaching the eye was greater than when the long diagonal was vertical.

When a plate of tourmaline was held between the eye and the bluish cloud, the quantity of light reaching the eye when the axis of the prism was perpendicular to the axis of the illuminating beam, was greater than when the axes of the crystal and of the beam were parallel to each other.

This was the result all round the experimental tube. Causing the crystal of tourmaline to revolve round the tube, with its axis perpendicular to the illuminating beam, the quantity of light that reached the eye was in all its positions a maximum. When the crystallographic axis was parallel to the axis of the beam, the quantity of light transmitted by the crystal was a minimum.

From the illuminated bluish cloud, therefore, polarized light was discharged, the direction of maximum polarization being at right angles to the illuminating beam; the *plane of vibration* of the polarized light, moreover, was that to which the beam was perpendicular.*

Thin plates of selenite or of quartz, placed between the Nicol and the bluish cloud, displayed the colours of polarized light, these colours being most vivid when the line of vision was at right angles to the experimental tube. The plate of selenite usually employed was a circle, thinnest at the centre, and augmenting uniformly in thickness from the centre outwards. When placed in its proper position between the Nicol and the cloud, it exhibited a system of splendidly coloured rings.

The cloud here referred to was the first operated upon in the manner described. It may, however, be greatly improved upon by the choice of proper substances, and by the application in proper quantities of the substances chosen. Benzol, bisulphide of carbon, nitrite of amyl, nitrite of butyl, iodide of allyl, iodide of isopropyl, and many other substances may be employed. I will take the nitrite of butyl as illustrative of the means adopted to secure the best result with reference to the present question.

And here it may be mentioned that a vapour, which when alone, or mixed with air in the experimental tube, resists the action of light, or shows but a feeble result of this action, may, by placing it in proximity with another gas or vapour, be caused to exhibit under light vigorous, if not violent, action. The case is similar to that of carbonic acid gas, which diffused in the atmosphere resists the decomposing action of solar light, but when placed in contiguity with the chlorophyl in the leaves of plants, has its molecules shaken asunder.

* I assume here that the plane of vibration is perpendicular to the plane of polarization. This is still an undecided point; but the probabilities are so much in its favour, and it is in my opinion so much preferable to have a physical image on which the mind can rest, that I do not hesitate to employ the phraseology in the text. Even should the assumption prove to be incorrect, no harm will be done by the provisional use of it.

Dry air was permitted to bubble through the liquid nitrite of butyl until the experimental tube, which had been previously exhausted, was filled with the mixed air and vapour. The visible action of light upon the mixture after fifteen minutes' exposure was slight. The tube was afterwards filled with half an atmosphere of the mixed air and vapour, and another half atmosphere of air which had been permitted to bubble through fresh commercial hydrochloric acid. On sending the beam through this mixture, the action paused barely sufficiently long to show that at the moment of commencement the tube was optically empty. But the pause amounted only to a small fraction of a second, a dense cloud being immediately precipitated upon the beam which traversed the mixture.

This cloud began *blue*, but the advance to whiteness was so rapid as almost to justify the application of the term instantaneous. The dense cloud, looked at perpendicularly to its axis, showed scarcely any signs of polarization. Looked at obliquely the polarization was strong.

The experimental tube being again cleansed and exhausted, the mixture of air and nitrite-of-butyl vapour was permitted to enter it until the associated mercury column was depressed $\frac{1}{10}$ of an inch. In other words, the air and vapour, united, exercised a pressure not exceeding $\frac{1}{300}$ of an atmosphere. Air passed through a solution of hydrochloric acid was then added till the mercury column was depressed three inches. The condensed beam of the electric-light passed for some time in darkness through this mixture. There was absolutely nothing within the tube competent to scatter the light. Soon, however, a superbly blue cloud was formed along the track of the beam, and it continued blue sufficiently long to permit of its thorough examination. The light discharged from the cloud at right angles to its own length was *perfectly* polarized. By degrees the cloud became of whitish-blue, and for a time the selenite colours obtained by looking at it normally were exceedingly brilliant. The direction of maximum polarization was distinctly at right angles to the illuminating beam. This continued to be the case as long as the cloud maintained a decided blue colour, and even for some time after the pure blue had changed to whitish-blue. But as the light continued to act the cloud became coarser and whiter, particularly at its centre, where it at length ceased to discharge polarized light in the direction of the perpendicular, while it continued to do so at both its ends.

But the cloud which had thus ceased to polarize the light emitted normally, showed vivid selenite colours when looked at *obliquely*. The direction of maximum polarization changed with the texture of the cloud. This point shall receive further illustration subsequently.

A blue, equally rich and more durable, was obtained by employing the nitrite-of-butyl vapour in a still more attenuated condition. Now the instance here cited is *representative*. In all cases, and with all substances, the cloud formed at the commencement, when the precipitated particles are sufficiently fine, is *blue*, and it can be made to display a colour rivalling that of the purest Italian sky. In all cases, moreover, this fine blue cloud polarizes *perfectly* the beam which illuminates it, the direction of polarization enclosing an angle of 90° with the axis of the illuminating beam.

It is exceedingly interesting to observe both the perfection and the decay of this polarization. For ten or fifteen minutes after its first appearance the light from a vividly illuminated incipient cloud, looked at horizontally, is absolutely quenched by a Nicol's prism with its longer diagonal vertical. But as the sky-blue is gradually rendered impure by the introduction of particles of too large a size, in other words, as *real* clouds begin to be formed, the polarization begins to deteriorate, a portion of the light passing through the prism in all its positions. It is worthy of note that for some time after the cessation of perfect polarization the *residual* light which passes, when the Nicol is in its position of minimum transmission, is of a gorgeous blue, the whiter light of the cloud being extinguished.* When the cloud texture has become sufficiently coarse to approximate to that of ordinary clouds, the rotation of the Nicol ceases to have any sensible effect on the quality of the light discharged normally.

The perfection of the polarization in a direction perpendicular to the illuminating beam is also illustrated by the following experiment. A Nicol's prism large enough to embrace the entire beam of the electric lamp was placed between the lamp and the experimental tube. A few bubbles of air carried through the liquid nitrite of butyl were introduced into the tube, and they were followed by about 3 inches (measured by the mercurial gauge) of air which had been passed through aqueous hydrochloric acid. Sending the polarized beam through the tube, I placed myself in front of it, my eye being on a level with its axis, my assistant, Mr. Cottrell, occupying a similar position behind the tube. The short diagonal of the large Nicol was in the first instance vertical, the plane of vibration of the emergent beam being therefore also vertical. As the light continued to act, a superb blue cloud, visible to both my assistant and myself, was slowly formed. But this cloud, so deep and rich when looked at from the positions mentioned, *utterly disappeared when looked at vertically downwards, or vertically upwards.* Reflexion from the cloud was not possible in these directions. When the large Nicol was slowly turned round its axis, the eye of the observer being on the level of the beam, and the line of vision perpendicular to it, entire extinction of the light emitted horizontally occurred where the longer diagonal of the large Nicol was vertical. But now a vivid blue cloud was seen when looked at downwards or upwards. This truly fine experiment was first definitely suggested by a remark addressed to me in a letter by Professor Stokes.

Now, as regards the polarization of sky-light, the greatest stumblingblock has hitherto been that, in accordance with the law of Brewster, which makes the index of refraction the tangent of the polarizing angle, the reflexion which produces perfect polarization would require to be made *in* air *upon* air; and indeed this led many of our most eminent men, Brewster himself among the number, to entertain the idea of *molecular reflexion.* I have, however, operated upon substances of widely different refractive indices, and therefore of very different polarizing angles as ordinarily defined, but the polarization of the beam by the incipient cloud has thus far proved itself to be *absolutely independent of the polarizing angle.* The law of Brewster does not apply to matter in this condition, and it rests with the undulatory theory to explain why. Whenever the precipitated particles are sufficiently fine, no matter what the substance forming the particles may be, the direction of maximum polarization is at

* This seems to prove that particles too large to polarize the blue, polarize perfectly light of lower refrangibility.

right angles to the illuminating beam, the polarizing angle for matter in this condition being invariably 45°.*

Suppose our atmosphere surrounded by an envelope impervious to light, but with an aperture on the sunward side through which a parallel beam of solar light could enter and traverse the atmosphere. Surrounded on all sides by air not directly illuminated, the track of such a beam through the air would resemble that of the parallel beam of the electric lamp through an incipient cloud. The sunbeam would be *blue*, and it would discharge laterally light in precisely the same condition as that discharged by the incipient cloud. In fact, the azure revealed by such a beam would be to all intents and purposes that which I have called a 'blue cloud.'

But, as regards the polarization of the sky, we know that not only is the direction of maximum polarization at right angles to the track of the solar beams, but that at certain angular distances, probably variable ones, from the sun, 'neutral points,' or points of no polarization, exist, on both sides of which the planes of atmospheric polarization are at right angles to each other.

The parallel beam employed in these experiments tracked its way through the laboratory air exactly as sunbeams are seen to do in the dusty air of London. This air showed, though far less vividly, all the effects of polarization obtained with the incipient clouds. The light discharged laterally from the track of the illuminating beam was polarized, though not perfectly, the direction of maximum polarization being at right angles to the beam.

The horizontal column of air thus illuminated was 18 feet long, and could therefore be looked at very obliquely without any disturbance from a solid envelope. At all points of the beam throughout its entire length the light emitted normally was in the same state of polarization. Keeping the positions of the Nicol and the selenite constant, the same colours were observed throughout the entire beam when the line of vision was perpendicular to its length.

* The difficulty referred to above is thus expressed by Sir John Herschel :—'The cause of the polarization is evidently a reflexion of the sun's light upon *something*. The question is on what? Were the angle of maximum polarization 76°, we should look to water or ice as the reflecting body, however inconceivable the existence in a cloudless atmosphere, and a hot summer's day of unevaporated molecules (? particles) of water. But though we were once of this opinion, careful observation has satisfied us that 90°, or thereabouts, is a correct angle, and that therefore, whatever be the body on which the light has been reflected, *if polarized by a single reflexion*, the polarizing angle must be 45°, and the index of refraction, which is the tangent of that angle, unity; in other words, the reflexion would require to be made *in* air *upon* air!' (*Meteorology*, par. 233).

Any particles, if small enough, will produce both the colour and the polarization of the sky. But is the existence of small water-particles on a hot summer's day *in the higher regions of our atmosphere* inconceivable? It is to be remembered that the oxygen and nitrogen of the air behave as a vacuum to radiant heat, the exceedingly attenuated vapour of the higher atmosphere being therefore in practical contact with the cold of space.

The opinion of Sir John Herschel, connecting the polarization and the blue colour of the sky is verified by the foregoing results. 'The more the subject [the polarization of sky-light] is considered,' writes this eminent philosopher, 'the more it will be found beset with difficulties, and its explanation when arrived at will probably be found to carry with it that of the blue colour of the sky itself and of the great quantity of light it actually does send down to us.' 'We may observe, too,' he adds, 'that it is only where the purity of the sky is most absolute that the polarization is developed in its highest degree, and that where there is the slightest perceptible tendency to cirrus it is materially impaired.' This applies word for word to the 'incipient clouds.'

I then placed myself near the end of the beam as it issued from the electric lamp, and looking through the Nicol and selenite more and more obliquely at the beam, observed the colours fading until they disappeared. Augmenting the obliquity, the colours appeared once more, *but they were now complementary to the former ones.*

Hence this beam, like the sky, exhibited its neutral point, at opposite sides of which the light was polarized in planes at right angles to each other.

Thinking that the action observed in the laboratory might be caused in some way by the vaporous fumes diffused in its air, I had a battery and an electric lamp carried to a room at the top of the Royal Institution. The track of the beam was seen very finely in the air of this room, a length of 14 or 15 feet being attainable. This beam exhibited all the effects observed with the beam in the laboratory. Even the uncondensed electric-light falling on the floating matter showed, though faintly, the effects of polarization.*

When the air was so sifted as to entirely remove the visible floating matter, it no longer exerted any sensible action upon the light, but behaved like a vacuum.

I had varied and confirmed in many ways those experiments on neutral points, operating upon the fumes of chloride of ammonium, the smoke of brown paper, and tobacco smoke, when my attention was drawn by Sir Charles Wheatstone to an important observation communicated to the Paris Academy in 1860 by Professor Govi, of Turin.† His observations on the light of comets had led M. Govi to examine a beam of light sent through a room in which was diffused the smoke of incense. He also operated on tobacco smoke. His first brief communication stated the fact of polarization by such smoke, but in his second communication he announced the discovery of a neutral point in the beam, at the opposite sides of which the light was polarized in planes at right angles to each other.

But, unlike my observations on the laboratory air, and unlike the action of the sky, the direction of maximum polarization in M. Govi's experiments enclosed a very small angle with the axis of the illuminating beam. The question was left in this condition, and I am not aware that M. Govi or any other investigator has pursued it farther.

I had noticed, as before stated, that as the clouds formed in the experimental tube became denser, the polarization of the light discharged at right angles to the beam became weaker, the direction of maximum polarization becoming oblique to the beam. Experiments on the fumes of chloride of ammonium gave me also reason to suspect that the position of the neutral point *was not constant,* but that it varied with the density of the illuminating fumes.

The examination of these questions led to the following new and remarkable results.—The laboratory being well filled with the fumes of incense, and sufficient time being allowed for their uniform diffusion, the electric beam was sent through the smoke. From the track of the beam polarized light was discharged, but the direction of maximum polarization, instead of being along the normal, now enclosed an angle of 12° or 13° with the axis of the beam.

A neutral point, with complementary effects at opposite sides of it, was also exhibited by the beam. The angle enclosed by the axis of the beam, and a

* I hope to try Alpine air next summer.

† *Comptes Rendus*, tome li. pp. 360 and 669.

line drawn from the neutral point to the observer's eye, measured in the first instance 66°.

The windows of the laboratory were now opened for some minutes, a portion of the incense smoke being permitted to escape. On again darkening the room and turning on the beam, the line of vision to the neutral point was found to enclose with the axis of the beam an angle of 63°.

The windows were again opened for a few minutes, more of the smoke being permitted to escape. Measured as before the angle referred to was found to be 54°.

This process was repeated three additional times; the neutral point was found to recede lower and lower down the beam, the angle between a line drawn from the eye to the neutral point and the axis of the beam falling successively from 54°, to 49°, 43°, and 33°.

The distances, roughly measured, of the neutral point from the lamp, corresponding to the foregoing series of observations, were these:—

1st observation	2	feet	2	inches.
2nd "	2	"	6	"
3rd "	2	"	10	"
4th "	3	"	2	"
5th "	3	"	7	"
6th "	4	"	6	"

At the end of this series of experiments the direction of maximum polarization had again become normal to the beam.

The laboratory was next filled with the fumes of gunpowder. In five successive experiments, corresponding to five different densities of the gunpowder smoke, the angles enclosed between the line of vision to the neutral point and the axis of the beam were 63°, 50°, 47°, 42°, and 38° respectively.

After the clouds of gunpowder had cleared away, the laboratory was filled with the fumes of common resin, rendered so dense as to be very irritating to the lungs. The direction of maximum polarization enclosed in this case an angle of 12°, or thereabouts, with the axis of the beam. Looked at, as in the former instances, from a position near the electric lamp *no neutral point* was observed throughout the entire extent of the beam.

When this beam was looked at normally through the selenite and Nicol, the ring system, though not brilliant, was distinct. Keeping the eye upon the plate of selenite and the line of vision normal, the windows were opened, the blinds remaining undrawn. The resinous fumes slowly diminished, and as they did so the ring system became paler. It finally disappeared. Continuing to look along the perpendicular, the rings revived, but now the colours were complementary to the former ones. *The neutral point had passed me in its motion down the beam consequent upon the attenuation of the fumes of resin.*

In the fumes of chloride of ammonium substantially the same results were obtained as those just described. Sufficient, I think, has been here stated to illustrate the variability of the position of the neutral point.*

* Brewster has proved the variability of the position of the neutral point for sky-light with the sun's altitude. Is not the proximate cause of this revealed by the foregoing experiments?

Before quitting the question of the reversal of the polarization by cloudy matter, I will make one or two additional observations. Some of the clouds formed in the experiments on the chemical action of light are astonishing as to form. The experimental tube is often divided into segments of dense cloud, separated from each other by nodes of finer matter. Looked at normally, as many as four reversals of the plane of polarization have been found in the tube in passing from node to segment, and from segment to node. With the fumes diffused in the laboratory, on the contrary, there was no change in the polarization along the normal, for here the necessary differences of cloud texture did not exist.

Further. By a puff of tobacco smoke or of condensed steam blown into the illuminated beam, the brilliancy of the colours may be greatly augmented. But with different clouds two different effects are produced. For example, let the ring system observed in the common air be brought to its maximum strength, and then let an attenuated cloud of chloride of ammonium be thrown into the beam at the point looked at: the ring system flashes out with augmented brilliancy, and the character of the polarization remains unchanged. This is also the case when phosphorus or sulphur is burned underneath the beam, so as to cause the fine particles of phosphoric acid or of sulphur to rise into the light. With the sulphur-fumes the brilliancy of the colours is exceedingly intensified; but in none of these cases is there any change in the character of the polarization.

But when a puff of aqueous cloud, or of the fumes of hydrochloric acid, hydriodic acid, or nitric acid is thrown into the beam, there is a complete reversal of the selenite tints. Each of these clouds twists the plane of polarization 90°. On these and kindred points experiments are still in progress.*

The idea that the colour of the sky is due to the action of finely divided matter, rendering the atmosphere a turbid medium, through which we look at the darkness of space, dates as far back as Leonardo da Vinci. Newton conceived the colour to be due to exceedingly small water particles acting as thin plates. Goethe's experiments in connexion with this subject are well known and exceedingly instructive. One very striking observation of Goethe's referred to what is technically called 'chill' by painters, which is due no doubt to extremely fine varnish particles interposed between the eye and a dark background. Clausius, in two very able memoirs, endeavoured to connect the colours of the sky with suspended water-vesicles, and to show that the important observations of Forbes on condensing steam could also be thus accounted for. Bruecke's experiments on precipitated mastic were referred to in my last abstract. Helmholtz has ascribed the blueness of the eyes to the action of suspended particles. In an article written nearly nine years ago by myself, the colours of the peat smoke of the cabins of Killarney † and the colours of the sky were referred to one and the same cause, while a chapter of the 'Glaciers of the Alps,' published in 1860, is also devoted to this question. Roscoe, in

* Sir John Herschel has suggested to me that this change of the polarization from positive to negative may indicate a change from polarization by reflexion to polarization by refraction. This thought repeatedly occurred to me while looking at the effects; but it will require much following up before it emerges into clearness.

† I have sometimes quenched almost completely, by a Nicol, the light discharged normally from burning leaves in Hyde Park. The blue smoke from the *ignited end* of a cigar polarizes also, but not perfectly.

connexion with his truly beautiful experiments on the photographic power of sky-light, has also given various instances of the production of colour by suspended particles. In the foregoing experiments the azure was produced in *air*, and exhibited a depth and purity far surpassing anything that I have ever seen in mote-filled liquids. Its polarization, moreover, was *perfect*.

In his experiments on fluorescence Professor Stokes had continually to separate the light reflected from the motes suspended in his liquids, the action of which he named 'false dispersion,' from the fluorescent light of the same liquids, which he ascribed to 'true dispersion.' In fact, it is hardly possible to obtain a liquid without motes, which polarize by reflexion the light falling upon them, truly dispersed light being unpolarized. At p. 530 of his celebrated memoir 'On the Change of the Refrangibility of Light,' Professor Stokes adduces some significant facts, and makes some noteworthy remarks, which bear upon our present subject. He notices more particularly a specimen of plate-glass which, seen by reflected light, exhibited a blue which was exceedingly like an effect of fluorescence, but which, when properly examined, was found to be an instance of false dispersion. 'It often struck me,' he writes, 'while engaged in these observations, that when the beam had a continuous appearance, the polarization was more nearly perfect than when it was sparkling, so as to force on the mind the conviction that it arose merely from motes.* Indeed in the former case the polarization has often appeared perfect, or all but perfect. It is possible that this may in some measure have been due to the circumstance, that when a given quantity of light is diminished in a given ratio, the illumination is perceived with more difficulty when the light is diffused uniformly than when it is spread over the same space, but collected into specks. Be this as it may, there was at least no tendency observed towards polarization in a plane perpendicular to the plane of reflexion, when the suspending particles became finer, and therefore the beam more nearly continuous.'

Through the courtesy of its owner, I have been permitted to see and to experiment with the piece of plate-glass above referred to. Placed in front of the electric lamp, whether edgeways or transversely, it discharges bluish polarized light laterally, the colour being by no means a bad imitation of the blue of the sky.

Professor Stokes considers that this deportment may be invoked to decide the question of the direction of the vibrations of polarized light. On this point I would say, if it can be demonstrated that when the particles are small in comparison to the length of a wave of light, the vibrations of a ray reflected by such particles cannot be perpendicular to the vibrations of the incident light; then assuredly the experiments recorded in the foregoing communication decide the question in favour of Fresnel's assumption.

As stated above, almost all liquids have motes in them sufficiently numerous to polarize sensibly the light, and very beautiful effects may be obtained by simple artificial devices. When, for example, a cell of distilled water is placed in front of the electric lamp, and a slice of the beam permitted to pass through it, scarcely any polarized light is discharged, and scarcely any colour produced

* The azure may be produced in the midst of a field of motes. By turning the Nicol, the interstitial blue may be completely quenched, the shining, and apparently unaffected, motes remaining masters of the field. A blue cloud, moreover, may be precipitated in the midst of the azure. An aqueous cloud thus precipitated reverses the polarization; but on the melting away of the cloud the azure and its polarization remain behind.

with a plate of selenite. But while the beam is passing through it, if a bit of soap be agitated in the water above the beam, the moment the infinitesimal particles reach the beam the liquid sends forth laterally almost perfectly polarized light; and if the selenite be employed, vivid colours flash into existence. A still more brilliant result is obtained with mastic dissolved in a great excess of alcohol.

The selenite rings constitute an extremely delicate test as to the quantity of motes in a liquid. Commencing with distilled water, for example, a thickish beam of light is necessary to make the polarization of its motes sensible. A much thinner beam suffices for common water; while with Brücke's precipitated mastic, a beam too thin to produce any sensible effect with most other liquids, suffices to bring out vividly the selenite colours.

XVI.

ON COMETARY THEORY.*

On the 8th of May 1869, in a lecture delivered before the Cambridge Philosophical Society, I ventured to enunciate a speculation regarding the origin and deportment of visible cometary matter. I had been led to reflect on the subject by my experiments on the decomposition of vapours by light. The speculation was introduced and communicated to the Philosophical Society in the following words:—

'In the course of my experiments on actinic action I have been often astonished at the body of light which a perfectly infinitesimal amount of matter, when diffused in the form of a cloud, can discharge from it by reflexion. I have been repeatedly perplexed and led into error by the action of residues so minute as to be simply inconceivable. In order to get rid of these residues, my experimental tubes, after having been employed for any vapour, are flooded with alcohol, sponged-out with soap and hot water, and finally flooded with pure water. Let me give you some idea of the quantities of matter that here come into play. The tube before you, which is 3 feet long and 3 inches wide, was so thoroughly cleansed that when filled with air, or with the vapour of aqueous hydrochloric acid, no amount of exposure to an intense light produced the least cloudiness. Having thus assured myself of the perfect purity of the tube, I took a small bit of bibulous paper, rolled up into a pellet not the fourth part of the size of a small pea, and moistened it with a liquid possessing a higher boiling-point than that of water. I held the pellet with my fingers until it had become almost dry, then introduced it into a connecting-piece, and allowed dry air to pass over it into this tube. The air charged with the modicum of vapour thus taken up was subjected to the action of light. A blue actinic cloud began to form immediately, and in five minutes the blue colour had extended quite through the experimental tube. For some minutes this cloud continued blue, and could be completely quenched by a Nicol's prism, no trace of its light reaching the eye when the Nicol was in its proper position. But its particles augmented gradually in magnitude, and at the end of fifteen minutes a dense white cloud filled the tube. Considering the amount of the vapour carried in by the air, the appearance of a cloud so massive and luminous seemed like the creation of a world out of nothing.

'But this is not all; the pellet of bibulous paper was removed, and the experimental tube was cleared out by sweeping a current of dry air through it. *This current passed also through the connecting-piece in which the pellet of bibulous paper had rested.* The air was at length cut off and the experimental tube

* *Philosophical Magazine* for April 1869.

exhausted. Fifteen inches of hydrochloric acid were then sent into the tube through the same connecting-piece. Now it is here to be noted: (1) that the total quantity of liquid absorbed by the pellet in the first instance was exceedingly small; (2) that nearly the whole of this small quantity had been allowed to evaporate between my fingers before the pellet was placed in the connecting-piece; (3) that the pellet had been ejected and the tube in which it rested rendered for some minutes the conduit of a strong current of pure air. It was part of such a residue as could linger in the connecting-piece after this process, that was carried into the experimental tube by the hydrochloric acid and subjected there to the action of light.

'One minute after the ignition of the electric lamp a faint cloud showed itself; in two minutes it had filled all the anterior portion of the tube and stretched a considerable way down it; it developed itself afterwards into a very beautiful cloud-figure; and at the end of fifteen minutes the body of light discharged by the cloud, considering the amount of matter involved in its production, was simply astounding. But though thus luminous, the cloud was far too fine to dim in any appreciable degree objects placed behind it. The flame of a candle seemed no more affected by it than it would be by a vacuum. Placing a page of print so that it might be illuminated by the cloud itself, it could be read *through* the cloud without any sensible enfeeblement. Nothing could more perfectly illustrate that 'spiritual texture' which Sir John Herschel ascribes to a comet than these actinic clouds. Indeed, experiments proved that matter of almost infinite tenuity is competent to shed forth light far more intense than that of the tails of comets. The weight of the matter which sent this body of light to the eye, would probably have to be multiplied by millions to bring it up to the weight of the air in which it hung.

'And now will you bear with me for five minutes will I endeavour to apply these results to cometary theory? I am encouraged to do so by a remark of Bessel's, who said that had any theory preceded his observations on Halley's comet, by fixing his attention either upon its verification or its confutation, it would have enabled him to return from his observations with a greater store of knowledge than he had actually derived from them.* If time permitted, I should like to lead you by an easy gradient up to the view that I wish to submit to you; but time does not permit of this, and therefore the speculation must suffer from the baldness arising from the absence of such preparation.

'You are doubtless aware of the tremendous difficulties which beset cometary theory. The comet examined by Newton in 1680 shot out a tail sixty millions of miles in length in two days. The comet of 1843, if I remember aright, shot out in a single day a tail which covered 100 degrees of the heavens. This enormous reach of cloudy matter is supposed to be generated in the head of the comet and driven backwards by some mysterious force of repulsion exerted by the sun. Bessel devised a kind of magnetic polarity and repulsion to account for it. "It is clear," says Sir John Herschel, "that *if we have to deal here with matter such as we conceive it, viz. possessing inertia, at all*, it must be under the dominion of forces incomparably more energetic than gravitation, and quite

* The remark of Bessel here referred to I found in Poggendorff's *Annalen*, vol. xxxviii. p. 499. These are his words:—'Ich glaube nämlich, dass wir weit brauchbarere Beobachtungen über die Beschaffenheit der Kometen besitzen würden, als wir wirklich besitzen, wenn eine Erklärung der Beobachtungen vorhanden gewesen wäre, an welcher sich der Widerspruch oder die Bestätigung hätten halten können.'

of a different nature." And in another place he states the difficulties of the subject in the following remarkable words:—

'"There is beyond question some profound secret and mystery of nature concerned in the phenomenon of their tails. Perhaps it is not too much to hope that future observation, borrowing every aid from rational speculation, grounded on the progress of physical science generally (especially those branches of it which relate to the æthereal or imponderable elements), may ere long enable us to penetrate this mystery, and to declare whether it is really *matter*, in the ordinary acceptation of the term, which is projected from their heads with such extravagant velocity, and if not impelled, at least *directed* in its course by a reference to the sun at its point of avoidance. In no respect is this question as to the materiality of the tail more forcibly pressed on us for consideration than in that of the enormous sweep which it makes round the sun *in perihelio*, in the manner of a straight and rigid rod, in defiance of the law of gravitation, nay, even of the received laws of motion, extending (as we have seen in the comets of 1680 and 1843) from near the sun's surface to the earth's orbit, yet whirled round unbroken—in the latter case through an angle of 180° in little more than two hours. It seems utterly incredible that in such a case it is one and the same material object which is thus brandished. [I would especially invite the reader's attention to these words in reference to the following theory.—J. T.] If there could be conceived such a thing as a *negative shadow*, a momentary impression made upon the luminiferous æther behind the comet, this would represent in some degree the conception such a phenomenon irresistibly calls up."

'I now ask for permission to lay before you a speculation which seems to do away with all these difficulties, and which, whether it represents a physical verity or not, ties together the phenomena exhibited by comets in a remarkably satisfactory way.

'1. The theory is, that a comet is composed of vapour decomposable by the solar light, the visible head and tail being an actinic cloud resulting from such decomposition; the texture of actinic clouds is demonstrably that of a comet.

'2. The tail, according to this theory, is not projected matter, but matter precipitated on the solar beams traversing the cometary atmosphere. It can be proved by experiment that this precipitation may occur either with comparative slowness along the beam, or that it may be practically momentary throughout the entire length of the beam. The amazing rapidity of the development of the tail would be thus accounted for without invoking the incredible motion of translation hitherto assumed.

'3. As the comet wheels round its perihelion, the tail is not composed throughout of the same matter, but of new matter precipitated on the solar beams, which cross the cometary atmosphere in new directions. The enormous whirling of the tail is thus accounted for without invoking a motion of translation.

'4. The tail is always turned from the sun for this reason:—Two antagonistic powers are brought to bear upon the cometary vapour,—the one an *actinic* power, tending to produce precipitation; the other a *calorific* power, tending to effect vaporization. Where the former prevails, we have the cometary cloud; where the latter prevails, we have the transparent cometary vapour. As a matter of fact, the sun emits the two agents here invoked. There is nothing whatever hypothetical in the assumption of their existence. That precipitation should occur behind the head of the comet, or in the space occupied by the head's shadow, it is only necessary to assume that the sun's calorific rays are absorbed more copiously by the head and nucleus than the actinic rays. This augments the relative superiority of the actinic rays behind the head and nucleus, and enables them to bring down the cloud which constitutes the comet's tail.

'5. The old tail, as it ceases to be screened by the nucleus, is dissipated by the solar heat; but its dissipation is not instantaneous. The tail leans towards that portion of space last shaded by the comet's head, a general fact of observation being thus accounted for.

'6. In the struggle for mastery of the two classes of rays a temporary advantage, owing to variations of density or some other cause, may be gained by the actinic rays even in parts of the cometary atmosphere which are unscreened by the nucleus. Occasional lateral streamers, and the apparent emission of feeble tails towards the sun, would be thus accounted for.

'7. The shrinking of the head in the vicinity of the sun is caused by the beating against it of the calorific waves, which dissipate its attenuated fringe and cause its apparent contraction.

'Throughout this theory I have dealt exclusively with true causes, and no agency has been invoked which does not rest on the sure basis either of observation or experiment. It remains with you to say whether in venturing to enunciate it I have transgressed the limits of "rational speculation."

'If I have done so, surely I could not have come to a place more certain to ensure my speedy correction. If the theory be a mere figment of the mind, your Adams and your Stokes (both happily here present), to whom I submit the speculation with the view of having it instantly annihilated by astronomy and physics, if it merit no better fate, will, I doubt not, effectually discharge that duty, and thus save both you and me from error before it has had time to lay any serious hold on our imagination.'*

* There may be comets whose vapour is undecomposable by the sun, or which, if decomposed, is not precipitated. This view opens out the possibility of invisible comets wandering through space, perhaps sweeping over the earth and affecting its sanitary condition without our being otherwise conscious of their passage. As regards tenuity, I entertain a strong persuasion that out of a few ounces (the possible weight assigned by Sir John Herschel to certain comets) of iodide-of-allyl vapour, an actinic cloud of the magnitude and luminousness of Donati's comet might be manufactured.

XVII.

ON THE FORMATION AND PHENOMENA OF CLOUDS.*

It is well known that when a receiver filled with ordinary undried air is exhausted, a cloudiness, due to the precipitation of the aqueous vapour diffused in the air, is produced by the first few strokes of the pump. It is, as might be expected, possible to produce clouds in this way with the vapours of other liquids than water.

In the course of the experiments on the chemical action of light which have been already communicated in abstract to the Royal Society, I had frequent occasion to observe the precipitation of such clouds in the experimental tubes employed; indeed several days at a time have been devoted solely to the generation and examination of clouds formed by the sudden dilatation of the air in the experimental tubes.

The clouds were generated in two ways: one mode consisted in opening the passage between the filled experimental tube and the air-pump, and then simply dilating the air by working the pump. In the other, the experimental tube was connected with a vessel of suitable size, the passage between which and the experimental tube could be closed by a stopcock. This vessel was first exhausted; on turning the cock the air rushed from the experimental tube into the vessel, the precipitation of a cloud within the tube being a consequence of the transfer. Instead of a special vessel, the cylinders of the air-pump itself were usually employed for this purpose.

It was found possible, by shutting off the residue of air and vapour after each act of precipitation, and again exhausting the cylinders of the pump, to obtain with some substances, and without refilling the experimental tube, fifteen or twenty clouds in succession.

The clouds thus precipitated differed from each other in luminous energy, some shedding forth a mild white light, others flashing out with sudden and surprising brilliancy. This difference of action is, of course, to be referred to the different reflective energies of the particles of the clouds, which were produced by substances of very different refractive indices.

Different clouds, moreover, possess very different degrees of stability; some melt away rapidly, while others linger for minutes in the experimental tube, resting upon its bottom as they dissolve like a heap of snow. The particles of other clouds are trailed through the experimental tube as if they were moving through a viscous medium.

Nothing can exceed the splendour of the diffraction-phenomena exhibited by some of these clouds; the colours are best seen by looking along the experi-

* *Proceedings of the Royal Society*, No. 110, 1869.

mental tube from a point above it, the face being turned towards the source of illumination. The differential motions introduced by friction against the interior surface of the tube often cause the colours to arrange themselves in distinct layers.

The difference in texture exhibited by different clouds caused me to look a little more closely than I had previously done into the mechanism of cloud-formation. A certain expansion is necessary to bring down the cloud; the moment before precipitation the mass of cooling air and vapour may be regarded as divided into a number of polyhedra, the particles along the bounding surfaces of which move in opposite directions when precipitation actually sets in. Every cloud-particle has consumed a polyhedron of vapour in its formation; and it is manifest that the size of the particle must depend, not only on the size of the vapour polyhedron, but also on the relation of the density of the vapour to that of its liquid. If the vapour were light, and the liquid heavy, other things being equal, the cloud-particle would be smaller than if the vapour were heavy and the liquid light. There would evidently be more shrinkage in the one case than in the other: these considerations were found valid throughout the experiments; the case of toluol may be taken as representative of a great number of others. The specific gravity of this liquid is 0·85, that of water being unity; the specific gravity of its vapour is 3·26, that of aqueous vapour being 0·6. Now, as the size of the cloud-particle is directly proportional to the specific gravity of the vapour, and inversely proportional to the specific gravity of the liquid, an easy calculation proves that, assuming the size of the vapour polyhedra in both cases to be the same, the size of the particle of toluol cloud must be more than six times that of the particle of aqueous cloud. It is probably impossible to test this question with numerical accuracy; but the comparative coarseness of the toluol cloud is strikingly manifest to the naked eye. The case is, as I have said, representative.

In fact, aqueous vapour is without a parallel in these particulars; it is not only the lightest of all vapours, in the common acceptation of that term, but the lightest of all gases, except hydrogen and ammonia. To this circumstance the soft and tender beauty of the clouds of our atmosphere is mainly to be ascribed.

The *sphericity* of the cloud-particles may be immediately inferred from their deportment under the luminous beam. The light which they shed when spherical is *continuous*: but clouds may also be precipitated in solid flakes; and then the incessant sparkling of the cloud shows that its particles are *plates*, and not spheres. Some portions of the same cloud may be composed of spherical particles, others of flakes, the difference being at once manifested through the *calmness* of the one portion of the cloud, and the *uneasiness* of the other. The sparkling of such flakes reminded me of the plates of mica in the River Rhone at its entrance into the lake of Geneva, when shone upon by a strong sun.

LONDON: PRINTED BY
SPOTTISWOODE AND CO., NEW-STREET SQUARE
AND PARLIAMENT STREET

Printed in the United States
By Bookmasters